NUMBER THEORY

Sailing on the Sea of Number Theory

Vol. 1 Arithmetic Geometry and Number Theory
edited by Lin Weng & Iku Nakamura

Vol. 2 Number Theory: Sailing on the Sea of Number Theory
edited by S. Kanemitsu & J.-Y. Liu

Series on Number Theory and Its Applications Vol. 2

NUMBER THEORY

Sailing on the Sea of Number Theory

Proceedings of the 4th China–Japan Seminar

Weihai, China 30 August – 3 September 2006

Editors

S. Kanemitsu *(Kinki University, Japan)*

J.-Y. Liu *(Shandong University, China)*

World Scientific

NEW JERSEY • LONDON • SINGAPORE • BEIJING • SHANGHAI • HONG KONG • TAIPEI • CHENNAI

Published by

World Scientific Publishing Co. Pte. Ltd.

5 Toh Tuck Link, Singapore 596224

USA office: 27 Warren Street, Suite 401-402, Hackensack, NJ 07601

UK office: 57 Shelton Street, Covent Garden, London WC2H 9HE

British Library Cataloguing-in-Publication Data
A catalogue record for this book is available from the British Library.

Series on Number Theory and Its Applications — Vol. 2
NUMBER THEORY
Sailing on the Sea of Number Theory

ISBN-13 978-981-270-810-6
ISBN-10 981-270-810-3

Printed in Singapore by World Scientific Printers (S) Pte Ltd

PREFACE

The present volume is a collection of survey papers of the talks presented at the very successful fourth China-Japan Seminar on number theory. The seminar was held in Shandong University Academic Center, in Weihai, Shandong, People's Republic of China, during August 30 – September 3, 2006, under the support of Shandong University, the Japan Society for the Promotion of Science (JSPS) and the National Natural Science Foundation of China (NSFC). The organizers Shigeru Kanemitsu and Jianya Liu would like to express their hearty thanks to Shandong University for their generosity and support. The title of the seminar reads "Sailing on the sea of number theory" which suggests that we are supposed to sail freely both on the sea of number theory and the sea of Weihai. The talks were given in a wide conference hall looking over the sea all round and the atmosphere was superb, and all the participants really got the feeling of sailing on the sea of number theory and the sea of Weihai.

Weihai Sea

It should be mentioned that Ms. Nan Luo, together with other staff members of Shandong University, helped out the organizers throughout in preparing and conducting the seminar very successfully. The organizers also would like to thank two of their Chinese colleagues, Professors T. -X. Cai and Z. -W. Sun for their attractions. The former read his own poem which has a passage referring to Pythagoras and the latter made a spot announcement of some recent results in a live fashion.

Shandong University Academic Center (Weihai)

Traditionally, we have been publishing the collection of papers presented at the seminars, but from this volume onwards, we are going to publish survey papers which give good perspectives of recent progress of research in number theory and related fields. We would like to thank the authors of these papers for their kind cooperation regarding the preparation of excellent manuscript and the subsequent unification process of style, which is to make the volume of a readable book quality.

In this volume we assemble the following papers:

1. S. Egami and K. Matsumoto, "Convolutions of the von Mangoldt Function and Related Dirichlet Series" has the theme of analytic continuation of multiple Dirichlet series of various types and their possible natural boundaries. Analytic continuation is furnished largely by means of the Mellin-Barnes integrals which is a form of the definition of the beta-function. The authors consider Φ_2 with the coefficients of the form of the Abel convolution G_2 of the von Mangoldt function as an example of a multiple Dirichlet series which may have a natural boundary. They also consider the Riesz sum of G_2 and obtain an asymptotic formula.

2. K.-Q. Feng and Y. Xue, "Constructing New Non-congruent Numbers by Graph Theory" is regarding the application of graph theory (combined with the results on elliptic curves) to finding non-congruent numbers with arbitrarily many prime factors. The main tools they use from elliptic curve theory is that if the rank of the rational points of a certain elliptic curve (with n in the coefficients) is 0, then n is a non-congruent number and this condition in turn is realized if the corresponding Selmer groups have

the minimal sizes. This last condition is then checked on the basis of the oddness of the suitable graphs. The reader can learn these kinds of various facts about elliptic curves and graph theory.

3. Y. Kitaoka, "Distribution of Units of an Algebraic Number Field Modulo an Ideal" is a massive work expounding the author's new investigation on the distribution of units modulo an ideal. Here the author combines algebraic structures with an analytic output, i.e. the density etc. and should give rise to a new fertile uncultivated land for the coming younger generation. Uncultivated because, compared with extensive study on other ingredients of an algebraic number field like class numbers, the units have not been paid much attention. Not only the field is promising but the problem setting will be very beneficial to younger scientists: Setting a problem in algebraic aspects and incorporate analytic tools to arrive at some statistical data. In the paper the reader can learn really all notions and tools in algebraic number theory, the Frobenius automorphism, Hilbert's ramification theory, Galois action, the Čebotarëv density theorem, and the Artin conjecture.

4. W. Kohnen, "Sign Changes of Fourier Coefficients and Eigenvalues of Cusp Forms". In the paper the recent results are summarized on the sign change problem of the Fourier coefficients $a(n)$ of cusp forms f (Siegel modular forms). The Theorem in §1 about infinitely many sign changes is proved using the Hecke L-series for f and the Rankin-Selberg zeta-function, which motivates the study of the first occurrence of the sign change. For a normalized Hecke eigenform that is a new form of even integral weight k and level N, sharp bounds for the occurrence of the first sign change is obtained, with or without the symmetric square L-function and the Hecke relations, (sub-)convexity bounds, prime number theorem, etc. The same ideas give results including sign changes in short intervals. Sign changes for Siegel modular forms of genus 2 are also summarized. Infinitely many sign changes may not occur in general, yet there are theorems which have similar flavor as those for the elliptic modular case. The proof uses the spinor zeta-function in place of Hecke L-series.

5. Y. -K. Lau, J. -Y. Liu and Y. -B. Ye, "Shifted Convolution Sums of Fourier Coefficients of Cusp Forms". Let α denote the infimum of the exponent of t in the estimation of the Riemann zeta-function on the critical line $\sigma = 1/2$. The GLH (Generalized Lindelöf Hypothesis) which follows from the GRH (Generalized Riemann Hypothesis) implies $\alpha = 0$. The convexity bound which is obtained with the aid of the Phragmén-Lindelöf convexity principle is $\alpha = 1/4$ and the authors call any improvement on the convex-

ity bound a subconvexity bound, while the bound $\alpha = 1/6$ which is 2/3-rd power of convexity and is due to Weyl, is called a Weyl-like bound. The paper is concerned with summarizing the research made hitherto toward these subconvexity and Weyl-like bounds for automorphic L-functions. The authors refer to the L-functions of degree n according to the degree of the generic polynomial factor in p^{-s} in their Euler product and after mentioning the degree 2 case, they present their newest Weyl-like bound for the Rankin-Selberg L-function $L(1/2+it, f\times g)$ formed from f, a holomorphic Hecke eigenform for $\Gamma_0(N)$ of weight k (or a Maass Hecke eigenform for $\Gamma_0(N)$ with Laplace eigenvalue $1/4 + k^2$) and g a fixed holomorphic or Maass cusp form for $\Gamma_0(N)$, or for $\Gamma_0(N')$ with $(N, N') = 1$. The Weyl-like bound obtained is 2/3 in the weight aspect, which is expounded in the authors' most recent article of 78 pages. The method hinges on the meromorphic continuation of the shifted convolution sum formed from the Fourier coefficients. In the paper, the intermediate developments toward the above bound are explained in some detail and the reader can learn a quite rich mixture of many new methods which have their origin in analytic number theory and spectral theory.

6. K. Miyake, "Two Expositions on Arithmetic of Cubics". The paper consists of two parts. Part I is devoted to the study of cubic generic polynomials $R = R(t;x)$ and $Q = Q(s;x)$ with parameters t and s for the symmetric group of degree 3 and the cyclic group of order 3, respectively. As an application, the divisibility of the class numbers of quadratic fields by 3 is obtained. For the algebraic tools used, the reader is referred to the recent books by G. Gras or J. Neukirch. Thus Part I motivates the study of cubic and more generally, non-abelian fields. In Part II, two types of families of elliptic curves are considered whose sets of rational points over $\mathbb{Q}$ are described by certain subsets of cubic fields. One type is given by $u^3 = u^3 + au^2 + bu + c$ whose short forms are Mordell curves, while the other family consists of the twists of Hessian family of elliptic curves over the splitting field of the cubic polynomial R and also over the quadratic field contained in the splitting field. Part II therefore introduces the reader to the world of arithmetic on elliptic curves. Basic knowledge is to be found in the textbooks of J. Silverman and of Silverman and Tate.

7. I. Shparlinski, "Distribution of Points on Modular Hyperbolas". For a positive integer m and an arbitrary integer a with $\gcd(a, m) = 1$, the author defines the modular hyperbola as $xy \equiv a \pmod m$ and designates all the points (x, y) on it by $\mathcal{H}_{a,m}$. The author considers the distribution and geometric properties of points of $\mathcal{H}_{a,m}$, denoted $\mathcal{H}_{a,m}(\mathcal{X}, \mathcal{Y})$ whose coordinates

x and y lie in prescribed sets of integers $\mathcal{X}$ and $\mathcal{Y}$, respectively. Acquiring precise asymptotic formulas and establishing positivity of $\#\mathcal{H}_{a,m}(\mathcal{X},\mathcal{Y})$ for various interesting sets have been the central theme in this area. However, as the author writes, "there is a large number of papers which rather routinely study various problems related to $\mathcal{H}_{a,m}$ on the case by case basis. Here, we explain some standard principles which can be used to derive these and many other results of similar spirit about the points on $\mathcal{H}_{a,m}$ as simple corollaries of just one general result about the uniformity of distribution of point on $\mathcal{H}_{a,m}$ in certain domains. In §3.1, this result, Theorem 10, is derived in a very straightforward fashion from (1.1)—the known bound of Kloosterman sums—using some standard arguments", and we learn one important lesson that if one keeps doing routine work, eventually all the result obtained will be relinquished into history's dustbox and will not be referred to as illustrating examples in the subsequent publications. §§2.1–2.3 give a nice survey on a diversity of methods used to study the point on $\mathcal{H}_{a,m}$, and especially, it is remarked that multiplicative character sums can sometimes give better results than Kloosterman sums.

Various applications are described as well. In particular in §5, one can find some interesting results, especially an unexpected result on torsion of elliptic curves in §5.4.

8. Z. -W. Sun, "A Survey of Problems and Results on Restricted Sumsets". The paper gives a useful survey on recent results in additive number theory, now being exciting, related to combinatorics. A restricted sumset has the form $\{a_1 + \cdots + a_n : \ a_1 \in A_1, \ldots, a_n \in A_n, \ P(a_1, ..., a_n) \neq 0\}$, where $A_1, \ldots, A_n$ are finite subsets of $\mathbb{Z}$, or a field or an abelian group, and P is a suitable polynomial. The book "Additive Combinatorics" (2006) by T. Tao and V. Vu, mainly summarizes important results on sumsets without restrictions, obtained by various different tools.

The author is concerned with the lower bounds for cardinalities of various sumsets obtained by algebraic methods. §1 gives rather enlightening discussion on the powerful "polynomial method" via Alon's "Combinatorial Nullstellensatz" the proof of which is given and which implies the Alon-Nathanson-Ruzsa theorem, generalizing Dias da Silva and Hamidoune's extension of the Erdős-Heilbronn conjecture. In the remaining sections one can find recent developments by the author and his school. The survey also contains some open conjectures.

9. H. Tsukada, "A General Modular Relation in Analytic Number Theory" presents equivalent forms of the functional equation with multiple gamma factors satisfied by two sets of Dirichlet series, in the form of mod-

ular relation between the corresponding two sets of H-function series, where H-function means the Fox H-function. As far as we have checked, there is no formula which does not come under this big umbral theorem. In the paper, there are some illustrating examples with multiple gamma factors. Hopefully, the theorem covers also those zeta-functions studied in the papers of Kohnen and Lau-Liu-Ye.

10. D. -Q. Wan, "L-functions of Function Fields". Let $\mathbb{F}_q$ denote a finite field of q elements with characteristic $p > 0$. As a generalization of the case of C= the projective plane, U= projective plane with the origin and infinity removed and $K = \mathbb{F}_q(t)$, the function field, the following general situation is considered: Let C denote a smooth projective geometrically connected curve defined over $\mathbb{F}_q$ with function field K, with its absolute Galois group denoted by G_K and let $j : U \to C$ be a Zariski open dense subset of C.

Let F_ℓ be a finite extension of $\mathbb{Q}_\ell$, ℓ being a prime number. With V a finite dimensional vector space over $\mathbb{F}_\ell$, let $\rho : G_K \to GL(V)$ be a continuous representation of G_K unramified on U. There are two L-functions introduced, the L-function $L(U, \rho)$ and the complete L-function of ρ on C, $L(C, \rho)$ which differ by a finite number of Euler factors. The paper gives a concise survey of recent results on analyticity of these L-functions, especially, for those representations arising from geometry. It is stated that in the ℓ-adic case, with $\ell \neq p$, all ℓ-adic representations are essentially geometric from L-function point of view including ℓ-adic function field analogue of Artin's entireness conjecture. In the more complicated $\ell = p$ case, the location of zeros and poles on the compact unit disc is determined and the author's result says that if the p-adic representation is geometric, the L-function $L(U, \rho)$ is p-adic meromorphic everywhere. The interested reader can go on by reading the references given.

All in all, the interested reader can really get benefited by going through this volume and we hope we will keep doing this work of putting together the recent developments in handy volumes.

The editors would like to express their hearty thanks to Professor Haruo Tsukada, Dr. Jing Ma and Ms. Nan Luo for their devoted help toward the completion of this volume. Dr. Jing Ma, especially, spent an immense amount of time to edit the final versions of all the papers; without her help, the volume could not have been completed so well and in time. Professor Haruo Tsukada kindly checked the final version of the manuscript and made essential improvement therein to whom the editors would like to express their hearty gratitude. Last but not least, thanks are due to the editor of the World Scientific, Mrs. Ji Zhang, for her kind and timely help throughout the preparation of the proceedings.

As usual, we end up with recording a poem describing our feeling, and this time it is the following.

威海連天碧
宇內盡光華
把酒歌聚散
一躬向天涯

The fifth China-Japan Seminar will be held in Osaka, at Kinki University and we do hope we'll meet there again. Wo men xi wang zai jian dao ni men.

Jin Gunagzi and Liu Jianya—the editors-organizers

PROGRAM

Wednesday, August 30, 2006

9:00–9:40 The Organizers,
Opening Address

Morning Session (Chair: Jianya Liu)
10:00–10:40 Andrzej Schinzel,
The Number of Solutions in a Box of a Linear Homogeneous Congruence
11:00–11:40 Shou-Wu Zhang,
Periods Integrals and Special Values of L-series

Afternoon Session (Chair: Krishnaswami Alladi)
14:30–15:00 Zhi-Wei Sun,
Curious Identities and Congruences Involving Bernoulli Polynomials
15:20–15:40 Shigeki Akiyama,
Rational Based Number System and Mahler's Problem
15:50–16:00 Masayuki Toda,
On Gauss' Formula for ψ and Finite Expressions for the L-series at 1
16:20–16:40 Xiumin Ren,
Estimates of Exponential Sums over Primes and Applications in Number Theory
16:50–17:05 Haruo Tsukada,
A General Modular Relation Associated with the Riemann Zeta-function
17:10–17:20 Takako Kuzumaki Kobayashi,
A Transformation Formula for Certain Lambert Series

Thursday, August 31, 2006

Morning Session (Chair: Vladimir N. Chubarikov)
9:00–9:40 Krishnaswami Alladi,
New Approaches to Jacobi's Triple Product Identity and a Quadruple Product Extension
10:00–10:40 Trevor D. Wooley,
Waring's Problem in Function Fields
11:00–11:40 Yangbo Ye,
A New Bound for Rankin-Selberg L-functions

Afternoon Session (Chair: Yoshio Tanigawa)
14:30–15:00 Wenpeng Zhang
The Mean Value of Dedekind Sum and Cochrane Sum in Short Intervals
15:20–15:50 Isao Wakabayashi,
Some Thue Equations and Continued Fractions
16:10–16:30 Yuk-Kam Lau,
The Error Terms in Dirichlet's Divisor Problem, the Circle Problem and the Mean Square Formula of the Riemann Zeta-function
16:40–17:00 Wenguang Zhai,
On the Fourth Power Moments of $\Delta(x)$ and $E(t)$
17:10–17:20 Tianxin Cai,
A Generalization of a Curious Congruence of Harmonic Numbers
17:40–17:50 Kentaro Ihara,
On the Structure of Algebra of Multiple Zeta Values

Saturday, September 2, 2006

Morning Session (Chair: Kohji Matsumoto)
9:00–9:40 Vladimir N. Chubarikov,
Trigonometric Sums in Number Theory, Analysis and Probability Theory
10:00–10:40 Winfried Kohnen,
Sign Changes of Fourier Coefficients and Hecke Eigenvalues of Cusp Forms
11:00–11:40 Igor Shparlinski,
Exponential and Character Sums with Combinatorial Sequences

Afternoon Session (Chair: Wenpeng Zhang)
14:30–15:00 Chaohua Jia,
On a Conjecture of Yiming Long
15:10–15:40 Koichi Kawada,
On the Sum of Five Cubes of Primes
15:50–16:10 Honggang Xia,
On Zeros of Cubic L-functions
16:20–16:40 Zhiguo Liu,
A Theta Function Identity to the Quintic Base
16:50–17:05 Masaki Sudo,
On the Exponential Equations $a^x - b^y = c$ $(1 \leq c \leq 300)$
17:25–17:40 Yoshinobu Nakai,
On a Function in the Biquadratic Theta-Weyl Sums
17:45–17:55 Deyu Zhang,
Zero Density Estimates for Automorphic L-functions

Sunday, September 3, 2006

Morning Session (Chair: Winfried Kohnen)
9:00–9:40 Daqing Wan,
L-functions of Infinite Symmetric Powers
10:00–10:30 Yoshiyuki Kitaoka,
Distribution of Units of an Algebraic Number Field
10:50–11:20 Kohji Matsumoto,
The Riesz Mean of the Convolution Product of Von Mangoldt Functions and the Related Zeta-function
11:30–12:00 Katsuya Miyake,
Twists of Hessian Elliptic Curves and Cubic Fields

Afternoon Session (Chair: Shigeru Kanemitsu)
14:30–15:00 Leo Murata,
On a Property of the Multiplicative Order of a (mod p)
15:20–15:50 Yonggao Chen,
On the Prime Power Factorization of $n!$
16:00–16:20 Yoshio Tanigawa,
Kronecker's Limit Formula and the Hypergeometric Function
16:25–16:35 Guangshi Lü,
Some Results in Classical Analytic Number Theory
16:40–16:50 Huaning Liu,
Mean Value of Dirichlet L-functions and Applications to Pseudorandom Binary Sequences in Cryptography
16:55–17:05 Hailong Li,
The Structural Elucidation of Eisenstein's Formula

Advisory Committee:
Professor Tao Zhan, Shandong University, China

Organizing Committee:
Professor Shigeru Kanemitsu, Kinki University, Japan
Professor Jianya Liu, Shandong University, China

Guest Speakers:
Professor Krishnaswami Alladi, University of Florida, USA
Professor Vladimir N. Chubarikov, Moscow State University, Russia
Professor Winfried Kohnen, Universitaet Heidelberg, Germany
Professor Andrzej Schinzel, Polish Academy of Sciences, Poland
Professor Igor Shparlinski, Macquarie University, Australia
Professor Daqing Wan, University of California, USA
Professor Trevor D. Wooley, University of Michigan, USA
Professor Yangbo Ye, University of Iowa, USA
Professor Shou-Wu Zhang, Columbia University, USA

Peripatetic Mathemagicians

Speakers:
Professor Shigeki Akiyama, Niigata University, Japan
Professor Tianxin Cai, Zhejiang University, China
Professor Yonggao Chen, Nanjing Normal University, China
Professor Chaohua Jia, The Chinese Academy of Sciences, China
Professor Koichi Kawada, Iwate University, Japan
Professor Yoshiyuki Kitaoka, Meijo University, Japan
Professor Yuk-Kam Lau, The University of Hong Kong, Hong Kong
Professor Zhiguo Liu, East China Normal University, China
Professor Kohji Matsumoto, Nagoya University, Japan
Professor Katsuya Miyake, Waseda University, Japan
Professor Leo Murata, Meiji-Gakuin University, Japan
Professor Yoshinobu Nakai, Yamanashi University, Japan
Professor Xiumin Ren, Shandong University, China
Professor Masaki Sudo, Seikei University, Japan
Professor Zhi-Wei Sun, Nanjing University, China
Professor Yoshio Tanigawa, Nagoya University, Japan
Professor Haruo Tsukada, Kinki University, Japan
Professor Isao Wakabayashi, Seikei University, Japan
Professor Honggang Xia, Shandong University, China
Professor Wenguang Zhai, Shandong Normal University, China
Professor Wenpeng Zhang, Northwest University, China

A Dead Heat Match

Short Communications:
Dr. Kentaro Ihara, Kinki University, Japan
Professor Takako Kuzumaki Kobayashi, Gifu University, Japan
Professor Hailong Li, Weinan Teacher's College, China
Dr. Huaning Liu, Northwest University, China
Professor Guangshi Lü, Shandong University, China
Mr. Masayuki Toda, Kinki University, Japan
Dr. Deyu Zhang, Shandong University, China

The Fourth China-Japan Conference on Number Theory

CONTENTS

CONVOLUTIONS OF THE VON MANGOLDT FUNCTION AND RELATED DIRICHLET SERIES

SHIGEKI EGAMI

Faculty of Engineering, Toyama University,
Gofuku, Toyama 930-8555, Japan
E-mail: megami@eng.u-toyama.ac.jp

KOHJI MATSUMOTO

Graduate School of Mathematics, Nagoya University,
Chikusa-ku, Nagoya 464-8602, Japan
E-mail: kohjimat@math.nagoya-u.ac.jp

In this paper, we first give a brief survey on the theory of meromorphic continuation and natural boundaries of multiple Dirichlet series. Then we consider the double Dirichlet series $\Phi_2(s)$ defined by the convolution of logarithmic derivatives of the Riemann zeta-function. Especially we propose the conjecture that $\Phi_2(s)$ would have the natural boundary on $\Re s = 1$, and give a supportive evidence. We further present an application of $\Phi_2(s)$ to the Riesz mean, and discuss its multiple analogues.

1. The analytic continuation of multiple Dirichlet series

Let $s = \sigma + it$ be a complex variable, and $P(X_1, \ldots, X_r)$ a polynomial of complex coefficients. The multiple zeta-function

$$\zeta_r(s; P) = \sum_{m_1=1}^{\infty} \cdots \sum_{m_r=1}^{\infty} P(m_1, \ldots, m_r)^{-s} \tag{1.1}$$

was first studied by Mellin [29,30], and independently by Barnes [5,6] for P a linear form, at the beginning of the 20th century. Mellin proved the meromorphic continuation of (1.1) to the whole complex plane $\mathbb{C}$ if all the coefficients of P have positive real parts. Several mathematicians after Mellin proved the meromorphic continuation of (1.1) under weaker assumptions. At present, the assumption (H_0S) introduced by Essouabri [12] is the weakest. Essouabri [11] also pointed out that the multi-variable generalization

$$\zeta_r(s_1,\ldots,s_n;P_1,\ldots,P_n)=\sum_{m_1=1}^{\infty}\cdots\sum_{m_r=1}^{\infty}P_1(m_1,\ldots,m_r)^{-s_1}\times\cdots\times P_n(m_1,\ldots,m_r)^{-s_n} \tag{1.2}$$

of (1.1), where $s_1,\ldots,s_n\in\mathbb{C}$ and $P_1,\ldots,P_n\in\mathbb{C}[X_1,\ldots,X_r]$, can be continued meromorphically to the whole space $\mathbb{C}^n$ under the same type of assumption.

A special type of multi-variable multiple series

$$\zeta_{EZ,r}(s_1,\ldots,s_r)=\sum_{m_1=1}^{\infty}\cdots\sum_{m_r=1}^{\infty}m_1^{-s_1}(m_1+m_2)^{-s_2}\times\cdots\times(m_1+\cdots+m_r)^{-s_r}, \tag{1.3}$$

which is called the Euler-Zagier r-fold sum, has been studied extensively in recent years. The meromorphic continuation of (1.3) to $\mathbb{C}^r$ is included in the above theorem of Essouabri [11], but [11] is unpublished. Various different proofs of the continuation were published by Arakawa and Kaneko [3], Zhao [37], Akiyama, Egami and Tanigawa [1], and the second-named author [27]. The method in [27] is based on the Mellin-Barnes integral formula

$$(1+\lambda)^{-s}=\frac{1}{2\pi i}\int_{(c)}\frac{\Gamma(s-z)\Gamma(z)}{\Gamma(s)}\lambda^{-z}dz \tag{1.4}$$

(where $s,\lambda\in\mathbb{C}$, $\lambda\neq 0$, $|\arg\lambda|<\pi$, $\Re s>0$, $0<c<\Re s$, and the path of integration is the vertical line from $c-i\infty$ to $c+i\infty$), which was already used in Mellin's papers [29,30].

For arithmetical applications, it is important to consider various multiple Dirichlet series with arithmetical coefficients. Peter [32] discussed the analytic continuation of the series

$$\sum_{m_1=1}^{\infty}\cdots\sum_{m_r=1}^{\infty}\frac{a_1(m_1)\cdots a_r(m_r)}{P(m_1,\ldots,m_r)^s}, \tag{1.5}$$

where $a_k(m_k)$ $(1\leq k\leq r)$ are complex numbers. Actually he treated the more general situation that $P(m_1,\ldots,m_r)$ in the denominator is replaced by $P(\lambda_1(m_1),\ldots,\lambda_r(m_r))$, where $\lambda_k(m)$ are complex numbers in a certain fixed cone on $\mathbb{C}$ satisfying $\lim_{m\to\infty}|\lambda_k(m)|=\infty$ $(1\leq k\leq r)$. The multi-variable series

$$\sum_{m_1=1}^{\infty}\cdots\sum_{m_r=1}^{\infty}\frac{f(m_1,\ldots,m_r)}{m_1^{s_1}\cdots m_r^{s_r}}, \tag{1.6}$$

where $f(m_1, \ldots, m_r)$ is a non-negative arithmetical function, was studied by de la Bretèche [8].

In connection with sums of the Euler-Zagier type, multiple L-series defined by twisting (1.3) by Dirichlet characters have been investigated by Goncharov [19], Arakawa and Kaneko [3,4], Akiyama and Ishikawa [2], and Ishikawa [21,22].

More generally, we may claim that if Dirichlet series

$$\varphi_k(s) = \sum_{m=1}^{\infty} \frac{a_k(m)}{m^s}, \qquad 1 \leq k \leq r \tag{1.7}$$

behave nicely, then we can show that the multiple Dirichlet series of the form

$$\begin{aligned}\Phi_r(s_1, \ldots, s_r; \varphi_1, \ldots, \varphi_r) = \sum_{m_1=1}^{\infty} \cdots \sum_{m_r=1}^{\infty} \frac{a_1(m_1)}{m_1^{s_1}} \frac{a_2(m_2)}{(m_1+m_2)^{s_2}} \\ \times \cdots \times \frac{a_r(m_r)}{(m_1+\cdots+m_r)^{s_r}}\end{aligned} \tag{1.8}$$

also behaves nicely. In fact, the following theorem was proved in Matsumoto and Tanigawa [28].

Theorem 1.1 ([28]). *Assume that $\varphi_k(s)$ $(1 \leq k \leq r)$ are absolutely convergent for $\sigma > \alpha_k (> 0)$, can be continued meromorphically to the whole plane $\mathbb{C}$, holomorphic except for a possible pole (of order at most 1) at $s = \alpha_k$, and of polynomial order in any fixed strip $\sigma_1 \leq \sigma \leq \sigma_2$. Then $\Phi_r(s_1, \ldots, s_r; \varphi_1, \ldots, \varphi_r)$ can be continued meromorphically to the whole space $\mathbb{C}^r$, and the location of its possible singularities can be described explicitly. In particular, if all $\varphi_k(s)$ are entire, then $\Phi_r(s_1, \ldots, s_r; \varphi_1, \ldots, \varphi_r)$ is also entire.*

The proof of the above theorem is an analogue of the second-named author's proof of the meromorphic continuation of (1.3) given in [27], whose basic tool is the Mellin-Barnes formula (1.4). The idea of applying formula (1.4) in such a situation had been already mentioned by the first-named author [10] in the one-variable case.

The authors express their sincere gratitude to Professor Gautami Bhowmik for pointing out an error in the original manuscript, and useful suggestions. In particular, the form of Conjecture (B) below was first suggested by her.

2. An example of double Dirichlet series with a natural boundary

In Theorem 1.1, there is the condition that each $\varphi_k(s)$ is holomorphic except for only one possible pole. Actually it is possible to prove a result of similar type under the weaker condition that each $\varphi_k(s)$ has finitely many poles.

However, if some of $\varphi_k(s)$ has infinitely many poles, the behaviour of the multiple series $\Phi_r(s_1,\ldots,s_r;\varphi_1,\ldots,\varphi_r)$ may be quite different. The following simple example illustrates this phenomenon. Let $\Lambda(n)$ be the von Mangoldt function, and

$$M(s) = -\frac{\zeta'}{\zeta}(s) = \sum_{n=1}^{\infty}\frac{\Lambda(n)}{n^s}, \tag{2.1}$$

where $\zeta(s)$ is the Riemann zeta-function. Then $M(s)$ is meromorphic in the whole plane, and has infinitely many poles because all zeros of $\zeta(s)$ are the poles of $M(s)$. In fact it is known that

$$N(T) \sim \frac{1}{2\pi}T\log T, \qquad T \geq 2 \tag{2.2}$$

(Theorem 9.4 of Titchmarsh [36]), where $N(T)$ is the number of zeros (counted with multiplicity) of $\zeta(s)$ in the region $0 < \sigma < 1$, $0 < t \leq T$, which is expected to be equal to the number of poles of $M(s)$ in the same region because all zeros of $\zeta(s)$ are conjectured to be simple.

Let

$$\Phi_2(s) = \Phi_2(0,s;M,M) = \sum_{k=1}^{\infty}\sum_{m=1}^{\infty}\frac{\Lambda(k)\Lambda(m)}{(k+m)^s}. \tag{2.3}$$

This can be rewritten as

$$\Phi_2(s) = \sum_{n=1}^{\infty}\frac{G_2(n)}{n^s}, \tag{2.4}$$

where

$$G_2(n) = \sum_{k+m=n}\Lambda(k)\Lambda(m). \tag{2.5}$$

The series on the right-hand side of (2.3), (2.4) is absolutely convergent for $\Re s > 2$, because

$$G_2(n) \leq \sum_{k=1}^{n-1}\log k\cdot\log(n-k) \leq n(\log n)^2. \tag{2.6}$$

In the present paper we will show, under the assumption of certain conjectures, that $\Phi_2(s)$ has the natural boundary on the line $\Re s = 1$ (Theorem

2.2 below). Therefore it seems that the behaviour of $\Phi_2(s)$ is completely different from that of multiple series studied in [28].

The history of the investigation of natural boundaries of Dirichlet series also goes back to the beginning of the 20th century. The analytic continuation and the natural boundary of the function $\sum_p p^{-s}$ (p runs over primes) were studied by Kluyver [23], Landau [25], and Landau and Walfisz [26]. In 1928, Estermann published two papers [13], [14] on natural boundaries of Dirichlet series. In the former paper [13], he considered a certain class of Dirichlet series which have Euler products, and gave a criterion when the series can be continued to the whole plane and when it has the natural boundary. The continuation and natural boundaries of Euler products were further studied in more general situations by several mathematicians such as Dahlquist [9], Kurokawa [24]. A multi-variable generalization was recently discussed by Bhowmik, Essouabri and Lichtin [7].

The results in the present paper give a different direction of research on natural boundaries of Dirichlet series. A part of the present work was already announced on the occasion of a conference on number theory (in honour of Professor Akio Fujii) held at Rikkyo University, Tokyo, in January 2005. On the other hand, independently of the present work, Tanigawa and Zhai [35] have considered Dirichlet series which are more general than ours, and have discussed the same type of problems (except for the Riesz mean). Their proof of the claim on natural boundaries (Theorem 1.3 of [35]) seems incomplete; some condition similar to our (B) below seems to be necessary to verify their argument.

We mention here the number-theoretic motivation of the study of $\Phi_2(s)$. The function $G_2(n)$ defined by (2.5) is a classical subject matter of number theory, because it is connected with the famous conjecture of C. Goldbach (that is, any even integer (≥ 4) can be expressed as a sum of two primes); in fact, the conjecture implies that $G_2(n) > 0$ for all even $n \geq 4$. Fujii [15] studied the mean value of $G_2(n)$ and proved that, if we assume the Riemann hypothesis (RH) for $\zeta(s)$, then

$$\sum_{n \leq X} G_2(n) = \frac{1}{2}X^2 + O(X^{3/2}) \tag{2.7}$$

for any large positive X. In [16], Fujii improved his result to obtain

$$\sum_{n \leq X} G_2(n) = \frac{1}{2}X^2 - H(X) + O((X \log X)^{4/3}) \tag{2.8}$$

under RH. Here

$$H(X) = 2\sum_{\rho} \frac{X^{1+\rho}}{\rho(1+\rho)},$$

where ρ runs over the non-trivial zeros of $\zeta(s)$, counted with multiplicity.

From the work [20] of Hardy and Littlewood it is expected that $G_2(n)$ for even n is approximated by $nS_2(n)$, where

$$S_2(n) = \prod_{p|n} \left(1 + \frac{1}{p-1}\right) \prod_{(p,n)=1} \left(1 - \frac{1}{(p-1)^2}\right). \tag{2.9}$$

Moreover it follows from Lemma 1 of Montgomery and Vaughan [31] that

$$\sum_{n \le X} nS_2(n) = \frac{1}{2}X^2 + O(X \log X). \tag{2.10}$$

From this viewpoint, Fujii [16] reformulated his formula (2.8) into

$$\sum_{n \le X} (G_2(n) - nS_2(n)) = -H(X) + O((X \log X)^{4/3}). \tag{2.11}$$

Hence the term $H(X)$ represents the main oscillation in the above formulation of Goldbach's problem. Some properties of $H(X)$ have been studied in Fujii [17].

By (2.4) and Perron's formula we have

$$\sum_{n \le X} G_2(n) = \frac{1}{2\pi i} \int_{c-iT}^{c+iT} \Phi_2(s) \frac{X^s}{s} ds + O(T^{-1}X^{2+\varepsilon}) \tag{2.12}$$

with $c > 2$. Therefore the study of $\Phi_2(s)$ will be useful to understand the behaviour of $G_2(n)$. In the next section we will prove the following:

Theorem 2.1 (under RH). *The function $\Phi_2(s)$ can be continued meromorphically to the half-plane $\Re s > 1$, and holomorphic except for the simple poles at $s = 2$ (with residue 1) and $s = 1 + \rho$ (with residue $-2n(\rho)/\rho$) for any non-trivial zero ρ of $\zeta(s)$, where $n(\rho)$ is the multiplicity of ρ.*

By this theorem, we can shift (under RH) the path of integration on the right-hand side of (2.12) to $\Re s = 1 + \varepsilon$. We encounter the poles $s = 2$ and $s = 1+\rho$, and the sum of their residues is $(1/2)X^2 - H(X)$, which coincides with the explicit terms on the right-hand side of (2.8). In particular, we find that the properties of $H(X)$ are closely connected with the behaviour of $\Phi_2(s)$ on the line $\Re s = 3/2$.

Next we consider the behaviour of $\Phi_2(s)$ on the line $\Re s = 1$. We propose the following:

Conjecture 2.1. *The line $\Re s = 1$ is the natural boundary of $\Phi_2(s)$.*

In the present paper we will show an evidence which supports the above conjecture. Let $\mathcal{I}$ be the set of all imaginary parts of non-trivial zeros of $\zeta(s)$. A well-known conjecture speculates that the positive elements of $\mathcal{I}$ would be linearly independent over the rationals. The following statement is a special case of this conjecture:

(A) If $\gamma_j \in \mathcal{I}$ $(1 \le j \le 4)$ and $\gamma_1 + \gamma_2 = \gamma_3 + \gamma_4 (\neq 0)$, then (γ_3, γ_4) equals (γ_1, γ_2) or (γ_2, γ_1).

These conjectures were mentioned on p.50 of Fujii [18]. In that paper Fujii made an extensive study on additive properties of the zeros of $\zeta(s)$. For instance he proved that the set

$$\{\gamma_1 + \gamma_2 \mid \gamma_1, \gamma_2 \in \mathcal{I}, \gamma_1 > 0, \gamma_2 > 0\}$$

is uniformly distributed mod 1 (Corollary 3 of [18]).

Here we introduce the following quantitative version of (A):

(B) There exists a constant α, with $0 < \alpha < \pi/2$, such that if $\gamma_j \in \mathcal{I}$ $(1 \le j \le 4)$, $\gamma_1 + \gamma_2 \neq 0$, and (γ_3, γ_4) is neither equal to (γ_1, γ_2) nor to (γ_2, γ_1), then

$$|(\gamma_1 + \gamma_2) - (\gamma_3 + \gamma_4)| \ge \exp\left(-\alpha(|\gamma_1| + |\gamma_2| + |\gamma_3| + |\gamma_4|)\right). \tag{2.13}$$

Clearly (B) implies (A).

In §4 of the present paper we will prove that, under RH, the set

$$\mathcal{K} = \{\kappa \mid \kappa = \gamma_1 + \gamma_2 \ \text{for some} \ \gamma_1, \gamma_2 \in \mathcal{I}\} \setminus \{0\} \tag{2.14}$$

is dense in the whole set of real numbers $\mathbb{R}$. This result will yield the following theorem.

Theorem 2.2 (under RH). *If we assume that* (B) *is true, then Conjecture* 1 *is true.*

Hence the continuation achieved by Theorem 1.1 seems to be best-possible. It is therefore not rash to propose the following

Conjecture 2.2. *The error term on the right-hand side of* (2.8) *is to be $O(X^{1+\varepsilon})$ and $\Omega(X)$, where $\Omega(X)$ means that it is not $o(X)$.*

3. Proof of Theorem 2.1

In this section we prove Theorem 2.1. First we assume $\Re s > 2 + 2\varepsilon$. Then we have

$$\begin{aligned}\Phi_2(s) =&\sum_{k=1}^{\infty}\sum_{m=1}^{\infty}\frac{\Lambda(k)\Lambda(m)}{(k+m)^s}\\ =&\sum_{k=1}^{\infty}\sum_{m=1}^{\infty}\frac{\Lambda(k)\Lambda(m)}{k^s}\left(1+\frac{m}{k}\right)^{-s}.\end{aligned}\tag{3.1}$$

We apply the Mellin-Barnes formula (1.4) with $\lambda = m/k$ to (3.1) to obtain

$$\begin{aligned}\Phi_2(s) =&\sum_{k=1}^{\infty}\sum_{m=1}^{\infty}\frac{\Lambda(k)\Lambda(m)}{k^s}\frac{1}{2\pi i}\int_{(c)}\frac{\Gamma(s-z)\Gamma(z)}{\Gamma(s)}\left(\frac{m}{k}\right)^{-z}dz\\ =&\frac{1}{2\pi i}\int_{(c)}\frac{\Gamma(s-z)\Gamma(z)}{\Gamma(s)}\sum_{k=1}^{\infty}\Lambda(k)k^{-s+z}\sum_{m=1}^{\infty}\Lambda(m)m^{-z}dz.\end{aligned}\tag{3.2}$$

Two infinite series in the integrand are convergent when $\sigma - c > 1$ and $c > 1$. These conditions, and also the condition $0 < c < \sigma$ (which is necessary to apply (1.4)), are satisfied by the choice $c = 1 + \varepsilon$. Under this choice of c, we have

$$\Phi_2(s) = \frac{1}{2\pi i}\int_{(c)}\frac{\Gamma(s-z)\Gamma(z)}{\Gamma(s)}M(s-z)M(z)dz.\tag{3.3}$$

The next step is to shift the path of integration from $\Re z = c = 1+\varepsilon$ to $\Re z = -\varepsilon$. First we have to show that this shifting is possible. It is known that

$$N(T+1) - N(T) \ll \log T\tag{3.4}$$

for any $T \geq 2$, where $f \ll g$ means $f = O(g)$ (Theorem 9.2 of Titchmarsh [36]). Hence we can find an arbitrarily large T such that

$$|T-\gamma| \gg (\log T)^{-1}\tag{3.5}$$

for any $\gamma \in \mathcal{I}$. Combining (3.5) with the formula

$$M(z) = -\sum_{|y-\gamma|<1}\frac{1}{z-\rho} + O(\log(|y|+2)), \quad y = \Im z, \gamma = \Im \rho,\tag{3.6}$$

which is known to hold uniformly for $-1 \leq x = \Re z \leq 2$ (Theorem 9.6(A) of [36]), we see that, if T satisfies (3.5), then

$$M(z) \ll (\log T)^2\tag{3.7}$$

for $z = x + iT$, $-1 \le x \le 2$. Also, $M(s-z) = O(1)$ for $-\varepsilon \le x \le 1+\varepsilon$. Hence, using Stirling's formula, we have

$$\begin{aligned}\int_{-\varepsilon+iT}^{1+\varepsilon+iT} \frac{\Gamma(s-z)\Gamma(z)}{\Gamma(s)} M(s-z)M(z)dz \ll e^{-(\pi/2)(T+|t-T|-|t|)} \\ \times \frac{(|t-T|+1)^{\sigma-1/2}(\log T)^2}{(|t|+1)^{\sigma-1/2}T^{1/2}} \int_{-\varepsilon}^{1+\varepsilon} (|t-T|+1)^{-x}T^x dx\end{aligned} \tag{3.8}$$

for any T satisfying (3.5), and (3.8) tends to 0 as T tends to infinity. This implies that the above shifting is possible.

In the course of this shifting, we encounter the poles $z = 1$, $z = \rho$ for any non-trivial zero, and $z = 0$. The residues of the integrand at those poles are

$$\frac{M(s-1)}{s-1}, \qquad -n(\rho)\frac{\Gamma(s-\rho)\Gamma(\rho)}{\Gamma(s)} M(s-\rho)$$

and

$$M(s)M(0) = -M(s)\log 2\pi,$$

respectively. Hence we obtain

$$\begin{aligned}\Phi_2(s) = \frac{M(s-1)}{s-1} - \sum_\rho \frac{\Gamma(s-\rho)\Gamma(\rho)}{\Gamma(s)} M(s-\rho) - M(s)\log 2\pi \\ + \frac{1}{2\pi i}\int_{(-\varepsilon)} \frac{\Gamma(s-z)\Gamma(z)}{\Gamma(s)} M(s-z)M(z)dz.\end{aligned} \tag{3.9}$$

Now we continue $\Phi_2(s)$ meromorphically by using (3.9). The first and the third terms on the right-hand side of (3.9) are clearly meromorphic on the whole plane. The poles of the third term coincide with the poles of $M(s)$, which are $s = 1$ and $s = \rho$ (non-trivial zeros). The poles of the first term are $s = 2$, $s = \rho + 1$, and $s = 1$. The residues of the first term at $s = 2$ and $s = \rho + 1$ are 1 and $-n(\rho)/\rho$, respectively.

The integral on the right-hand side of (3.9) is convergent uniformly in any compact subset of the half-plane $\Re s > 1 - \varepsilon$, and hence holomorphic in that half-plane. Actually it is possible to continue this integral meromorphically to the whole plane, by shifting the path further to the left.

The most difficult part is the second term

$$B_2(s) = -\sum_\rho \frac{\Gamma(s-\rho)\Gamma(\rho)}{\Gamma(s)} M(s-\rho). \tag{3.10}$$

The factor $\Gamma(s-\rho)$ has poles at $s = \rho - \ell$ ($\ell = 0, 1, 2, \ldots$), while the factor $M(s-\rho)$ has poles at $s = \rho + 1$ and at $s = \rho + \rho'$, where ρ' denotes the non-trivial zeros of $\zeta(s)$. In order to control this situation, we now assume RH (to the end of this section). Then the only poles of $B_2(s)$ in the region $\Re s > 1$ are $s = \rho + 1$ for non-trivial zeros ρ, and the residue there is

$$-n(\rho)\frac{\Gamma(1)\Gamma(\rho)}{\Gamma(\rho+1)} = -\frac{n(\rho)}{\rho}.$$

These poles are isolated singularities, and hence $B_2(s)$ can be continued to $\Re s > 1$. This implies the meromorphic continuation of $\Phi_2(s)$ to $\Re s > 1$. The residue of $\Phi_2(s)$ at $s = 2$ is 1, and at $s = 1 + \rho$ is

$$-\frac{n(\rho)}{\rho} - \frac{n(\rho)}{\rho} = -\frac{2n(\rho)}{\rho}.$$

Now the proof of Theorem 2.1 is complete.

4. Proof of Theorem 2.2

To prove Theorem 2.2, we use the classical explicit formula

$$M(s) = b + \frac{1}{s-1} + \frac{1}{2}\frac{\Gamma'}{\Gamma}\left(\frac{s}{2}+1\right) - \sum_{\rho}\left(\frac{1}{s-\rho} + \frac{1}{\rho}\right), \tag{4.1}$$

where $b = 1 + (C_0/2) - \log 2\pi$ and C_0 is Euler's constant (formula (2.12.7) of [36]). Substituting this into (3.10), for $\Re s > 1$ we obtain

$$\begin{aligned} B_2(s) &= -\sum_{\rho}\frac{\Gamma(s-\rho)\Gamma(\rho)}{\Gamma(s)}\left\{b + \frac{1}{s-\rho-1} + \frac{1}{2}\frac{\Gamma'}{\Gamma}\left(\frac{s-\rho}{2}+1\right)\right\} \\ &\quad + \sum_{\rho}\sum_{\rho'}\frac{\Gamma(s-\rho)\Gamma(\rho)}{\Gamma(s)}\left(\frac{1}{s-\rho-\rho'} + \frac{1}{\rho'}\right) \\ &= B_{21}(s) + B_{22}(s), \end{aligned} \tag{4.2}$$

say. Clearly $B_{21}(s)$ is meromorphic on the whole plane, and has no pole on the line $\Re s = 1$. To investigate $B_{22}(s)$, we assume RH (to the end of this section), and rewrite $\rho = \rho_1 = 1/2 + i\gamma_1$ and $\rho' = \rho_2 = 1/2 + i\gamma_2$ to obtain

$$B_{22}(s) = \frac{1}{\Gamma(s)}\sum_{\rho_1}\sum_{\rho_2}\frac{\Gamma(s+1-\rho_1)\Gamma(\rho_1)}{(s-\rho_1-\rho_2)\rho_2}, \quad \Re s > 1. \tag{4.3}$$

Therefore $B_{22}(s)$ may behave singularly as s tends to $\rho_1 + \rho_2$, that is, any point of the form $1 + i\kappa$ with $\kappa \in \mathcal{K}$ (where $\mathcal{K}$ is the set defined by (2.14)). Before studying this phenomenon closely, we first prove

Lemma 4.1 (under RH). *The set $\mathcal{K}$ is dense in $\mathbb{R}$.*

Proof. It is classically known that

$$N(T) = \frac{1}{2\pi}T\log T - C_1T + O(\log T), \qquad C_1 = \frac{1+\log 2\pi}{2\pi}$$

(Theorem 9.4 of [36]), and, under RH, the error term in the above formula can be replaced by $O(\log T/\log\log T)$ (Theorem 14.13 of [36]). Therefore, for any fixed $h \in \mathbb{R}$, the number of zeros on the interval $(1/2+iT, 1/2+i(T+h)]$ is

$$\begin{aligned} &\frac{1}{2\pi}(T+h)\log(T+h) - C_1(T+h) \\ &\qquad - \frac{1}{2\pi}T\log T + C_1T + O\left(\frac{\log T}{\log\log T}\right) \\ &= \frac{1}{2\pi}T\left\{\log T + \log\left(1+\frac{h}{T}\right)\right\} + \frac{1}{2\pi}h\left\{\log T + \log\left(1+\frac{h}{T}\right)\right\} \\ &\qquad - C_1h - \frac{1}{2\pi}T\log T + O\left(\frac{\log T}{\log\log T}\right) \\ &= \frac{h}{2\pi}\log T + O\left(\frac{\log T}{\log\log T}\right). \end{aligned} \tag{4.4}$$

There exists a sufficiently large $T_0 = T_0(h)$, such that the right-hand side of (4.4) is positive for any $T \geq T_0$. Let α be any non-zero real number, and ε be arbitrarily small. Then, by using this positivity, we can find a sufficiently large $T = T(\alpha, \varepsilon)$ and $\gamma_1, \gamma_2 \in \mathcal{I}$, satisfying

$$\gamma_1 \in (T+\alpha-\varepsilon/2, T+\alpha+\varepsilon/2], \qquad \gamma_2 \in (-T-\varepsilon/2, -T+\varepsilon/2].$$

Hence $|\alpha-(\gamma_1+\gamma_2)| < \varepsilon$. Moreover, if $\varepsilon < |\alpha|$, then $\gamma_2 \neq -\gamma_1$, so $\gamma_1+\gamma_2 \in \mathcal{K}$. Thus we conclude the assertion of the lemma. □

In view of the above lemma we now know that the points of the form $1+i\kappa$ ($\kappa \in \mathcal{K}$) are dense on the line $\Re s = 1$. Now we assume (B), and prove the following

Lemma 4.2 (under RH and (B)). *For any $\kappa \in \mathcal{K}$, the function $B_{22}(s)$ tends to infinity as s tends to $1+i\kappa$ from the right.*

Proof. By (A) we see that there is only one pair (γ_1^0, γ_2^0) (and its reverse-ordered pair (γ_2^0, γ_1^0)) satisfying $\gamma_1^0 + \gamma_2^0 = \kappa$. Put $\rho_1^0 = (1/2) + i\gamma_1^0$, $\rho_2^0 =$

$(1/2)+i\gamma_2^0$. Then

$$\begin{aligned}B_{22}(s)=\frac{n(\rho_1^0)n(\rho_2^0)}{\Gamma(s)}&\left\{\frac{\Gamma(s+(1/2)-i\gamma_1^0)\Gamma((1/2)+i\gamma_1^0)}{(s-1-i\gamma_1^0-i\gamma_2^0)((1/2)+i\gamma_2^0)}\right.\\&\left.+\frac{\Gamma(s+(1/2)-i\gamma_2^0)\Gamma((1/2)+i\gamma_2^0)}{(s-1-i\gamma_1^0-i\gamma_2^0)((1/2)+i\gamma_1^0)}\right\}\\&+\frac{1}{\Gamma(s)}\sum_{\gamma_1}\sum_{\gamma_2}{}^{*}\frac{\Gamma(s+1-\rho_1)\Gamma(\rho_1)}{(s-\rho_1-\rho_2)\rho_2}\\&=B_{22}^{*}(s)+B_{22}^{**}(s),\end{aligned}$$

say, where the symbol $\sum\sum^*$ means the sum over all (γ_1,γ_2) satisfying $(\gamma_1,\gamma_2)\neq(\gamma_1^0,\gamma_2^0),(\gamma_2^0,\gamma_1^0)$. Then $B_{22}^*(s)$ is meromorphic on the whole plane, and its residue at $s=1+i\kappa=1+i(\gamma_1^0+\gamma_2^0)$ is

$$\begin{aligned}\frac{n(\rho_1^0)n(\rho_2^0)}{\Gamma(1+i\kappa)}&\left\{\frac{\Gamma((3/2)+i(\kappa-\gamma_1^0))\Gamma((1/2)+i\gamma_1^0)}{(1/2)+i\gamma_2^0}\right.\\&\left.+\frac{\Gamma((3/2)+i(\kappa-\gamma_2^0))\Gamma((1/2)+i\gamma_2^0)}{(1/2)+i\gamma_1^0}\right\}\\=\frac{2n(\rho_1^0)n(\rho_2^0)}{\Gamma(1+i\kappa)}&\Gamma(\rho_1^0)\Gamma(\rho_2^0),\end{aligned}\tag{4.5}$$

which does not vanish. That is, $B_{22}^*(s)\to\infty$ as $s\to 1+i\kappa$. Therefore the remaining task is to show that $B_{22}^{**}(s)$ remains finite as $s\to 1+i\kappa$. Putting $s=1+\eta+i\kappa$ ($\eta\geq 0$, small), we have

$$\begin{aligned}B_{22}^{**}(1+\eta+i\kappa)&=\frac{1}{\Gamma(1+\eta+i\kappa)}\\&\times\sum_{\gamma_1}\sum_{\gamma_2}{}^{*}\frac{\Gamma((3/2)+\eta+i(\kappa-\gamma_1))\Gamma((1/2)+i\gamma_1)}{(\eta+i(\kappa-\gamma_1-\gamma_2))((1/2)+i\gamma_2)}.\end{aligned}\tag{4.6}$$

To prove the lemma, it is enough to show that the right-hand side of (4.6) is absolutely convergent, uniformly in η. By using Stirling's formula we have

$$\begin{aligned}B_{22}^{**}(1+\eta+i\kappa)\ll\frac{1}{\Gamma(1+\eta+i\kappa)}&\sum_{\gamma_1}(|\kappa-\gamma_1|+1)^{1+\eta}\\&\times e^{-(\pi/2)(|\kappa-\gamma_1|+|\gamma_1|)}\sum_{\gamma_2}{}^{*}\frac{1}{|\kappa-\gamma_1-\gamma_2|(1+|\gamma_2|)}.\end{aligned}\tag{4.7}$$

The inner sum on the right-hand side of (4.7) can be divided into

$$\sum_{0<|\gamma_2-\lambda|\leq 1}+\sum_{|\gamma_2-\lambda|>1}=\Sigma_1+\Sigma_2,$$

say, where $\lambda = \kappa - \gamma_1$. If $\lambda = 0$, then obviously $\Sigma_2 = O(1)$. If $\lambda > 0$, we divide Σ_2 as

$$\Sigma_2 = \sum_{\gamma_2 > \lambda+1} + \sum_{o<\gamma_2<\lambda-1} + \sum_{\gamma_2<0} = \Sigma_{21} + \Sigma_{22} + \Sigma_{23},$$

say. The sum Σ_{23} is clearly $O(1)$, while by using partial summation and (2.2) we can easily show that Σ_{21}, Σ_{22} are $O(\log(\lambda+1))$. The case $\lambda < 0$ can be treated similarly. The conclusion is that

$$\Sigma_2 = O(\log(|\kappa - \gamma_1| + 1)). \tag{4.8}$$

Next consider Σ_1. Since $0 < |\gamma_2 - \lambda| \leq 1$, we have $1 + |\gamma_2| \gg 1 + |\lambda|$. Hence

$$\Sigma_1 \ll \frac{1}{1+|\lambda|} \sum_{0<|\gamma_2-\lambda|\leq 1} \frac{1}{|\kappa - \gamma_1 - \gamma_2|}.$$

Then, since $\kappa = \gamma_1^0 + \gamma_2^0$, applying assumption (B) we have

$$\begin{aligned}\Sigma_1 &\ll \frac{1}{1+|\lambda|} \sum_{0<|\gamma_2-\lambda|\leq 1} \exp\left(\alpha(|\gamma_1| + |\gamma_2| + |\gamma_1^0| + |\gamma_2^0|)\right) \\ &\ll \frac{1}{1+|\lambda|} \exp\left(\alpha(|\gamma_1| + |\lambda| + |\gamma_1^0| + |\gamma_2^0|)\right) \sum_{0<|\gamma_2-\lambda|\leq 1} 1,\end{aligned}$$

where we have used $|\gamma_2| \leq |\lambda| + 1$. Applying (3.4) we obtain

$$\Sigma_1 \ll \frac{\log(1+|\lambda|)}{1+|\lambda|} \exp\left(\alpha(|\gamma_1| + |\lambda| + |\gamma_1^0| + |\gamma_2^0|)\right). \tag{4.9}$$

Substituting (4.8) and (4.9) into (4.7), we find that

$$\begin{aligned}B_{22}^{**}(1+\eta+i\kappa) \ll \frac{\exp\left(\alpha(|\gamma_1^0| + |\gamma_2^0|)\right)}{\Gamma(1+\eta+i\kappa)} &\sum_{\gamma_1} (|\kappa-\gamma_1|+1)^{1+\eta} \\ \times \log(|\kappa-\gamma_1|+1) \exp&\left(\left(\alpha - \frac{\pi}{2}\right)(|\gamma_1| + |\kappa - \gamma_1|)\right),\end{aligned} \tag{4.10}$$

which is absolutely convergent because $\alpha < \pi/2$. Hence the assertion of Lemma 4.2 follows. □

From Lemma 4.1 and Lemma 4.2 we can conclude that singular points of $\Phi_2(s)$ are distributed densely on the line $\Re s = 1$. In fact, (4.5) implies

$$\Phi_2(s) \sim \frac{2}{\Gamma(1+i\kappa)} n(\rho_1^0) n(\rho_2^0) \Gamma(\rho_1^0) \Gamma(\rho_2^0) \frac{1}{s - (1+i\kappa)} \tag{4.11}$$

as $s \to 1 + i\kappa$. This completes the proof of Theorem 2.2.

5. An application to the Riesz mean

At present the authors have no idea how to prove the desired estimate $O(X^{1+\varepsilon})$ of Conjecture 2.2. Instead, in this section we consider the Riesz mean of $G_2(n)$, that is,

$$\mathcal{D}_{2,a}(X) = \frac{1}{\Gamma(a+1)} \sum_{n \le X} (X-n)^a G_2(n), \tag{5.1}$$

where $a > 0$. The treatment of $\mathcal{D}_{2,a}(X)$ becomes easier when a becomes larger. The aim of this section is to prove the following theorem.

Theorem 5.1 (under RH). *For any $a > 1/2$, the asymptotic formula*

$$\begin{aligned}\mathcal{D}_{2,a}(X) = & \frac{1}{\Gamma(3+a)} X^{2+a} - \sum_{\rho} \frac{2\Gamma(1+\rho)}{\rho\Gamma(2+\rho+a)} X^{1+\rho+a} \\ & + O(X^{1+a+\varepsilon})\end{aligned} \tag{5.2}$$

holds.

If (5.2) would hold for $a = 0$, then it would give (2.8) with the desired error estimate $O(X^{1+\varepsilon})$.

The basic tool for the proof of Theorem 5.1 is the Mellin transformation formula

$$\mathcal{D}_{2,a}(X) = \frac{1}{2\pi i} \int_{(2+\varepsilon)} \frac{\Gamma(s)}{\Gamma(s+a+1)} \Phi_2(s) X^{s+a} ds. \tag{5.3}$$

In order to obtain (5.2), we shift the path of integration on the right-hand side of (5.3) to the left. By Theorem 2.1 we know that $\Phi_2(s)$ can be continued to $\Re s > 1$. Therefore we shift the path to $\Re s = 1 + \varepsilon$. To check that this shifting procedure is possible, we need the following lemma.

Lemma 5.1 (under RH). *There exists an arbitrarily large T for which the estimate*

$$\Phi_2(s) \ll T^{1/2} (\log T)^2 \tag{5.4}$$

holds for $s = \sigma + iT$, $1 + \varepsilon \le \sigma \le 2 + \varepsilon$.

Proof. We estimate each term on the right-hand side of (3.9) for $1+\varepsilon \le \sigma \le 2+\varepsilon$.

First we do not assume RH, and let $T = \Im s$ be any sufficiently large positive number. Since $M(s)$ is absolutely convergent for $\sigma \ge 1+\varepsilon$, we have

$$M(s) \log 2\pi = O(1), \qquad 1+\varepsilon \le \sigma \le 2+\varepsilon. \tag{5.5}$$

Next, denote the integral term on the right-hand side of (3.9) by $I_2(s)$. From (3.6) we have

$$M(-\varepsilon+iy) \ll \sum_{|y-\gamma|<1} 1 + O(\log(|y|+2)) \ll \log(|y|+2), \tag{5.6}$$

where in the second inequality we have used (3.4). It is clear that $M(s-(-\varepsilon+iy)) = O(1)$ for $1+\varepsilon \le \sigma \le 2+\varepsilon$. Using these estimates and Stirling's formula, we find that

$$\begin{aligned} I_2(s) &\ll e^{(\pi/2)T}T^{1/2-\sigma}\int_{-\infty}^{\infty} e^{-(\pi/2)(|y|+|T-y|)} \\ &\qquad \times (|T-y|+1)^{\sigma+\varepsilon-1/2}(|y|+1)^{-\varepsilon-1/2}\log(|y|+1)dy \\ &= e^{(\pi/2)T}T^{1/2-\sigma}\left(\int_{-\infty}^{0}+\int_{0}^{T}+\int_{T}^{\infty}\right) \\ &= e^{(\pi/2)T}T^{1/2-\sigma}(J_1+J_2+J_3), \end{aligned} \tag{5.7}$$

say. By changing y by $-y$, we see that

$$J_1 = e^{-(\pi/2)T}\int_0^{\infty} e^{-\pi y}(T+y+1)^{\sigma+\varepsilon-1/2}(y+1)^{-\varepsilon-1/2}\log(y+1)dy.$$

Divide the last integral into two parts at $y = T$. In the interval $[0,T]$ we use $T+y+1 \asymp T$ to conclude that the integral from 0 to T is $O(T^{\sigma+\varepsilon-1/2})$. The integral from T to ∞ is of exponential decay. Hence $J_1 \ll e^{-(\pi/2)T}T^{\sigma+\varepsilon-1/2}$. As for J_3, changing $y-T$ by y we obtain

$$J_3 = e^{-(\pi/2)T}\int_0^{\infty} e^{-\pi y}(y+1)^{\sigma+\varepsilon-1/2}(T+y+1)^{-\varepsilon-1/2}\log(T+y+1)dy.$$

Dividing the integral at $y = T$ and proceeding similarly to the case of J_1, we find that $J_3 \ll e^{-(\pi/2)T}T^{-\varepsilon-1/2}\log T$. Lastly, dividing at $y = T/2$ we find that

$$\begin{aligned} J_2 &\ll e^{-(\pi/2)T}\left\{\int_0^{T/2} T^{\sigma+\varepsilon-1/2}(y+1)^{-\varepsilon-1/2}\log(y+1)dy\right. \\ &\qquad \left.+\int_{T/2}^{T}(T-y+1)^{\sigma+\varepsilon-1/2}T^{-\varepsilon-1/2}\log T dy\right\} \\ &\ll e^{-(\pi/2)T}T^{\sigma}\log T. \end{aligned}$$

Substituting these estimates into (5.7), we obtain

$$I_2(s) = O(T^{1/2}\log T). \tag{5.8}$$

In order to treat the second term $B_2(s)$ on the right-hand side of (3.9), we now assume RH (to the end of this section). Then from (3.6) we have

$$\begin{aligned} M(s-\rho) &= M\left(\sigma - \frac{1}{2} + I(T-\gamma)\right) \qquad (5.9)\\ &= -\sum_{\substack{\gamma' \\ |T-\gamma-\gamma'|<1}} \frac{1}{\sigma - 1 + i(T-\gamma-\gamma')} + O(\log(|T-\gamma|+1)) \\ &\ll \sum_{\substack{\gamma' \\ |T-\gamma-\gamma'|<1}} 1 + O(\log(|T-\gamma|+1)) \\ &\ll \log(|T-\gamma|+1) \end{aligned}$$

(using (3.4) for the last inequality) for $1+\varepsilon \le \sigma \le 2+\varepsilon$, where γ' runs over all imaginary parts of non-trivial zeros of $\zeta(s)$ satisfying $|T-\gamma-\gamma'|<1$. By using Stirling's formula and (5.9), we have

$$\begin{aligned} B_2(s) &\ll e^{(\pi/2)T} T^{1/2-\sigma} \qquad (5.10)\\ &\quad \times \sum_{\rho} e^{-(\pi/2)(|\gamma|+|T-\gamma|)} (|T-\gamma|+1)^{\sigma-1} \log(|T-\gamma|+1) \\ &= e^{(\pi/2)T} T^{1/2-\sigma} \left(\sum_{\gamma<0} + \sum_{0<\gamma\le T} + \sum_{\gamma>T} \right) \\ &= e^{(\pi/2)T} T^{1/2-\sigma} (C_1 + C_2 + C_3), \end{aligned}$$

say. We can estimate C_j analogously to the case of J_j $(j=1,2,3)$. As for C_1, changing γ by $-\gamma$, we obtain

$$C_1 = e^{-(\pi/2)T} \sum_{\gamma>0} e^{-\pi\gamma} (T+\gamma+1)^{\sigma-1} \log(T+\gamma+1).$$

Divide the last sum into two parts corresponding to $0<\gamma\le T$ and $\gamma>T$, respectively. Using partial summation and (2.2), we find that the first sum is $O(T^{\sigma-1}\log T)$ and the second sum is of exponential decay. Hence $C_1 \ll e^{-(\pi/2)T} T^{\sigma-1} \log T$. The treatment of C_3 is even simpler than that of J_3; by partial summation and (2.2) we have

$$\begin{aligned} C_3 &= e^{(\pi/2)T} \sum_{\gamma>T} e^{-\pi\gamma} (\gamma - T + 1)^{\sigma-1} \log(T-\gamma+1) \\ &\ll e^{(\pi/2)T} \int_T^\infty \xi \log\xi \cdot e^{-\pi\xi} (\xi - T + 1)^{\sigma-1} \log(\xi - T + 1) d\xi \\ &\ll e^{-(\pi/2)T} T^{\sigma} (\log T)^2. \end{aligned}$$

Lastly, dividing C_2 into two parts corresponding to $0 < \gamma \le T/2$ and $T/2 < \gamma \le T$, and applying partial summation and (2.2), we obtain $C_2 \ll e^{-(\pi/2)T} T^{\sigma} (\log T)^2$. Therefore

$$B_2(s) = O(T^{1/2}(\log T)^2). \tag{5.11}$$

Now the only remaining term on the right-hand side of (3.9) is the first term $M(s-1)/(s-1)$. For the purpose of estimating this term suitably, we now specify T; we choose the same T as in (3.5), and put $s = \sigma + iT$. Then from (3.6) we obtain

$$M(s-1) = M(\sigma - 1 + iT) = O((\log T)^2). \tag{5.12}$$

From (5.5), (5.8), (5.11) and (5.12), the assertion of Lemma 5.1 follows. □

In the above proof, the special choice of T is necessary only for obtaining (3.5). But (3.5) is required only when $\sigma - 1$ is near $1/2$. In fact, if $\sigma \notin (3/2-\varepsilon, 3/2+\varepsilon)$, then just using (3.6) we obtain $M(s-1) \ll \log T$ for any T (under RH). Therefore we obtain

Lemma 5.2 (under RH). *The estimate* (5.4) *of Lemma 5.1 is valid for any* T *if* $1+\varepsilon \le \sigma \le 3/2-\varepsilon$ *or* $3/2+\varepsilon \le \sigma \le 2+\varepsilon$.

Remark 5.1. The estimate of Lemma 5.1 can be improved for $3/2+\varepsilon \le \sigma \le 2+\varepsilon$. In fact, from Lemma 5.2 we have

$$\Phi_2((3/2)+\varepsilon+iT) \ll T^{1/2}(\log T)^2$$

for any T. It is clear that $\Phi_2(2+\varepsilon+iT) = O(1)$. Moreover from Theorem 2.1 we know that $\Phi_2(s)$ is holomorphic (except for $s=2$) in the region $3/2+\varepsilon \le \sigma \le 2+\varepsilon$. Therefore by the Phragmén-Lindelöf convexity principle we obtain

$$\Phi_2(\sigma+iT) \ll \left(T^{1/2}(\log T)^2\right)^{2(2+\varepsilon-\sigma)} \ll T^{2-\sigma+\varepsilon} \tag{5.13}$$

for any T when $3/2+\varepsilon \le \sigma \le 2+\varepsilon$ (under RH).

Now we complete the proof of Theorem 5.1. Choose T as in Lemma 5.1, and write (5.3) as

$$\mathcal{D}_{2,a}(X) = \lim_{T\to\infty} \frac{1}{2\pi i} \int_{2+\varepsilon-iT}^{2+\varepsilon+iT} \frac{\Gamma(s)}{\Gamma(s+a+1)} \Phi_2(s) X^{s+a} ds. \tag{5.14}$$

Deform the path into the oriented polygonal path joining $2+\varepsilon-iT$, $1+\varepsilon-iT$, $1+\varepsilon+iT$, and $2+\varepsilon+iT$. By using Lemma 5.1 and Stirling's formula we see that the integrand is $O(T^{-a-1/2}(\log T)^2 X^{\sigma+a})$ on the horizontal segments, and hence the integrals on these segments vanish as $T \to \infty$. On the line $\Re s = 1+\varepsilon$, estimate (5.4) is valid for any T by Lemma 5.2. Hence, as $T \to \infty$, the integral on the line $\Re s = 1+\varepsilon$ is

$$\ll \int_{-\infty}^{\infty} T^{-a-1/2}(\log T)^2 X^{1+\varepsilon+a} dT \ll X^{1+\varepsilon+a}$$

if $a > 1/2$. Therefore we can shift the path of integration on the right-hand side of (5.3) to $\Re s = 1+\varepsilon$ if $a > 1/2$. The relevant poles are at $s = 2$ and $s = 1+\rho$, the residues at which are

$$\frac{1}{\Gamma(3+a)} X^{2+a}, \qquad -\frac{\Gamma(1+\rho)}{\Gamma(2+\rho+a)} \cdot \frac{2n(\rho)}{\rho} \cdot X^{1+\rho+a},$$

respectively. The assertion of Theorem 5.1 now follows.

Remark 5.2. From the case $a = 1$ of Theorem 5.1, with the aid of (2.6), we can deduce

$$\sum_{n \le X} G_2(n) = \frac{1}{2} X^2 - H(X) + O(X^{3/2+\varepsilon}) \tag{5.15}$$

(under RH) by the standard difference-operator argument. However this is weaker than Fujii's result (2.8).

6. The multiple case

So far we have mainly discussed the double Dirichlet series $\Phi_2(s)$, but it is possible to consider the multiple case in an analogous manner. Let

$$G_r(n) = \sum_{k_1+\cdots+k_r=n} \Lambda(k_1)\cdots\Lambda(k_r), \qquad r \ge 2 \tag{6.1}$$

and

$$\Phi_r(s) = \sum_{n=1}^{\infty} \frac{G_r(n)}{n^s} = \sum_{k_1=1}^{\infty} \cdots \sum_{k_r=1}^{\infty} \frac{\Lambda(k_1)\cdots\Lambda(k_r)}{(k_1+\cdots+k_r)^s}. \tag{6.2}$$

Similarly to (2.6) we have $G_r(n) \le n^{r-1}(\log n)^r$, and hence the series (6.2) is absolutely convergent for $\Re s > r$. In this section we prove some analytic properties of $\Phi_r(s)$.

It is to be noted that in the paper [32] quoted in §1, Peter considered the more general series

$$S_r(s) = \sum_{k_1=1}^{\infty} \cdots \sum_{k_r=1}^{\infty} \frac{\Lambda(k_1)\cdots\Lambda(k_r)}{P(k_1,\ldots,k_r)^s}$$

for the purpose of evaluating

$$\sum_{P(k_1,\ldots,k_r)\leq X} \Lambda(k_1)\cdots\Lambda(k_r). \tag{6.3}$$

The reason why Peter's method can treat $S_r(s)$ is that his method is based on the idea of Sargos [33,34], which can be applied to the case when the associated single series has infinitely many poles. By using the Tauberian theorem of Ikehara, Peter proved a certain asymptotic formula for (6.3).

The first purpose of this section is to show the following:

Theorem 6.1 (under RH). *The function $\Phi_r(s)$ $(r \geq 2)$ can be continued meromorphically to $\Re s > r-1$, and holomorphic there except for the simple poles at $s = r$ and $s = r-1+\rho$ for all non-trivial zeros ρ of $\zeta(s)$. The residues at $s = r$ and $s = r-1+\rho$ are*

$$\frac{1}{(r-1)!}, \qquad -\frac{r\cdot n(\rho)}{\rho(1+\rho)\cdots(r-2+\rho)},$$

respectively.

Proof. We prove this theorem by induction on r. When $r = 2$, this theorem is exactly Theorem 2.1. Assume that the theorem is true for $r-1$. Applying (1.4) to (6.2), we obtain

$$\Phi_r(s) = \frac{1}{2\pi i}\int_{(c)} \frac{\Gamma(s-z)\Gamma(z)}{\Gamma(s)}\Phi_{r-1}(s-z)M(z)dz \tag{6.4}$$

for $\Re s > r$, where $1 < c < \Re s - (r-1)$. Shift the path of integration to $\Re z = -\varepsilon$. By using the same T as in (3.5), we can show that this shifting is possible. (Note that in the strip $-\varepsilon \leq \Re z \leq c$ the factor $\Phi_{r-1}(s-z)$ is in the domain of its absolute convergence, hence is $O(1)$.) The result is that

$$\begin{aligned}\Phi_r(s) = {}& \frac{\Phi_{r-1}(s-1)}{s-1} - \sum_{\rho}\frac{\Gamma(s-\rho)\Gamma(\rho)}{\Gamma(s)}\Phi_{r-1}(s-\rho) \\ & - \Phi_{r-1}(s)\log 2\pi + \frac{1}{2\pi i}\int_{(-\varepsilon)}\frac{\Gamma(s-z)\Gamma(z)}{\Gamma(s)}\Phi_{r-1}(s-z)M(z)dz.\end{aligned} \tag{6.5}$$

Under the induction assumption, this expression gives the continuation of $\Phi_r(s)$ to $\Re s > r-1$. Moreover, the residues of $\Phi_{r-1}(s-1)/(s-1)$ at $s = r$, $s = r-1+\rho$ are

$$\frac{1}{(r-1)!}, \qquad -\frac{(r-1)n(\rho)}{\rho(1+\rho)\cdots(r-2+\rho)},$$

respectively, while the residue of

$$B_r(s) = -\sum_\rho \frac{\Gamma(s-\rho)\Gamma(\rho)}{\Gamma(s)}\Phi_{r-1}(s-\rho)$$

at $s = r-1+\rho$ is

$$-n(\rho)\cdot\frac{\Gamma(r-1)\Gamma(\rho)}{\Gamma(r-1+\rho)}\cdot\frac{1}{(r-2)!} = -\frac{n(\rho)}{\rho(1+\rho)\cdots(r-2+\rho)}.$$

Hence the assertion of Theorem 6.1 follows. □

The function $\Phi_{r-1}(s-\rho)$ is singular at $s = r-2+\rho+\rho'$ for any non-trivial zero ρ'. Hence, in view of Lemma 4.1, it is natural to raise the following:

Conjecture 6.1. *The line $\Re s = r-1$ is the natural boundary of $\Phi_r(s)$.*

In fact, under a certain assumption, we can show that

$$\Phi_r(s) \sim \frac{r(r-1)}{\Gamma(r-1+i\kappa)} n(\rho_1^0)n(\rho_2^0)\Gamma(\rho_1^0)\Gamma(\rho_2^0)\frac{1}{s-(r-1+i\kappa)} \tag{6.6}$$

as $s \to r-1+i\kappa$ for any $\kappa = \gamma_1^0+\gamma_2^0 \in \mathcal{K}$. This implies, as in the proof of Theorem 2.2, that Conjecture 6.1 is true.

When $r = 2$, (6.6) is nothing but (4.11), which has been shown under RH and (B).

We prove (6.6) for general r by induction. When $s \to r-1+i\kappa$, we have

$$\begin{aligned}\frac{\Phi_{r-1}(s-1)}{s-1} &\sim \frac{(r-1)(r-2)}{\Gamma(r-2+i\kappa)} n(\rho_1^0)n(\rho_2^0)\Gamma(\rho_1^0)\Gamma(\rho_2^0) \\ &\qquad\times\frac{1}{r-2+i\kappa}\frac{1}{(s-1)-(r-2+i\kappa)} \\ &= \frac{(r-1)(r-2)}{\Gamma(r-1+i\kappa)} n(\rho_1^0)n(\rho_2^0)\Gamma(\rho_1^0)\Gamma(\rho_2^0)\frac{1}{s-(r-1+i\kappa)}\end{aligned} \tag{6.7}$$

by induction assumption. Next, we divide $B_r(s)$ as

$$\begin{aligned}B_r(s) = &-n(\rho_1^0)\frac{\Gamma(s-\rho_1^0)\Gamma(\rho_1^0)}{\Gamma(s)}\Phi_{r-1}(s-\rho_1^0) \\ &-n(\rho_2^0)\frac{\Gamma(s-\rho_2^0)\Gamma(\rho_2^0)}{\Gamma(s)}\Phi_{r-1}(s-\rho_2^0) \\ &-\sum_{\rho\neq\rho_1^0,\rho_2^0}\frac{\Gamma(s-\rho)\Gamma(\rho)}{\Gamma(s)}\Phi_{r-1}(s-\rho) \\ = &\, B_{r1}(s)+B_{r2}(s)+B_{r3}(s),\end{aligned} \tag{6.8}$$

say. The factor $\Phi_{r-1}(s-\rho_1^0)$ has a pole at $s = r-1+i\kappa = r-2+\rho_1^0+\rho_2^0$, whose residue is given by Theorem 6.1. Therefore

$$\begin{aligned} B_{r1}(s) &\sim -n(\rho_1^0)\frac{\Gamma(r-2+\rho_2^0)\Gamma(\rho_1^0)}{\Gamma(r-1+i\kappa)} \\ &\quad \times \left(-\frac{(r-1)n(\rho_2^0)}{\rho_2^0(1+\rho_2^0)\cdots(r-3+\rho_2^0)}\right)\frac{1}{s-(r-1+i\kappa)} \\ &= \frac{r-1}{\Gamma(r-1+i\kappa)}n(\rho_1^0)n(\rho_2^0)\Gamma(\rho_1^0)\Gamma(\rho_2^0)\frac{1}{s-(r-1+i\kappa)} \end{aligned}$$

as $s \to r-1+i\kappa$. The asymptotic behaviour of $B_{r2}(s)$ when $s \to r-1+i\kappa$ is exactly the same. Therefore, if we assume

$(C)_r$ The sum $B_{r3}(s)$ remains finite when $s \to r-1+i\kappa$,

then we have

$$B_r(s) \sim \frac{2(r-1)}{\Gamma(r-1+i\kappa)}n(\rho_1^0)n(\rho_2^0)\Gamma(\rho_1^0)\Gamma(\rho_2^0)\frac{1}{s-(r-1+i\kappa)} \tag{6.9}$$

as $s \to r-1+i\kappa$. From (6.5), (6.7) and (6.9), we obtain (6.6), which implies the following:

Theorem 6.2 (under RH). *If we assume that* (B) *and* $(C)_k$ $(k \le r)$ *are true, then Conjecture 6.1 is true.*

References

1. S. Akiyama, S. Egami and Y. Tanigawa, Analytic continuation of multiple zeta functions and their values at non-positive integers, *Acta Arith.*, **98** (2001), 107–116.
2. S. Akiyama and H. Ishikawa, On analytic continuation of multiple L-functions and related zeta-functions, in *Analytic Number Theory*, C. Jia and K. Matsumoto (eds.), Dev. Math., Vol.6, Kluwer, 2002, pp.1–16.
3. T. Arakawa and M. Kaneko, Multiple zeta values, poly-Bernoulli numbers, and related zeta functions, *Nagoya Math. J.*, **153** (1999), 189–209.
4. T. Arakawa and M. Kaneko, On multiple L-values, *J. Math. Soc. Japan*, **56** (2004), 967–991.
5. E. W. Barnes, The theory of the double gamma function, *Philos. Trans. Roy. Soc.* (A), **196** (1901), 265–387.
6. E. W. Barnes, On the theory of the multiple gamma function, *Trans. Cambridge Philos. Soc.*, **19** (1904), 374–425.
7. G. Bhowmik, D. Essouabri and B. Lichtin, Meromorphic continuation of multivariable Euler products, *Forum Math.*, to appear.
8. R. de la Bretèche, Estimation de sommes multiples de fonctions arithmétiques, *Compositio Math.*, **128** (2001), 261–298.

9. G. Dahlquist, On the analytic continuation of Eulerian products, *Ark. Mat.*, **1** (1952), 533–555.
10. S. Egami, Some curious Dirichlet series, *Sūrikaisekikenkyūsho Kōkyūroku*, **1091**, RIMS Kyoto Univ. (1999), 172–174.
11. D. Essouabri, *Singularités des séries de Dirichlet associées à des polynômes de plusieurs variables et applications à la théorie analytique des nombres*, Thése, Univ. Henri Poincaré - Nancy I, 1995.
12. D. Essouabri, Singularités des séries de Dirichlet associées à des polynômes de plusieurs variables et applications en théorie analytique des nombres, *Ann. Inst. Fourier*, **47** (1997), 429–483.
13. T. Estermann, On certain functions represented by Dirichlet series, *Proc. London Math. Soc.*, (2)**27** (1928), 435–448.
14. T. Estermann, On a problem of analytic continuation, ibid., 471–482.
15. A. Fujii, An additive problem of prime numbers, *Acta Arith.*, **58** (1991), 173–179.
16. A. Fujii, An additive problem of prime numbers II, *Proc. Japan Acad.*, **67A** (1991), 248–252.
17. A. Fujii, An additive problem of prime numbers III, ibid., 278–283.
18. A. Fujii, An additive theory of the zeros of the Riemann zeta function, *Comment. Math. Univ. St. Pauli*, **45** (1996), 49–116.
19. A. B. Goncharov, Multiple polylogarithms, cyclotomy and modular complexes, *Math. Res. Letters*, **5** (1998), 497–516.
20. G. H. Hardy and J. E. Littlewood, Some problems of "partitio numerorum" (V): A further contribution to the study of Goldbach's problem, *Proc. London Math. Soc.*, (2)**22** (1924), 46–56.
21. H. Ishikawa, On analytic properties of a multiple L-function, in *Analytic Extension Formulas and their Applications*, S. Saitoh et al. (eds.), Soc. Anal. Appl. Comput., Vol.9, Kluwer, 2001, pp.105–122.
22. H. Ishikawa, A multiple character sum and a multiple L-function, *Arch. Math.*, **79** (2002), 439–448.
23. J. C. Kluyver, Benaderingsformules betreffende de priemgetallen beneden eene gegeven grens, *Koninglijke Akad. Wet. Amsterdam, Versl. Gew. Verg. Wis- en Natuur. Afd.*, **8** (1900), 672–682.
24. N. Kurokawa, On the meromorphy of Euler products (I), *Proc. London Math. Soc.*, (3)**53** (1986), 1–47; (II), ibid., 209–236.
25. E. Landau, Üder die Multiplikation Dirichlet'scher Reihen, *Rend. Circ. Mat. Palermo*, **24** (1907), 81–160.
26. E. Landau and A. Walfisz, Üder die Nichtfortsetzbarkeit einiger durch Dirichletsche Reihen definierter Funktionen, ibid. **44** (1920), 82–86.
27. K. Matsumoto, Asymptotic expansions of double zeta-functions of Barnes, of Shintani, and Eisenstein series, *Nagoya Math. J.*, **172** (2003), 59–102.
28. K. Matsumoto and Y. Tanigawa, The analytic continuation and the order estimate of multiple Dirichlet series, *J. Théorie des Nombres de Bordeaux*, **15** (2003), 267–274.
29. H. Mellin, Eine Formel für den Logarithmus transcendenter Funktionen von endlichem Geschlecht, *Acta. Soc. Sci. Fenn.*, **29** (1900), no.4.

30. H. Mellin, Die Dirichlet'schen Reihen, die zahlentheoretischen Funktionen und die unendlichen Produkte von endlichem Geschlecht, *Acta Math.*, **28** (1904), 37–64.
31. H. L. Montgomery and R. C. Vaughan, Error terms in additive prime number theory, *Quart. J. Math. Oxford*, (2)**24** (1973), 207–216.
32. M. Peter, Dirichlet series associated with polynomials, *Acta Arith.*, **84** (1998), 245–278.
33. P. Sargos, Prolongement méromorphe des séries de Dirichlet associées à des fractions rationnelles de plusieurs variables, *Ann. Inst. Fourier*, **34**(3) (1984), 83–123.
34. P. Sargos, Croissance de certaines séries de Dirichlet et applications, *J. Reine Angew. Math.*, **367** (1986), 139–154.
35. Y. Tanigawa and W. Zhai, Dirichlet series associated with polynomials and applications, *J. Number Theory*, to appear.
36. E. C. Titchmarsh, *The Theory of the Riemann Zeta-function*, Oxford, 1951.
37. J. Zhao, Analytic continuation of multiple zeta-functions, *Proc. Amer. Math. Soc.*, **128** (2000), 1275–1283.

CONSTRUCTING NEW NON-CONGRUENT NUMBERS BY GRAPH THEORY

KEQIN FENG*

Department of Mathematical Science, Tsinghua University,
Beijing 100084, China
E-mail: kfeng@math.tsinghua.edu.cn

YAN XUE

Department of Mathematical Science, Tsinghua University,
Beijing 100084, China
E-mail: xueyan97@mails.tsinghua.edu.cn

This paper is a survey on recent results of new series of non-congruent numbers which can be acquired by traditional arithmetic theory of elliptic curves plus a result from algebraic graph theory. More precisely, we start from the following two facts: (1) A square-free positive integer n is a non-congruent number if and only if the rank of the group $E_n(\mathbb{Q})$ of rational points of the elliptic curve $E_n : y^2 = x^3 - n^2x$ is zero. (2) If the 2-Selmer groups S_n and $\hat{S_n}$ of the elliptic curve E_n and its dual curve $\hat{E_n} : y^2 = x^3 + 4n^2x$ have minimal sizes $|S_n| = 1$ and $|\hat{S_n}| = 4$, then $\text{rank}(E_n(\mathbb{Q})) = 0$. Selmer groups can be determined by the data of locally solvability conditions of the homogenous spaces of elliptic curves E_n and $\hat{E_n}$. The next step is to organize the data into carefully constructed graphs so that the Selmer groups have minimum if and only if the graphs have specific "odd" property. By a result in algebraic graph theory, an odd graph can be described by the rank of the Laplace matrix of the graph over $\mathbb{F}_2$. Thus, by computing the rank of a certain matrix over $\mathbb{F}_2$, we can determine all n such that S_n and $\hat{S_n}$ have minimum, in which case n is a non-congruent number. After explaining this method and related concepts, we describe the results on new series of non-congruent numbers obtained in this way and illustrate them by examples.

1. Introduction

A square-free positive integer n is called a congruent number if n is the area of a certain rational right triangle where "rational" means that the

*Supported by a National Scientific Research Project 973 of China No. 2004 CB 3180004 and a NSFC grant No. 60433050.

lengths of three sides of this triangle are rational numbers. Otherwise n is called a non-congruent number. To determine all congruent numbers is one of long-standing problems in number theory. Using the theory of modular forms, Tunnell [17] (also see [10]) presented an elementary criterion on congruent numbers under the Birch and Swinnerton-Dyer conjecture, but this problem has not been solved completely. Many congruent numbers and non-congruent numbers have been determined [6,7,11–15], but most of them have at most 4 prime divisors. In this paper we are concerned only with non-congruent numbers, and we list up several known results in the following (p, q, r, s are odd prime numbers, and $(\frac{p}{q})$ is the Legendre symbol.)

Lemma 1.1. *In the following cases, n is a non-congruent number.*
(1) (Genocchi [6])

$$\begin{aligned} &n = p \equiv 3 \pmod 8, \\ &n = pq, \quad p \equiv q \equiv 3 \pmod 8, \\ &n = 2p, \quad p \equiv 5 \pmod 8, \\ &n = 2pq, \quad p \equiv q \equiv 5 \pmod 8. \end{aligned}$$

(2) (Lagrange [12])

$$\begin{array}{lll} n = pq, & (p,q) \equiv (1,3) \pmod 8, & (\frac{p}{q}) = -1, \\ n = 2pq, & (p,q) \equiv (1,5) \pmod 8, & (\frac{p}{q}) = -1, \\ n = pqr, & (p,q,r) \equiv (1,1,3) \pmod 8, & \text{with the condition } (*), \\ n = 2pqr, & (p,q,r) \equiv (1,1,5) \pmod 8, & \text{with the condition } (*). \end{array}$$

Condition $(*)$: n can be written as $n = p_1p_2p_3$ or $2p_1p_2p_3$ such that

$$\left(\frac{p_1}{p_2}\right) = \left(\frac{p_1}{p_3}\right) = -1.$$

(3) (Serf [15])

$$\begin{array}{lll} n = pq, & (p,q) \equiv (5,7) \pmod 8, & (\frac{p}{q}) = -1, \\ n = pqr, & (p,q,r) \equiv (1,3,3) \pmod 8, & (\frac{p}{q}) = -(\frac{p}{r}), \\ n = pqr, & (p,q,r) \equiv (3,5,7) \pmod 8, & (\frac{q}{r}) = -1, \\ n = 2pqr, & (p,q,r) \equiv (1,5,5) \pmod 8, & (\frac{p}{q}) = -(\frac{p}{r}), \\ n = pqrs, & (p,q,r,s) \equiv (5,5,7,7) \pmod 8, & \textit{and} \end{array}$$

$$\begin{aligned} 1 &= \left(\frac{p}{r}\right) = -\left(\frac{p}{s}\right) = -\left(\frac{q}{r}\right); \quad \textit{or} \\ 1 &= -\left(\frac{p}{r}\right) = \left(\frac{p}{s}\right) = -\left(\frac{q}{s}\right); \quad \textit{or} \\ 1 &= -\left(\frac{p}{r}\right) = -\left(\frac{p}{s}\right), \quad \left(\frac{q}{r}\right) = -\left(\frac{q}{s}\right). \end{aligned}$$

Serf also claimed in [15] that he found more cases of non-congruent numbers n with 5 or 6 odd prime divisors, but "it is almost impossible to exhibit them in a reasonable way."

In this survey paper we introduce several works [2–4,18,19] trying to find a reasonable way to describe a remarkable portion of new non-congruent numbers, including many series of such n with arbitrarily large number of prime factors. The method starts from the following basic fact (see N. Koblitz' book [10], for example).

Lemma 1.2. *A square-free positive integer n is a non-congruent number if and only if the group $E_n(\mathbb{Q})$ of rational points of the elliptic curve E_n : $y^2 = x^3 - n^2x$ has rank zero.*

Traditional way to deal with $\mathrm{rank}(E_n(\mathbb{Q}))$ is the 2-descent method which we explain in §2. With this method, the rank problem of $E_n(\mathbb{Q})$ is reduced to determining the Selmer groups S_n and $\hat{S_n}$ of the elliptic curve E_n and its dual curve $\hat{E_n}$: $y^2 = x^3 + 4n^2x$. In particular, if the Selmer groups have minimal sizes $|S_n| = 1$ and $|\hat{S_n}| = 4$, then $\mathrm{rank}(E_n(\mathbb{Q})) = 0$ so that n is a non-congruent number by Lemma 1.2. The Selmer groups are defined by locally solvability of certain equations so that can be determined in principle by Hensel lemma. But the data of such conditions of locally solvability are rather complicated when the number of prime factors of n is larger. Next step is that the data can be organized into carefully constructed graphs so that the Selmer groups have minimum if and only if the graphs have specific "odd" property. By a result in algebraic graph theory, odd graphs can be described by the rank of the Laplace matrix of the graph over $\mathbb{F}_2$. Thus, all n such that S_n and $\hat{S_n}$ have minimum, so that $\mathrm{rank}(E_n(\mathbb{Q})) = 0$ and, a fortiori, n is a non-congruent number, can be determined by computing the rank of a certain matrix over $\mathbb{F}_2$. We explain our graph-theory tools in §3 and introduce the results on new series of non-congruent numbers obtained in this way in §4. Finally we mention the Birch and Swinnerton-Dyer conjecture for elliptic curve E_n briefly in §5.

2. Selmer groups

In this section we describe the 2-descent method for elliptic curve E_n. For details we refer to the last chapter of Silverman's book [16].

Let $n = p_1 \cdots p_t$ or $n = 2p_1 \cdots p_t$ where $t \geqslant 1$ and $p_1, \cdots, p_t$ are distinct odd prime numbers. We define a set of prime divisors of the rational number field $\mathbb{Q}$ by

$$S = \{\infty,\ 2,\ p_1, \cdots, p_t\},$$

and a subgroup M of the multiplicative group $\mathbb{Q}^*/(\mathbb{Q}^*)^2$ generated by $-1, 2, p_1, \cdots, p_t$

$$M = \langle -1,\ 2,\ p_1, \cdots, p_t\rangle \subseteq \mathbb{Q}^*/(\mathbb{Q}^*)^2.$$

For each $d \in M$ we have the homogenous spaces C_d and $\hat{C}_d$ of E_n and its 2-dual curve $\hat{E}_n:\ y^2 = x^3 + 4n^2x$ defined by

$$C_d:\ dw^2 = d^2t^4 + 4n^2z^4,$$
$$\hat{C}_d:\ dw^2 = d^2t^4 - n^2z^4.$$

For each prime divisor $v \in S$, we denote $C_d(\mathbb{Q}_v)$ and $\hat{C}_d(\mathbb{Q}_v)$ the set of non-trivial solutions $(w, t, z) \neq (0,0,0)$ of C_d and $\hat{C}_d$ in the local field $\mathbb{Q}_v$ respectively. The Selmer groups S_n and $\hat{S_n}$ of E_n and $\hat{E_n}$ are defined by locally solvability of C_d and $\hat{C}_d$:

$$S_n = \{d \in M:\ C_d(\mathbb{Q}_v) \neq \emptyset \text{ for all } v \in S\},$$
$$\hat{S_n} = \{d \in M:\ \hat{C}_d(\mathbb{Q}_v) \neq \emptyset \text{ for all } v \in S\}.$$

It is proved that S_n and $\hat{S_n}$ are subgroups of M, and

$$1 \in S_n, \quad \{\pm 1, \pm n\} \subseteq \hat{S_n},$$

since C_1 and $\hat{C}_d$ for $d = \pm 1, \pm n$ have global non-trivial solutions in $\mathbb{Q}$. The following result is a special consequence of 2-descent method.

Lemma 2.1. *If S_n and $\hat{S_n}$ have minimal sizes: $S_n = \{1\}$ and $\hat{S_n} = \{\pm 1, \pm n\}$, then* $\operatorname{rank}(E_n(\mathbb{Q})) = 0$, *so that n is a non-congruent number.*

Now the problem is reduced to finding an explicit criterion to describe $S_n = \{1\}$ and $\hat{S_n} = \{\pm 1, \pm n\}$. By the definition of Selmer groups, $S_n = \{1\}$ if and only if for all $d \in M$ and $d \neq 1$, there exists $v \in S$ such that $C_d(\mathbb{Q}_v) = \emptyset$. Similarly, $\hat{S_n} = \{\pm 1, \pm n\}$ can also be described by such kind of local solvabilities of $\hat{C}_d$. It is not difficult to give the following result by Hensel lemma and careful computation.

Lemma 2.2 ([3, Lemma 3.1, 3.2, 5.1, 5.2]). *Let $p_1, \cdots, p_t$ $(t \geqslant 1)$ be distinct odd prime numbers, $d \in M = \langle -1, 2, p_1, \cdots, p_t\rangle \subset \mathbb{Q}^*/(\mathbb{Q}^*)^2$, and p denotes an odd prime number.*

(A) *If $n = p_1 \cdots p_t$, then*

(A1) $C_d(\mathbb{Q}_\infty) = \emptyset \ \Leftrightarrow d < 0.$

(A2) *For* $p|d$, $C_d(\mathbb{Q}_p) = \emptyset \Leftrightarrow (\frac{-1}{p}) = -1$ *or* "$(\frac{-1}{p}) = 1$ *and* $(\frac{n/d}{p}) = -1$".

(A3) *For* $p|\frac{2n}{d}$, $C_d(\mathbb{Q}_p) = \emptyset \Leftrightarrow (\frac{d}{p}) = -1$.

(A4) *If* $n \equiv \pm 3 \pmod 8$ *and* $2|d$, *then* $C_d(\mathbb{Q}_2) = \emptyset$.

(A5) $d \equiv 1 \pmod 4 \Rightarrow C_d(\mathbb{Q}_2) \neq \emptyset$.

(A6) *If* $n \equiv \pm 1 \pmod 8$, $d = 2d'$ *where* $d'|n$ *and* $d' \equiv 1 \pmod 4$, *then* $C_d(\mathbb{Q}_2) \neq \emptyset$.

(A1′) $2|d \Rightarrow \hat{C}_d(\mathbb{Q}_2) = \emptyset$.

(A2′) *If* $2 \nmid d$, *then* $\hat{C}_d(\mathbb{Q}_2) = \emptyset \Leftrightarrow d \equiv \pm 3 \pmod 8$ *and* $\frac{n}{d} \equiv \pm 3 \pmod 8$.

(A3′) *If* $p|d$, *then* $\hat{C}_d(\mathbb{Q}_p) = \emptyset \Leftrightarrow (\frac{-1}{p}) = 1$ *and* $(\frac{n/d}{p}) = -1$.

(A4′) *If* $p|\frac{2n}{d}$, *then* $\hat{C}_d(\mathbb{Q}_p) = \emptyset \Leftrightarrow (\frac{-1}{p}) = 1$ *and* $(\frac{d}{p}) = -1$.

(B) *If* $n = 2p_1 \cdots p_t$, *then*

(B1) $C_d(\mathbb{Q}_\infty) = \emptyset \Leftrightarrow d < 0$.

(B2) $2|d \Rightarrow C_d(\mathbb{Q}_2) = \emptyset$.

(B3) *For* $p|d$, *if* $(\frac{-1}{p}) = -1$ *or* "$(\frac{-1}{p}) = 1$ *and* $(\frac{-1}{p})_4(\frac{2n/d}{p}) = -1$", *then* $C_d(\mathbb{Q}_p) = \emptyset$.

(B4) *For* $p|\frac{n}{d}$, *then* $(\frac{d}{p}) = -1 \Rightarrow C_d(\mathbb{Q}_p) = \emptyset$.

(B1′) *For* $p|d$, $\hat{C}_d(\mathbb{Q}_p) = \emptyset \Leftrightarrow (\frac{-1}{p}) = 1$ *and* $(\frac{n/d}{p}) = -1$.

(B2′) *For* $p|\frac{n}{d}$, $\hat{C}_d(\mathbb{Q}_p) = \emptyset \Leftrightarrow (\frac{-1}{p}) = 1$ *and* $(\frac{d}{p}) = -1$.

(B3′) $2 \nmid d \Rightarrow \hat{C}_d(\mathbb{Q}_2) \neq \emptyset$.

For the Genocchi cases in Lemma 1.1 (1), it can be seen by Lemma 2.2 that $S_n = \{1\}$ and $\hat{S_n} = \{\pm 1, \pm n\}$ so that n is a non-congruent number. For the first two cases of Lemma 1.1 (2), it can be seen that $S_n = \{1\}$ and $\hat{S_n} = \{\pm 1, \pm n\}$ if and only if $(\frac{p}{q}) = -1$. But in general cases, it seems no simple way to write down a necessary and sufficient condition for $S_n = \{1\}$ and $\hat{S_n} = \{\pm 1, \pm n\}$ if n has large number of prime divisors. What we did in next step is to find proper graphs such that the condition can be described by a specific property of graphs which we introduce in next section.

3. Oddness of graphs

We use standard terminology in graph theory (see [8] for example). Let $G = (V, A)$ be a (simple) directed graph where $V = V(G) = \{v_1, \cdots, v_m\}$ is the set of vertices of G, and $A = A(G)$ is the set of arcs in G. We denote an arc $(v_i, v_j) \in A$ by $\overrightarrow{v_iv_j}$. If both of $\overrightarrow{v_iv_j}$ and $\overrightarrow{v_jv_i}$ belong to A, we have a two-direction arc $\overline{v_i v_j}$ in G and call it an edge. If all arcs in $A(G)$ are two-directed, the graph G is called non-directed.

The adjacency matrix of G is defined by

$$M(G) = (a_{ij})_{1 \leqslant i,j \leqslant m}$$

where

$$a_{ij} = \begin{cases} 1 & \text{if } \overrightarrow{v_i v_j} \in A(G), \quad 1 \leqslant i \neq j \leqslant m, \\ 0 & \text{otherwise.} \end{cases}$$

Let

$$d_i = \sum_{j=1}^{m} a_{ij} \quad \text{(the outdegree of vertex } v_i\text{)}, \quad 1 \leqslant i \leqslant m.$$

The Laplace matrix of G is defined by

$$L(G) = \operatorname{diag}(d_1, \cdots, d_m) - M(G).$$

Since the sum of each row of $L(G)$ is zero, we know that $\operatorname{rank}_{\mathbb{Q}}(L(G)) \leqslant m-1$ and

$$L_{ij} = (-1)^{j+k} L_{ik}$$

where L_{ij} is the co-factor of $L(G)$ at the position (i,j). For non-directed graph G, the matrices $M(G)$ and $L(G)$ are symmetric and

$$L_{11} = (-1)^{i+k} L_{ik}, \quad 1 \leqslant i,k \leqslant m.$$

In this case, it is well known that the absolute value of $|L_{11}|$ is the number of spanning trees of the non-directed graph G (see [2] Section 1.2.4).

Definition 3.1. Let $G = (V, A)$ be a directed graph. A partition $\{V_1, V_2\}$ of V is called odd if either there exists a vertex $v_1 \in V_1$ such that $\#\{v_1 \to V_2\}$ (the total number of arcs from v_1 to vertices in V_2) is odd, or there exists $v_2 \in V_2$ such that $\#\{v_2 \to V_1\}$ is odd. Otherwise the partition $\{V_1, V_2\}$ is called even. The graph G is called odd if all non-trivial partitions $\{V_1, V_2\} \neq \{V, \emptyset\}$ of V are odd.

The following result presents a simple criterion for oddness of a graph G in terms of the rank of the Laplace matrix $L(G)$ over finite field $\mathbb{F}_2$. Remark that $\operatorname{rank}_{\mathbb{F}_2}(L(G)) \leqslant \operatorname{rank}_{\mathbb{Q}}(L(G)) \leqslant m-1$ and the total number of partitions of V is 2^{m-1} ($m = |V|$) since we view $\{V_1, V_2\}$=$\{V_2, V_1\}$.

Lemma 3.1 ([2, Lemma 2.2]). *Let $G = (V, A)$ be a directed graph, $m = |V|$ and $r = \mathrm{rank}_{\mathbb{F}_2}(L(G))$ $(\leqslant m-1)$. Then the total number of even partitions of V is 2^{m-r-1}. In particular, the graph G is odd if and only if $r = m-1$. For non-directed graph G, G is odd if and only if the number $t(G) = |L_{11}(G)|$ of spanning trees of G is odd.*

An odd graph should be connected. There exists plenty of odd graphs as shown in following examples:

(1) All directed cycles C'_m with $V = \{v_1, \cdots, v_m\}$ and $A = \{\overrightarrow{v_1v_2}, \cdots, \overrightarrow{v_{m-1}v_m}, \overrightarrow{v_mv_1}\}$.

For non-directed graphs:

(2) All trees T, since $t(T) = 1$.

(3) All complete graphs K_m with odd integer $m \geqslant 3$ defined by $V = \{v_1, \cdots, v_m\}$ and $A = \{\overline{v_iv_j} : 1 \leqslant i \neq j \leqslant m\}$ since we have Cayley formula $t(K_m) = m^{m-2}$.

(4) All cycles C_m with odd integer $m \geqslant 3$ defined by $V = \{v_1, \cdots, v_m\}$ and $A = \{\overline{v_1v_2}, \cdots, \overline{v_{m-1}v_m}, \overline{v_mv_1}\}$.

The concept of odd graph has been used in number theory to determine the 4-rank of the class group of imaginary quadratic number fields in 1930's by L. Rédei and H. Reichardt and to present a sufficient condition for the Pell's equation $x^2 - ny^2 = -1$ having no integral solution in 1970's. In the next section we show its new application in number theory to present series of new non-congruent numbers.

4. New non-congruent numbers

Let $n = p_1 \cdots p_t$ $(t \geqslant 1)$ be a product of distinct odd primes. We define a graph $G(n) = (V, A)$ by

$$V = \{p_1, \cdots, p_t\}, \quad A = \left\{\overrightarrow{p_ip_j} : \left(\frac{p_j}{p_i}\right) = -1, \quad 1 \leqslant i \neq j \leqslant t\right\}.$$

In 1996, the first author found that for specific n, the oddness of the graph $G(n)$ is a sufficient condition for $S_n = \{1\}$ and $\hat{S_n} = \{\pm 1, \pm n\}$ as described in Lemma 4.1.

Lemma 4.1 ([2, Theorem 3.1]). (1) *Suppose that $n = p_1 \cdots p_t$ $(t \geqslant 1)$, $p_1 \equiv 3 \pmod 8$ and $p_i \equiv 1 \pmod 8$ when $i \geqslant 2$. If $G(n)$ is an odd graph, then $S_n = \{1\}$ and $\hat{S_n} = \{\pm 1, \pm n\}$.*
(2) *Suppose that $n = 2p_1 \cdots p_t$ $(t \geqslant 1)$, $p_1 \equiv 5 \pmod 8$ and $p_i \equiv 1 \pmod 8$ when $i \geqslant 2$. If $G(\frac{n}{2})$ is an odd graph, then $S_n = \{1\}$ and $\hat{S_n} = \{\pm 1, \pm n\}$.*

This result presents series of non-congruent number n such that n can have arbitrarily lager number of prime divisors since it is easy to show by the Dirichlet theorem on primes in arithmetic progressions that for each non-directed graph G there exist infinitely many of n in form (1) of Lemma 4.1 and $2n$ in form (2) of Lemma 4.1 such that $G(n) = G$ and $G(\frac{n}{2}) = G$ respectively. Later, we find suitable graphs to do this for all cases of n with more careful consideration. Now we describe our results and omit technique in proofs.

Case $2|n$**.** Let $n = 2n'$, where

$$\begin{aligned} &n' = p_1 \cdots p_t q_1 \cdots q_s, \\ &p_i \equiv 1 \pmod 4, \quad q_j \equiv 3 \pmod 4, \quad 1 \leqslant i \leqslant t, \quad 1 \leqslant j \leqslant s, \end{aligned}$$

is a product of distinct prime numbers ($t + s \geqslant 1$). Let $P = \{p_1, \cdots, p_t\}$, $Q = \{q_1, \cdots, q_s\}$. We define a graph $\hat{G}(n) = (V, A)$ by

$$\begin{aligned} V &= \{2, p_1, \cdots, p_t, q_1, \cdots, q_s\}, \\ A &= \left\{\overline{p_i p_j} : \left(\frac{p_j}{p_i}\right) = -1,\ p_i, p_j \in P\right\} \\ &\cup \left\{\overrightarrow{pq} : \left(\frac{q}{p}\right) = -1,\ p \in P,\ q \in Q\right\} \\ &\cup \left\{\overrightarrow{p2} : \left(\frac{2}{p}\right) = -1\ (\Leftrightarrow p \equiv 5 \pmod 8),\ p \in P\right\}. \end{aligned}$$

From Lemma 2.2 (B) we obtain the following result.

Theorem 4.1 ([3, Lemma 5.3, 5.4]). *For $n = 2n'$, we have*

$$\hat{S_n} = \{\pm 1, \pm n\} \ \Leftrightarrow\ \hat{G}(n) \text{ is an odd graph} \ \Rightarrow\ S_n = \{1\}.$$

Then by the matrix characterization of odd graphs in Lemma 3.2, we get the following result.

Theorem 4.2 ([3, Theorem 2.6]). *Let $n = 2n'$ and n' has decomposition (4.1). Then $S_n = \{1\}$ and $\hat{S_n} = \{\pm 1, \pm n\}$ (so that $\mathrm{rank}(E_n(\mathbb{Q})) = 0$ and n is a non-congruent number) if and only if the following two conditions are satisfied.*

(1) $s = 0$ *so that* $n' = p_1 \cdots p_t$ ($t \geqslant 1$) *and* $p_i \equiv 1 \pmod 4$.

(2) *Define the following numbers*

$$a_{ij} = \begin{cases} 1 & \text{if } (\frac{p_j}{p_i}) = -1, \\ 0 & \text{if } (\frac{p_j}{p_i}) = 1, \end{cases} \qquad 1 \leqslant i \neq j \leqslant t,$$

$$c_i = \begin{cases} 1 & \text{if } p_i \equiv 5 \pmod 8, \\ 0 & \text{otherwise}, \end{cases} \qquad 1 \leqslant i \leqslant t,$$

$$a_{ii}^* = \sum_{j=1, j\neq i}^{t} a_{ij} + c_i, \qquad 1 \leqslant i \leqslant t.$$

Then

$$\begin{vmatrix} a_{11}^* & a_{12} & \cdots & a_{1t} \\ a_{21} & a_{22}^* & \cdots & a_{2t} \\ \vdots & \vdots & & \vdots \\ a_{t1} & a_{t2} & \cdots & a_{tt}^* \end{vmatrix} = 1 \in \mathbb{F}_2.$$

In particular, there exists at least one i ($1 \leqslant i \leqslant t$) such that $p_i \equiv 5 \pmod 8$.

As an example of new consequences, the following result can be derived from Theorem 4.3.

Corollary 4.1. *If* $n = 2p_1 \cdots p_t$ $(t \geqslant 1)$, $p_i \equiv 5 \pmod 8$ $(1 \leqslant i \leqslant t)$ *and* $(\frac{p_j}{p_i}) = 1$ *for all* $1 \leqslant i \neq j \leqslant t$, *then* $S_n = \{1\}$ *and* $\hat{S_n} = \{\pm 1, \pm n\}$ *so that* $\text{rank}(E_n(\mathbb{Q})) = 0$ *and n is a non-congruent number.*

Case $2 \nmid n$. For $n = p_1 \cdots p_t \equiv \pm 3 \pmod 8$ we have the following result where $G(n)$ is the (non-directed) graph $G(n) = (V, E)$ defined by

$$V = \{p_1, \cdots, p_t\},$$

$$E = \left\{ \overline{p_i p_j} : \left(\frac{p_j}{p_i}\right) = -1, \ 1 \leqslant i \neq j \leqslant t \right\}.$$

Theorem 4.3 ([3, Theorem 2.4]). *For* $n \equiv \pm 3 \pmod 8$, $S_n = \{1\}$ *and* $\hat{S_n} = \{\pm 1, \pm n\}$ *(so that* $\text{rank}(E_n(\mathbb{Q})) = 0$ *and n is a non-congruent number) if and only if the following three conditions are satisfied.*
(1) $n \equiv 3 \pmod 8$.
(2) $n = p_1 \cdots p_t$, $p_1 \equiv 3 \pmod 4$ *and* $p_i \equiv 1 \pmod 4$ *for* $2 \leqslant i \leqslant t$.
(3) $G(n)$ *is an odd graph.*

For the case $n \equiv \pm 1 \pmod 8$, we need to generalize the concept of odd graph a little more. Here we state the final result. Let $n = p_1 \cdots p_t q_1 \cdots q_s$ be a product of distinct prime numbers, where $p_i \equiv 1 \pmod 4$, $q_j \equiv 3 \pmod 4$,

$(1 \leqslant i \leqslant t,\ \ 1 \leqslant j \leqslant s)$, $(t+s \geqslant 1)$. Let $P = \{p_1, \cdots, p_t\}$, $Q = \{q_1, \cdots, q_s\}$. We define a graph $G^*(n) = (V, A)$ by

$$V = \{2, p_1, \cdots, p_t, q_1, \cdots, q_s\},$$

$$A = \left\{\overrightarrow{p_i p_j} : \left(\frac{p_j}{p_i}\right) = -1,\ p_i, p_j \in P\right\}$$

$$\cup \left\{\overrightarrow{pq} : \left(\frac{q}{p}\right) = -1,\ p \in P,\ q \in Q\right\}$$

$$\cup \left\{\overrightarrow{2r} : r \equiv \pm 3 \pmod 8,\ r \in P \cup Q\right\}.$$

Theorem 4.4 ([3, Theorem 2.5]). *For $n \equiv \pm 1 \pmod 8$, $S_n = \{1\}$ and $\hat{S}_n = \{\pm 1, \pm n\}$ if and only if the following three conditions are satisfied.*
(1) $n \equiv 1 \pmod 8$.
(2) *The decomposition of n has one of the following forms:*
(2.1) $n = p_1 \cdots p_r P_1 \cdots P_s Q_1 Q_2$;
(2.2) $n = p_1 \cdots p_r P_1 \cdots P_t q_1 q_2$;
(2.3) $n = p_1 \cdots p_r P_1 \cdots P_l Q_1 q_1$
where $p_i \equiv 1$, $P_j \equiv 5$, $Q_\lambda \equiv 3$, $q_\mu \equiv 7 \pmod 8$ and

$$r \geqslant 0, \quad 2|s \geqslant 0, \quad 2|t \geqslant 2, \quad 2 \nmid l \geqslant 1.$$

(3) *There exists only one non-trivial even partition $V_1 = \{2\}$ and $V_2 = V \backslash V_1$ for the graph $G^*(n)$. Namely, the rank of $L(G^*(n))$ over $\mathbb{F}_2$ is $|V| - 2$ where $L(G^*(n))$ is the Laplace matrix of $G^*(n)$ defined in §3.*

Many knowing results on non-congruent numbers in Lemma 1.1 are special cases of Theorem 4.2-6.

Now under the change of variable, the elliptic curve $E_n : y^2 = x^3 - n^2 x$ transforms into the elliptic curve

$$E'_n : \ y^2 = x^3 - 3nx^2 + 2n^2 x.$$

It is obvious that $\mathrm{rank}(E_n(\mathbb{Q})) = \mathrm{rank}(E'_n(\mathbb{Q}))$. The homogenous spaces of E'_n and its dual curve $\hat{E}'_n : \ y^2 = x^3 + 6nx^2 + n^2 x$ are

$$C'_d : \ dw^2 = d^2 t^4 + 6ndt^2 z^2 + n^2 z^4,$$

$$\hat{C}'_d : \ dw^2 = d^2 t^4 - 3ndt^2 z^2 + 2n^2 z^4 = (dt^2 - 2nz^2)(dt^2 - nz^2).$$

Let S'_n and $\hat{S}'_n$ be the Selmer groups of E'_n and $\hat{E}'_n$ respectively. Then $1 \in S'_n$ and $\{1, 2, n, 2n\} \subset \hat{S}'_n$. As a consequence of 2-descent method, we also have the following fact.

Lemma 2.1′. *If $S'_n = \{1\}$ and $\hat{S}'_n = \{1, 2, n, 2n\}$, then* $\mathrm{rank}(E'_n(\mathbb{Q})) = \mathrm{rank}(E_n(\mathbb{Q})) = 0$, *so that n is a non-congruent number.*

In 2002, Goto [6] obtained new non-congruent numbers n for n having at most 4 prime divisors by using Lemma 2.1′. With helping of odd graphs, we obtain following general results which present more non-congruent numbers.

Case $2 \nmid n$. Let

$$\begin{aligned} &n = p_1 \cdots p_t q_1 \cdots q_s, \quad t+s \geqslant 1, \\ &p_i \equiv \pm 1 \pmod 8, \ q_j \equiv \pm 3 \pmod 8, \quad 1 \leqslant i \leqslant t, \ 1 \leqslant j \leqslant s. \end{aligned} \tag{4.1}$$

Theorem 4.5 ([4, Theorem 3.1]). *Assume that* $n \equiv 3 \pmod 4$ *with decomposition* (4.2). *Then* $S'_n = \{1\}$ *and* $\hat{S}'_n = \{1, 2, n, 2n\}$ *if and only if the following two conditions are satisfied.*

(1) $n \equiv 3 \pmod 8$ *and* $s = 1$, *so that* $n = p_1 \cdots p_t q$ $(t \geqslant 0)$ *where* $p_i \equiv \pm 1 \pmod 8$ $(1 \leqslant i \leqslant t)$, $q \equiv \pm 3 \pmod 8$.

(2) *The graph* $\tilde{G}(n) = (\tilde{V}, \tilde{A})$ *defined by*

$$\tilde{V} = \{p_1, \cdots, p_t, \ q\},$$

$$\tilde{A} = \left\{\overrightarrow{p_i p_j} \,\middle|\, \left(\frac{p_j}{p_i}\right) = -1, \ 1 \leqslant i \neq j \leqslant t\right\} \cup \left\{\overrightarrow{p_i q} \,\middle|\, \left(\frac{q}{p_i}\right) = -1, \ 1 \leqslant i \leqslant t\right\}$$

is odd.

Theorem 4.6 ([4, Theorem 3.2]). *Assume that* $n \equiv 1 \pmod 4$ *with decomposition* (4.2). *Then* $S'_n = \{1\}$ *and* $\hat{S}'_n = \{1, 2, n, 2n\}$ *if and only if the following two conditions are satisfied.*

(1) $s = 2$ *and* $n \equiv 1 \pmod 8$ *so that* $n = p_1 \cdots p_t q_1 q_2$.

(2)

$$\begin{vmatrix} m_{11}^* & m_{12} & \cdots & m_{1t} & b_{11} \\ \vdots & \vdots & & \vdots & \vdots \\ m_{t1} & m_{t2} & \cdots & m_{tt}^* & b_{t1} \\ l_1 & l_2 & \cdots & l_t & k_1 \end{vmatrix} = 1 \in \mathbb{F}_2,$$

where

$$m_{ij} = \begin{cases} 1 & \text{if } (\frac{p_j}{p_i}) = -1, \\ 0 & \text{otherwise}, \end{cases} \qquad 1 \leqslant i \neq j \leqslant t,$$

$$b_{i\lambda} = \begin{cases} 1 & \text{if } (\frac{q_\lambda}{p_i}) = -1, \\ 0 & \text{otherwise}, \end{cases} \qquad 1 \leqslant i \leqslant t, \ 1 \leqslant \lambda \leqslant 2,$$

$$l_i = \begin{cases} 1 & \text{if } p_i \equiv 7 \pmod 8, \\ 0 & \text{if } p_i \equiv 1 \pmod 8, \end{cases} \qquad 1 \leqslant i \leqslant t,$$

$$k_1 = \begin{cases} 1 & \text{if } q_1 \equiv 3 \pmod 8, \\ 0 & \text{if } q_1 \equiv 5 \pmod 8, \end{cases}$$

and

$$m^*_{ii} = \sum_{j=1,j\neq i}^{t} m_{ij} + b_{i1} + b_{i2}, \quad 1 \leqslant i \leqslant t.$$

Theorem 4.7 and 4.8 present series of explicit non-congruent numbers as following.

Corollary 4.2 ([4, Corollary 3.3]). *Suppose that* $n = p_1 \cdots p_t q$ $(t \geqslant 1)$ *and*

$$p_i \equiv 1 \pmod 8,\ 1 \leqslant i \leqslant t-1, \quad p_t \equiv 7 \pmod 8, \quad q \equiv 5 \pmod 8,$$

$$\left(\frac{q}{p_i}\right) = -1, \quad 1 \leqslant i \leqslant t.$$

Then n *is a non-congruent number provided one of the following conditions is satisfied.*
(1) $\left(\frac{p_i}{p_j}\right) = 1$ *for all* $1 \leqslant i \neq j \leqslant t$.
(2) $\left(\frac{p_i}{p_j}\right) = -1$ *for all* $1 \leqslant i \neq j \leqslant t$ *and* t *is even.*

Corollary 4.3 ([4, Corollary 3.4]). *Suppose that* $n = p_1 \cdots p_t q$ $(t \geqslant 1)$ *and*

$$p_i \equiv 1 \pmod 8, 1 \leqslant i \leqslant t-1,\ p_t \equiv 7 \pmod 8, \quad q \equiv 5 \pmod 8.$$

Then n *is a non-congruent number if the following two conditions are satisfied.*
(1) *There is exactly one* i $(1 \leqslant i \leqslant t)$ *such that* $(\frac{q}{p_i}) = -1$.
(2) *The (non-directed) graph* $G = (V, A)$ *defined by*

$$V = \{p_1, \cdots p_t\}, \quad A = \left\{ \overline{p_i p_j} \,\middle|\, \left(\frac{p_j}{p_i}\right) = -1,\ 1 \leqslant i \neq j \leqslant t \right\}$$

is odd.

Corollary 4.4 ([4, Corollary 3.5]). *Suppose that* $n = p_1 \cdots p_t q_1 q_2$ $(t \geqslant 0)$ *and*
(1) $p_i \equiv 1 \pmod 8$, $q_1 \equiv q_2 \equiv 3 \pmod 8$, $(\frac{q_1}{p_i})(\frac{q_2}{p_i}) = -1$, $(1 \leqslant i \leqslant t)$.
(2) $(\frac{p_i}{p_j}) = 1$ $(1 \leqslant i \neq j \leqslant t)$ *or* "$(\frac{p_i}{p_j}) = -1$ $(1 \leqslant i \neq j \leqslant t)$ *and* t *is even*".
Then n *is a non-congruent number.*

Corollary 4.5 ([4, Corollary 3.6]). *Suppose that* $n = p_1 \cdots p_t q_1 q_2$ $(t \geqslant 1)$ *and*
(1) $p_i \equiv 1 \pmod 8$ $(1 \leqslant i \leqslant t-1)$, $(p_t, q_1, q_2) \equiv (7, 5, 3) \pmod 8$;

(2) *All* $(\frac{q_\lambda}{p_i}) = 1$, $(1 \leqslant i \leqslant t,\ 1 \leqslant \lambda \leqslant 2)$ *except* $(\frac{q_1}{p_t}) = -1$;
(3) *The non-directed graph* $G = (V, A)$ *defined in Corollary 4.10 is odd.*
Then n *is a non-congruent number.*

Case $2|n = 2n'$**.** We proved (see [5]) that there is no $n' \equiv 7 \pmod 8$ such that $S_n' = \{1\}$ and $\hat{S}_n' = \{1, 2, n', n\}$. We have not completed the case $n' \equiv 1$ or $3 \pmod 8$, but for $n' \equiv 5 \pmod 8$ we get the following result.

Theorem 4.7 ([5, Theorem 3.1]). *Suppose that* $n = 2n'$ *and* $n' = p_1 \cdots p_t q_1 \cdots q_s \equiv 5 \pmod 8$ *where* $t, s \geqslant 0$, $t + s \geqslant 1$, $p_i \equiv \pm 1 \pmod 8$ *and* $q_j \equiv \pm 3 \pmod 8$, $(1 \leqslant i \leqslant t,\ 1 \leqslant j \leqslant s)$. *Then the following two conditions are equivalent.*
(1) $S_n' = \{1\}$ *and* $\hat{S}_n' = \{1, 2, n', n\}$ *so that* $\operatorname{rank}(E_n'(\mathbb{Q})) = 0$ *and* n *is a non-congruent number;*
(2) $s = 1$ *so that* $n' = p_1 \cdots p_t q$, $q \equiv \pm 3 \pmod 8$; *and*

$$D = \begin{vmatrix} m_{11}^* & m_{12} & \cdots & m_{1t} \\ \vdots & \vdots & & \vdots \\ m_{t1} & m_{t2} & \cdots & m_{tt}^* \end{vmatrix} = 1 \in \mathbb{F}_2,$$

where

$$m_{ij} = \begin{cases} 1 & \text{if } (\frac{p_j}{p_i}) = -1, \\ 0 & \text{otherwise,} \end{cases} \qquad 1 \leqslant i \neq j \leqslant t,$$

$$b_i = \begin{cases} 1 & \text{if } (\frac{q}{p_i}) = -1, \\ 0 & \text{otherwise,} \end{cases} \qquad 1 \leqslant i \leqslant t,$$

and

$$m_{ii}^* = \sum_{j=1, j\neq i}^{t} m_{ij} + b_i, \quad 1 \leqslant i \leqslant t.$$

(We assume $D = 1$ *for* $t = 0$.*)*

From simple computations, we can derive the following consequences of Theorem 4.13.

Corollary 4.6 ([5, Corollary 3.2]). *Suppose that* $n = 2n'$ *and* n' *satisfies one of following conditions where* p, q, p_1, p_2 *are prime numbers and* $p_1 \neq p_2$.
(1) $n' = q \equiv 5 \pmod 8$;
(2) $n' = pq$, $(\frac{q}{p}) = -1$ *and* $(p, q) \equiv (1, 5)$ *or* $(7, 3) \pmod 8$;

(3) $n' = p_1p_2q$, $(p_1, p_2, q) \equiv (1, 1, 5)$ *or* $(1, 7, 3) \pmod 8$ *and there exists one of* p_1, p_2, q *which is quadratic non-residue of the other two prime numbers;*
(4) $n' = p_1p_2q$, $(p_1, p_2, q) \equiv (7, 7, 5) \pmod 8$ *and* $(\frac{p_2}{p_1}) = (\frac{q}{p_2}) = -(\frac{q}{p_1})$.
Then $S'_n = \{1\}$ *and* $\hat{S'_n} = \{1, 2, n', n\}$ *so that* $\mathrm{rank}(E'_n(\mathbb{Q})) = 0$ *and* n *is a non-congruent number.*

Corollary 4.7 ([5, Corollary 3.3]). *Let* $n = 2n'$, $n' = p_1 \cdots p_t q \equiv 5 \pmod 8$ *where* $p_1, \cdots, p_t$ *are distinct prime numbers,* $p_i \equiv \pm 1 \pmod 8$ $(1 \leqslant i \leqslant t)$ *and* $q \equiv \pm 3 \pmod 8$. *Let* $r_1, \cdots, r_l$ *be distinct prime numbers,* $r_\lambda \equiv \pm 1 \pmod 8$ $(1 \leqslant \lambda \leqslant l)$, $(\frac{r_i}{r_j}) = 1$ $(1 \leqslant i, j \leqslant l)$, $r_1 \cdots r_l \equiv 1 \pmod 8$ *and*

$$\left(\frac{p_i}{r_\lambda}\right) = \left(\frac{r_\lambda}{p_i}\right) = 1, \quad 1 \leqslant i \leqslant s,\ 1 \leqslant \lambda \leqslant l,$$

$$\left(\frac{q}{r_\lambda}\right) = -1, \quad 1 \leqslant \lambda \leqslant l.$$

If $D = 1 \in \mathbb{F}_2$ *where* D *is given by Theorem* 4.13 (2) *(so that* n *is a non-congruent number), then* $N = r_1 \cdots r_l n$ *is a non-congruent number.*

5. Birch and Swinnerton-dyer conjecture for E_n

The Birch and Swinnerton-Dyer conjecture says that for each elliptic curve E over $\mathbb{Q}$,

(BSD1) The order of zero of L-function $L(E, s)$ at $s = 1$ is equal to $\mathrm{rank}(E(\mathbb{Q}))$.

(BSD2) If $\mathrm{rank}(E(\mathbb{Q})) = 0$ (so that $L(E, 1) \neq 0$ by (BSD1)), then $L(E, 1)$ is equal to a certain conjectured value.

For several cases of E_n such that $\mathrm{rank}(E_n(\mathbb{Q})) = 0$, Chunlai Zhao [18,19] calculated $L(E_n, 1)$ by using Eisenstein series and odd-graph language and then verified conjecture (BSD1) and (BSD2). For the case that n has at most 2 odd prime divisors, this can be done by using Tunnell's elementary criterion, see [2].

References

1. N. Aoki, On the 2-Selmer groups of elliptic curves arising from the congruent number problems, *Comment. Math. Univ. St. Paul.*, **48** (1999), 77–101.
2. K. Feng, Non-congruent numbers, odd graphs and the Birch-Swinnerton-Dyer conjecture, *Acta Arith.*, **80** (1996), 71–83.
3. K. Feng and M. Xiong, On elliptic curves $y^2 = x^3 - n^2x$ with rank zero, *Jour. of Number Theory*, **109** (2004), 1–26.

4. K. Feng and Y. Xue, New series of odd non-congruent numbers, to appear in *Science in China (A)*, 2006.
5. K. Feng and Y. Xue, New series of non-congruent numbers $n \equiv 10 \pmod{16}$, preprint, 2006.
6. A. Genocchi, Sur l'impossibilité de quelques égalités doubles, *C.R. Acad. Sci. Paris*, **78** (1874), 423–436.
7. T. Goto, *A study on the Selmer groups of elliptic curves with a rational 2-torsion*, Kyushu University, Doctoral thesis, 2002.
8. J. M. Harris, J. L. Hirst, M. J. Mossignhoff, *Combinatorics and Graph Theory*, Springer-Verlag, Berlin, 2000.
9. B. Iskra, Non-congruent numbers with arbitrarily many prime factors congruent to 3 modulo 8, *Proc. Japan Acad.*, **72** (1996), 168–169.
10. N. Koblitz, Introduction to Elliptic Curves and Modular Forms, **GTM 97**, 2nd ed. Springer-Verlag, 1993.
11. J. Lagrange, Construction d'une table de nombres congruents, *Bull. Soc. Math. France, Suppl. Mem.*, 49–50 (1977), 125–130.
12. J. Lagrange, *Nombres congruents et courbes elliptiques, Sémin. Delange-Pisot-Poitou*, 1974/75, Fasc. 1, Exposé 16,17pp.
13. F. Lemmermeyer, Some families of non-congruent numbers, *Acta Arith.*, **110** (2003), 15–36.
14. F. R. Nemenzo, All congruent number less than 40000, *Proc. Japan Acad.*, **74** (1998) 29–31.
15. P. Serf, *Congruent numbers and elliptic curves, in Computational Number Theory* (Debrecen, 1989), de Gruyter, 1991, 227–238.
16. J. Silverman, *The Arithmetic of Elliptic Curves*, **GTM 106**, Springer-Verlag, 1986.
17. J. B. Tunnell, A classical Diophantine problem and modular forms of weight 3/2, *Invent. Math.*, **72** (1983), 323–334.
18. C. Zhao, A criterion for elliptic curves with lowest 2-power in $L(1)$, *Math. Proc. Cambridge Philos. Soc.*, **121** (1997), 385–400.
19. C. Zhao, A criterion for elliptic curves with lowest 2-power in $L(1)$ II, *Acta Math. Sinica (English ser.)*, **21** (2005), 961–976.

DISTRIBUTION OF UNITS OF AN ALGEBRAIC NUMBER FIELD MODULO AN IDEAL

YOSHIYUKI KITAOKA

Department of Mathematics, Meijo University,
Tenpaku, Nagoya, 468-8502, Japan
E-mail: kitaoka@ccmfs.meijo-u.ac.jp

Let F be an algebraic number field and o_F the maximal order of F. We are interested in how units of F distribute in $(o_F/\mathfrak{n})^\times$, where $\mathfrak{n}$ is an integral ideal. When $\mathfrak{n}$ is a prime ideal, we give the upper bound of the order of the subgroup represented by units in $(o_F/\mathfrak{n})^\times$, using new invariants. Prime ideals are ruled by an automorphism of an overfield of F, which is a Galois extension of the rationals. We give the expected density of the set of prime ideals which attain the upper bound, taking account of Chebotarev's density theorem. In the third section, we try to generalize the above to principal ideals generated by rational primes. On the contrary to the prime ideal case, there remain much to do even in order to complete the algebraic framework.

1. Introduction

Let F be an algebraic number field and o_F the ring of algebraic integers in F. The structure of the group $o_F^\times$ of units in F is well described by Dirichlet's Theorem, which says that there exist a primitive w-th root ζ_w of unity and units $\epsilon_1, \cdots, \epsilon_r$ so that they generate $o_F^\times$ and the equality $\zeta_w^{a_0} \prod_i \epsilon_i^{a_i} = 1$ implies $a_0 \equiv 0 \bmod w$, $a_1 = \cdots = a_r = 0$.

In this paper, we are interested in the distribution of units modulo an integral ideal. For an integral ideal $\mathfrak{n}$, we put

$$E(\mathfrak{n}) = \left\{ \epsilon \bmod \mathfrak{n} \,\middle|\, \epsilon \in o_F^\times \right\} \ \left(\subset (o_F/\mathfrak{n})^\times \right).$$

This is a finite group and therefore infinitely many multiplicative relations modulo $\mathfrak{n}$ arise among units. We would like to know these relations, the structures of $E(\mathfrak{n})$ and $(o_F/\mathfrak{n})^\times/E(\mathfrak{n})$. However, these are heavily dependent on the modulus ideal $\mathfrak{n}$, and we are to extract a property common to some appropriate set of ideals.

In §2, we take up the case that $\mathfrak{n}$ is a prime ideal. In this case, $(o_F/\mathfrak{n})^\times$ is cyclic and the structure being determined by the order, this case is easier, as already studied in [7]. Let us briefly explain its outline (see the text for details). Let K be a subsidiary extension field of F, and assume that K is a Galois extension of the rational number field $\mathbb{Q}$. We take an element $\eta \in \mathrm{Gal}(K/\mathbb{Q})$ to control prime ideals of F. $g(x) \in \mathbb{Z}[x]$ is the monic polynomial of minimal degree such that

$$W_1(g(x)) := \left\{ \epsilon^{g(\eta)} \,\middle|\, \epsilon \in o_F^\times \right\}$$

is a finite group, whose order we denote by δ_1 (cf. (2.1)) and put (cf. (2.3))

$$\delta_0 = \max \left\{ m \,\middle|\, \sqrt[m]{\epsilon^{\delta_1 g(\rho)}} = 1 \text{ for } \forall \epsilon \in o_F^\times, \forall \rho_{|K} = \eta \right\},$$

and we say that a prime ideal $\mathfrak{p}$ of F corresponds to η if there is a prime ideal of K lying above $\mathfrak{p}$ whose Frobenius automorphism is η. Then, for every prime ideal $\mathfrak{p}$ of F corresponding to η, we see that $\#E(\mathfrak{p})$ divides $\delta_1 g(p)/\delta_0$, where p denotes a rational prime number lying below $\mathfrak{p}$, and we conjectured in [7]

$$\begin{aligned} &\# \left\{ \mathfrak{p} \,\middle|\, p < x, p \nmid 2D_K, \#E(\mathfrak{p}) = \delta_1 g(p)/\delta_0 \text{ and } \mathfrak{p} \text{ corresponds to } \eta \right\} \\ &\sim \mathrm{den}(\eta) \mathrm{Li}(x), \end{aligned}$$

where

$$\begin{aligned} \mathrm{den}(\eta) = &\left[\{ \sigma \in \mathrm{Gal}(K/F) \,|\, \sigma\eta = \eta\sigma \} : \mathrm{Gal}(K/F) \cap \langle \eta \rangle \right]^{-1} \\ &\times \sum_{m=1}^{\infty} \frac{\mu(m) \# H_{\delta_0 m}(\eta)}{[K_{\delta_0 m} : K_\eta]}. \end{aligned}$$

Here, $\mu(m)$ denotes the Möbius function, $K_m = K\left(\sqrt[m]{o_K^\times}\right)$, K_η is the fixed subfield of K by η and

$$H_m(\eta) = \left\{ \rho \in \mathrm{Gal}(K_m/\mathbb{Q}) \,\middle|\, \rho_{|K} = \eta \text{ and } \sqrt[m]{\epsilon^{\delta_1 g(\rho)}} = 1 \text{ for } \forall \epsilon \in o_F^\times \right\}.$$

We showed that the expected density $\mathrm{den}(\eta)$ is indeed finite and positive. The conjecture is true under Generalized Riemann Hypothesis in a few cases [2,6,12–14]. We know the arithmetic frame-work, but as Artin's conjecture on primitive roots, the remaining problem is the estimation of the accumulation of error terms when we apply Chebotarev's density theorem to infinitely many algebraic number fields.

In §3, we go on to study the case where $\mathfrak{n}$ is a principal ideal generated by a rational prime number. We have already studied a few cases ([8,9]). In this paper, we deal with slightly wider classes; let F be a Galois extension of $\mathbb{Q}$ and η an element of the center of $\mathrm{Gal}(F/\mathbb{Q})$. These two assumptions play the essential role at present, although they should be loosened. The automorphism η controls rational primes through the Frobenius automorphism. Denote the group of roots of unity in F by W_F. Then, we see, as η-modules

$$\mathbb{Q} \otimes_{\mathbb{Z}} (o_F^\times / W_F) \cong \bigoplus_i \mathbb{Q}[d_i],$$

where $\mathbb{Q}[d_i]$ denotes the cyclotomic field $\mathbb{Q}(\zeta_{d_i})$ with η action given by $\alpha^\eta = \zeta_{d_i}\alpha$ for $\alpha \in \mathbb{Q}(\zeta_{d_i})$, and d_i is a divisor of the order d of η. Then, writing $E(p)$ for $E((p))$, we will see in Corollary 3.3 that

$$\#E(p) = \#W_F \prod_i |\Phi_{d_i}(p)| \Big/ \left[\ker \iota_p : W_F o_F^{\times\,\eta - p}\right]$$

for a rational prime $p\,(\nmid 2D_F)$ whose Frobenius automorphism is η. Here, $\Phi_m(x)$ is the cyclotomic polynomial of index m, and a canonical mapping $\iota_p : o_F^\times \to E(p)/W_F$ is defined by

$$\epsilon \mapsto \epsilon \bmod p.$$

Putting

$$\mathrm{Rel}_\eta = \gcd_p \left[\ker \iota_p : W_F o_F^{\times\,\eta - p}\right],$$

where p corresponds to η as above and is sufficiently large (cf. (3.7)), we have

$$\#E(p) \,\Big|\, \#W_F \prod_i |\Phi_{d_i}(p)| \Big/ \mathrm{Rel}_\eta$$

and hence

$$\mathrm{Rel}_\eta \,\Big|\, \#W_F \prod_i |\Phi_{d_i}(p)| \Big/ \#E(p).$$

Experimentally, we expect that there are infinitely many prime numbers p which correspond to η and satisfy $\mathrm{Rel}_\eta = \#W_F \prod_i |\Phi_{d_i}(p)|/\#E(p)$, replacing "|"by "=". We do not know how to evaluate Rel_η. But there is a plausible description for it with supporting experimental data, which we will explain below.

Putting together subgroups corresponding to the same d_i, we put, for a divisor m of d

$$U(m) = \left\{ u \in o_F^\times \,\Big|\, u^{\Phi_m(\eta)} \in W_F \right\}.$$

Let $g(x)$ be the polynomial defined in the case $K = F$ above, and define natural numbers $\tilde{\Delta}$ and $\tau_m(\tilde{\eta})$ for any extension $\tilde{\eta}$ of η as follows:

$$\tilde{\Delta} = \max\left\{t \,\middle|\, \zeta_t^{g(\rho)} = 1 \text{ for } \forall\rho \in \mathrm{Gal}(F(\zeta_t)/\mathbb{Q}) \text{ satisfying } \rho_{|F} = \eta\right\},$$

$$\tau_m = \tau_m(\tilde{\eta}) := \max\left\{t \,\middle|\, \zeta_t^{\Phi_m(\tilde{\eta})} = 1, t|\tilde{\Delta}\right\}.$$

And for a prime number p whose Frobenius class in $\mathrm{Gal}\left(F\left(\sqrt[\tilde{\Delta}]{o_F^\times}\right)/\mathbb{Q}\right)$ contains $\tilde{\eta}$, we put

$$\mathcal{R}(\tilde{\eta}) = \left\{\prod_{m|d} v_m \,\middle|\, \begin{array}{l}\text{(i) } v_m \in U(m),\\ \text{(ii) } \prod_{m|d} v_m^{\Phi_m(p)/\tau_m} \equiv \zeta \bmod p \text{ for } \exists\zeta \in W_F\end{array}\right\}.$$

Here $\Phi_m(p)/\tau_m$ is an integer, and the group $\mathcal{R}(\tilde{\eta})$ depends only on $\tilde{\eta}$ in spite of its definition. Obviously, it induces a subgroup of $\ker\iota_p$

$$\mathcal{R}(\tilde{\eta}, p) = \left\{\prod_{m|d} v_m^{\Phi_m(p)/\tau_m} \,\middle|\, \prod_{m|d} v_m \in \mathcal{R}(\tilde{\eta})\right\} W_F {o_F^\times}^{\eta - p},$$

and hence $[\ker\iota_p : W_F {o_F^\times}^{\eta-p}]$ is divisible by the index $[\mathcal{R}(\tilde{\eta}, p) : W_F {o_F^\times}^{\eta-p}]$, which depends only on $\tilde{\eta}$. Now we have

$$\kappa(\eta) := \gcd_{\tilde{\eta}}\left[\mathcal{R}(\tilde{\eta}, p) : W_F {o_F^\times}^{\eta-p}\right] \,\Big|\, \mathrm{Rel}_\eta.$$

After such preparations, we give $\kappa(\eta)$ explicitly for several types of algebraic number fields of low degree in §4, and making use of it, we confirm the existence of a prime number p satisfying $\kappa(\eta) = [\ker\iota_p : W_F {o_F^\times}^{\eta-p}]$ by computer experiment, which yields the expectation $\mathrm{Rel}_\eta = \kappa(\eta)$.

As referred to above, we have already studied several cases [8] where the rank of $o_F^\times$ is one. There, we have given the value $\kappa(\eta)$ by experiments and showed that the expected density of the set of primes p satisfying $\kappa(\eta) = \#W_F \prod |\Phi_{d_i}(p)|/\#E(p)$ is indeed positive. Our argument here elucidates the theoretical background of their values.

In the appendix, we give the structure of the Galois group of the field extended by roots of units, which is necessary to consider all extensions $\tilde{\eta}$ of η explicitly.

Notations : For an algebraic number field L, we denote by o_L, $o_L^\times$, W_L, D_L, L_m the ring of algebraic integers in L, the group of units in L, the group of roots of unity in L, the discriminant of L, and the field $L\left(\sqrt[m]{o_L^\times}\right)$ extended by all m-th roots of units in L, respectively. Assume that L is a

Galois extension over $\mathbb{Q}$. $\sigma_{L/\mathbb{Q}}(\mathfrak{p})$ denotes the Frobenius automorphism of a prime ideal of $\mathfrak{p}$ of L, and for a prime number p lying below $\mathfrak{p}$, $\sigma_{L/\mathbb{Q}}(p)$ denotes a conjugacy class $\{\rho\sigma_{L/\mathbb{Q}}(\mathfrak{p})\rho^{-1} \mid \rho \in \mathrm{Gal}(L/\mathbb{Q})\}$. For integers a, b we denote their greatest common divisor by (a, b) or by $\gcd(a, b)$. For a polynomial $f(x) = a_0 + a_1x + \cdots + a_nx^n \in \mathbb{Z}[x]$, $\rho \in \mathrm{Gal}(L/\mathbb{Q})$ and $u \in L^\times$, we write

$$u^{f(\rho)} = \prod_{t=0}^{n} u^{a_t\rho^t}.$$

We denote by F an algebraic number field, with which we are mainly concerned in this article, and for an integral ideal $\mathfrak{n}$ of F, we put

$$E(\mathfrak{n}) = \left\{\epsilon \bmod \mathfrak{n} \,\middle|\, \epsilon \in o_F^\times\right\} \left(\subset (o_F/\mathfrak{n})^\times\right).$$

We denote $\#W_F$ by w. For a natural number n, ζ_n is a primitive n-th root of unity and the polynomial

$$\Phi_n(x) = \prod_{(a,n)=1} (x - \zeta_n^a)$$

is the cyclotomic polynomial of index n. For a polynomial $h(x)$, we put

$$h(x, y) = \frac{h(x) - h(y)}{x - y}.$$

2. Case of prime ideals

Throughout this section, fields $K \supset F$ are fixed algebraic number fields and we assume that $\#o_F^\times = \infty$ and that K is a Galois extension of the rational number field $\mathbb{Q}$. We choose and fix an element $\eta \in \mathrm{Gal}(K/\mathbb{Q})$ to control prime ideals of F through the Frobenius automorphism.

2.1. *Polynomial $g(x)$*

First, we introduce a key polynomial $g(x)$, which plays a central role.

Lemma 2.1. *Let $g(x)$ be a non-zero polynomial in $\mathbb{Z}[x]$ such that*

$$W_1(g(x)) := \left\{ \epsilon^{g(\eta)} \,\middle|\, \epsilon \in o_F^\times \right\}$$

is a finite group. We fix a primitive polynomial $g(x)$ of minimal degree among them. Then it divides $x^d - 1$ in $\mathbb{Z}[x]$ for

$$d := \left[\langle\eta\rangle : \langle\eta\rangle \cap \mathrm{Gal}(K/F)\right].$$

Proof. By virtue of $\eta^d \in \mathrm{Gal}(K/F)$, $W_1(x^d-1)=\{1\}$ is clear and we may take a primitive polynomial $g(x)$ of minimal degree satisfying $\#W_1(g(x)) < \infty$. Then there exist an integer a and polynomials $q(x), r(x) \in \mathbb{Z}[x]$ so that $a(x^d-1) = q(x)g(x)+r(x)$ and $\deg r(x) < \deg g(x)$. The assumption $\#o_F^\times = \infty$ implies $\deg g(x) \geq 1$. For $\epsilon \in o_F^\times$, we have $\epsilon^{r(\eta)} = \epsilon^{a(\eta^d-1)-q(\eta)g(\eta)} = (\epsilon^{g(\eta)})^{-q(\eta)}$ and hence $W_1(r(x))$ is a finite group. Hence the minimality of $\deg g(x)$ implies $r(x)=0$ and then the primitiveness of both x^d-1 and $g(x)$ entails that $g(x)$ divides x^d-1. □

Since the polynomial $g(x)$ divides x^d-1, we may assume that $g(x)$ is monic. Hereafter the monic polynomial $g(x)$ means the one defined in Lemma 2.1 and put

$$\delta_1 := \#W_1(g(x)). \tag{2.1}$$

Example 2.1. If $\eta \in \mathrm{Gal}(K/F)$ holds, then obviously $\epsilon^\eta = \epsilon$ for any $\epsilon \in o_F^\times$, and so we have $g(x) = x-1$ and $\delta_1 = 1$. Although the determination of $g(x)$ in general is complicated, we know the following [7]:

> Suppose that $K = F$ is a Galois extension of $\mathbb{Q}$; then the polynomial $g(x)$ is given as follows:
>
> **(R)** The case where F is real.
>
> **(R1)** $g(x) = x^{d-1} + x^{d-2} + \cdots + 1$ if $\mathrm{Gal}(F/\mathbb{Q}) = \langle\eta\rangle$.
> **(R2)** $g(x) = x^d - 1$ otherwise.
>
> **(I)** The case where F is imaginary.
> We denote the complex conjugation by J.
>
> **(I1)** $g(x) = x^{d-1} + x^{d-2} + \cdots + 1$ if $[\mathrm{Gal}(F/\mathbb{Q}) : \langle\eta\rangle] = 2$ and $J \notin \langle\eta\rangle$.
> **(I2)** $g(x) = x^d - 1$ if $[\mathrm{Gal}(F/\mathbb{Q}) : \langle\eta\rangle] > 2$ and $J \notin \langle\eta\rangle$.
> **(I3)** $g(x) = x^{d/2-1} + x^{d/2-2} + \cdots + 1$ if $\mathrm{Gal}(F/\mathbb{Q}) = \langle\eta\rangle$.
> **(I4)** The case of $J \in \langle\eta\rangle \neq \mathrm{Gal}(F/\mathbb{Q})$.
>
> **(i)** If there is an element $u \in \mathrm{Gal}(F/\mathbb{Q})$ such that $JuJ^{-1}u^{-1} \notin \langle\eta\rangle$, then $g(x) = x^d - 1$.
> **(ii)** If $JuJ^{-1}u^{-1} \in \langle\eta\rangle$ holds for every element $u \in \mathrm{Gal}(F/\mathbb{Q})$, then $g(x) = x^{d/2} - 1$.
>
> **Remark.** *In the case of* (ii) *in* (I4), $g(x)$ *is* $(x^{d/2}-1)(\sum_i x^{a_i} - \sum_i x^{b_i})$ *in* [7], *but it turns out that it must be* $x^{d/2}-1$ *in virtue of* $J = \eta^{d/2} \in Z(\mathrm{Gal}(F/\mathbb{Q}))$.

In the general case, suppose that there is a real infinite place of F; then we can show

$$g(x) = \begin{cases} x^{d-1} + x^{d-2} + \cdots + 1 & \text{if } [F:\mathbb{Q}] = d, \\ x^d - 1 & \text{otherwise.} \end{cases}$$

But in the case of F being totally imaginary, the evaluation of $g(x)$ is not easy.

2.2. *Upper bound for $\#E(\mathfrak{p})$*

Let $\mathfrak{P}$ ($\nmid 2D_K$) be an unramified prime ideal of K whose Frobenius automorphism $\sigma_{K/\mathbb{Q}}(\mathfrak{P})$ is η; then we say a prime number p and a prime ideal $\mathfrak{p}$ of F lying below $\mathfrak{P}$ correspond to η. Note that the assumption yields that the condition $\zeta \equiv 1 \bmod \mathfrak{P}$ for a root of unity ζ in K implies $\zeta = 1$. By ramification theory, in particular, applying the assertion (3) for $i = 1$ in the following theorem to $L = K, M = F, N = \mathbb{Q}$ and $\mathfrak{Q} = \mathfrak{P}$ with $H = \mathrm{Gal}(K/F)$ and $Z = \langle\eta\rangle$, we derive that

$$d := \left[\langle\eta\rangle : \langle\eta\rangle \cap \mathrm{Gal}(K/F)\right] = \deg(\mathfrak{P} \cap F), \tag{2.2}$$

where $p^{\deg \mathfrak{p}}$ signifies the number of elements of the residue class field modulo a prime ideal $\mathfrak{p}$.

Hilbert's ramification theory for intermediate fields :
Let $L \supset M \supset N$ be algebraic number fields, and suppose that L/N is a Galois extension with Galois group G, and that H is the subgroup corresponding to M. For a prime ideal $\mathfrak{Q}$ of L, the decomposition group and the inertia group of $\mathfrak{Q}$ with respect to L/N are denoted by Z, T respectively. Then we have
(1) *For $\sigma, \tau \in G$, $\mathfrak{Q}^\sigma \cap M = \mathfrak{Q}^\tau \cap M$ if and only if $Z\sigma H = Z\tau H$.*
(2) *Let*

$$G = Z\sigma_1 H + \cdots + Z\sigma_s H \quad (\sigma_1 = \mathrm{id})$$

be the double coset decomposition. Then the ideals $\mathfrak{Q}^{\sigma_1} \cap M, \cdots, \mathfrak{Q}^{\sigma_s} \cap M$ are all distinct prime ideals of M lying above $\mathfrak{Q} \cap N$.
(3) *Let e_i, f_i be the ramification index and the relative degree of $\mathfrak{Q}^{\sigma_i} \cap M$ with respect to M/N respectively. Then we have*

$$e_i f_i = \left[\sigma_i^{-1} Z \sigma_i : \sigma_i^{-1} Z \sigma_i \cap H\right], \quad e_i = \left[\sigma_i^{-1} T \sigma_i : \sigma_i^{-1} T \sigma_i \cap H\right].$$

Lemma 2.2. *We put*

$$h(x) := (x^d - 1)/g(x) \, (\in \mathbb{Z}[x]).$$

If a prime number $p \, (\nmid 2D_K)$ *corresponds to* η, *then* δ_1 *divides* $h(p)$.

Proof. Take a unit $\epsilon \in o_F^\times$ such that $\epsilon^{g(\eta)}$ is a primitive δ_1-th root of unity, and let $\mathfrak{P}$ be an unramified prime ideal of K lying above p such that $\sigma_{K/\mathbb{Q}}(\mathfrak{P}) = \eta$, and put $\mathfrak{p} = \mathfrak{P} \cap F$. For a generator $\alpha \in o_F$ of $(o_F/\mathfrak{p})^\times$, we put $\epsilon \equiv \alpha^a \bmod \mathfrak{p}$ $(a \in \mathbb{Z})$. Then we have

$$\begin{aligned} 1 = \epsilon^{\delta_1 g(\eta)} &\equiv \epsilon^{\delta_1 g(p)} \mod \mathfrak{P} \\ &\equiv \alpha^{a\delta_1 g(p)} \mod \mathfrak{P}, \end{aligned}$$

which implies $1 \equiv \alpha^{a\delta_1 g(p)} \bmod \mathfrak{p}$. Since by (2.2) $d = \deg \mathfrak{p}$, there is an integer b such that $ag(p)\delta_1 = (p^d - 1)b$, or $a\delta_1 = h(p)b$. We have only to show that $(\delta_1, b) = 1$. Suppose that q is a prime number dividing (δ_1, b); then we have

$$\epsilon^{g(p)\delta_1/q} \equiv \alpha^{ag(p)\delta_1/q} \equiv \alpha^{(p^d-1)b/q} \equiv 1 \bmod \mathfrak{p},$$

which implies $\epsilon^{g(\eta)\delta_1/q} \equiv 1 \bmod \mathfrak{P}$. Since $\epsilon^{g(\eta)\delta_1/q}$ is a root of unity in K and $\mathfrak{P} \nmid 2D_K$, we have $\epsilon^{g(\eta)\delta_1/q} = 1$. This contradicts the fact that $\epsilon^{g(\eta)}$ is a primitive δ_1-th root of unity. □

Lemma 2.3. *Let* m *be a natural number and* $\mathfrak{p} \, (\nmid m)$ *a prime ideal of* F *corresponding to* η. *Let* $\mathfrak{P}_m (\mid \mathfrak{p})$ *be a prime ideal of* K_m *whose Frobenius automorphism* ρ *is an extension of* η. *Then we have*

$$m\#E(\mathfrak{p}) \mid \delta_1 g(p) \iff \sqrt[m]{\epsilon}^{\,\delta_1 g(\rho)} = 1 \text{ for } \forall \epsilon \in o_F^\times,$$

where p *is a prime number lying below* $\mathfrak{p}$ *and* $\sqrt[m]{\epsilon}$ *means all* m*-th roots of* ϵ.

Proof. The left-hand side assertion is equivalent to $m \mid \delta_1 g(p)$ and $\epsilon^{\delta_1 g(p)/m} \equiv 1 \bmod \mathfrak{p}$ for $\forall \epsilon \in o_F^\times$ as $(o_F/\mathfrak{p})^\times$ is cyclic. Since $\sqrt[m]{\epsilon}^{\,\delta_1 g(\rho)}$ is an m-th root of unity in K_m for $\epsilon \in o_F^\times$, the congruence $\epsilon^{\delta_1 g(p)/m} \equiv 1 \bmod \mathfrak{P}_m$ is equivalent to $\sqrt[m]{\epsilon}^{\,\delta_1 g(\rho)} = 1$ by $p \nmid 2mD_K$. Therefore the left-hand side assertion is equivalent to $m \mid \delta_1 g(p)$ and $\sqrt[m]{\epsilon}^{\,\delta_1 g(\rho)} = 1$ for $\forall \epsilon \in o_F^\times$. Noting that the condition $m \mid \delta_1 g(p)$ is contained in the second condition, taking 1 as ϵ, we complete the proof. □

In the lemma, the right-hand side assertion holds for $m = 1$ and therefore

$$\#E(\mathfrak{p}) \mid \delta_1 g(p).$$

We note that

$$m\#E(\mathfrak{p}) \mid \delta_1 g(p) \iff m \cdot h(p)/\delta_1 \mid [(o_F/\mathfrak{p})^\times : E(\mathfrak{p})],$$

where $h(p)/\delta_1$ is an integer by Lemma 2.2.

We put for a natural number m

$$H_m(\eta) := \left\{\rho \in \mathrm{Gal}(K_m/\mathbb{Q}) \,\middle|\, \rho_{|K} = \eta \text{ and } \sqrt[m]{\epsilon}^{\delta_1 g(\rho)} = 1 \text{ for } \forall\epsilon \in o_F^\times \right\}.$$

Here $\sqrt[m]{\epsilon}$ means all m-th roots of ϵ and so $\zeta_m^{\delta_1 g(\rho)} = 1$ for $\rho \in H_m(\eta)$. Now we introduce another constant

$$\delta_0 = \max\left\{m \,\middle|\, \sqrt[m]{\epsilon}^{\delta_1 g(\rho)} = 1 \text{ for } \forall\epsilon \in o_F^\times, \forall\rho \in \mathrm{Gal}(K_m/\mathbb{Q}) \text{ with } \rho_{|K} = \eta \right\}. \tag{2.3}$$

The maximum is assured to exist, by applying Proposition 5.1 in the appendix to $L = K$, $f(x) = \delta_1 g(x)$ with $\epsilon = 1$. Then, taking $m = \delta_0$ in Lemma 2.3, we have

$$\#E(\mathfrak{p}) \mid \delta_1 g(p)/\delta_0$$

for all prime ideals $\mathfrak{p}$ corresponding to η. The evaluation of δ_0 is not easy in general.

Proposition 2.1. *We have $(\delta_0, \delta_1) = 1$ and $\delta_0 \mid g(p)$ for a prime number p corresponding to η.*

Proof. Let m be a divisor of (δ_0, δ_1). Take $\epsilon = \epsilon_0 \in o_F^\times$ so that the order of $\epsilon_0^{g(\eta)}$ is δ_1; then $\sqrt[\delta_0]{\epsilon_0}^{\delta_1 g(\rho)} = 1$ for any extension ρ of η implies $\epsilon_0^{(\delta_1/m) g(\eta)} = 1$, which yields $m = 1$, i.e., $(\delta_0, \delta_1) = 1$. Then, Lemma 2.3 implies $\delta_0 \mid g(p)$. □

Proposition 2.2. *Let $\mathfrak{p}$ be a prime ideal corresponding to η. For a natural number m, which is not divisible by p, the condition $m \mid \delta_1 g(p)/\#E(\mathfrak{p})$ holds if and only if $\rho := \sigma_{K_m/\mathbb{Q}}(\mathfrak{P}_m) \in H_m(\eta)$, where $\mathfrak{P}_m$ is a prime ideal of K_m lying above $\mathfrak{p}$ and satisfies $\sigma_{K/\mathbb{Q}}(\mathfrak{P}_m \cap K) = \eta$.*

Proof. This is an immediate consequence of Lemma 2.3. □

2.3. *Conjecture*

The previous proposition means that for any given natural number m, the condition on $\mathfrak{p}$ that $\delta_1 g(p)/\#E(\mathfrak{p})$ is a multiple of m is characterized in terms of Frobenius automorphisms. Therefore, after some transformations, we can apply Chebotarev's density theorem (see [7] for details).

Theorem 2.1. *Let m be a natural number. If $\mathfrak{p}$ is a prime ideal of F and p is a prime number lying below it, then the density of the set*

$$\left\{\mathfrak{p} \,\middle|\, p \nmid 2m\delta_0 D_K,\ m\#E(\mathfrak{p}) \mid g(p)\delta_1/\delta_0,\ \text{and } \mathfrak{p} \text{ corresponds to } \eta \right\}$$

is equal to

$$\left[\{\sigma \in \mathrm{Gal}(K/F) \,|\, \sigma\eta = \eta\sigma\} : \mathrm{Gal}(K/F) \cap \langle\eta\rangle\right]^{-1} \frac{\#H_{m\delta_0}(\eta)}{[K_{m\delta_0} : K_\eta]}.$$

Here K_η is a subfield of K fixed by $\langle\eta\rangle$.

Then the usual procedure

$$\sum_{\mathfrak{p}\,:\#E(\mathfrak{p})=g(p)\delta_1/\delta_0} 1 = \sum_m \mu(m) \sum_{\mathfrak{p}\,:m\#E(\mathfrak{p})|g(p)\delta_1/\delta_0} 1$$

suggests

Conjecture 2.1. *Denoting by* $\mathrm{den}(\eta)$

$$\left[\{\sigma \in \mathrm{Gal}(K/F) \,|\, \sigma\eta = \eta\sigma\} : \mathrm{Gal}(K/F) \cap \langle\eta\rangle\right]^{-1} \sum_{m=1}^{\infty} \frac{\mu(m)\#H_{\delta_0 m}(\eta)}{[K_{\delta_0 m} : K_\eta]},$$

and denoting a prime number and a prime ideal of F by $p, \mathfrak{p}$ ($\mathfrak{p} \mid p$), we have

$$\#\left\{\mathfrak{p} \,\middle|\, p < x,\ p \nmid 2D_K,\ \#E(\mathfrak{p}) = g(p)\delta_1/\delta_0 \text{ and } \mathfrak{p} \text{ corresponds to } \eta \right\}$$
$$\sim \mathrm{den}(\eta)\mathrm{Li}(x).$$

That the infinite sum $\mathrm{den}(\eta)$ is convergent to a positive number is shown in [7]. We note that the condition $p < x$ is used instead of $N_{F/\mathbb{Q}}(\mathfrak{p}) < x$, and so this is a modification of the usual natural density. This conjecture is a generalization of [2,6,12–14] and hence the conjecture for a real quadratic field $K = F$ is true under the Generalized Riemann Hypothesis. When $\eta \in \mathrm{Gal}(K/F)$, we know that $d = 1$ and $g(x) = x - 1$ in Lemma 2.1 and hence both $\sigma_{K/\mathbb{Q}}(\mathfrak{P}) = \sigma_{K/F}(\mathfrak{P})$ and $h(x) = 1$ hold. Therefore the conjecture is true under G.R.H. by a result of [12]. The situation of [12] is as follows: Let K/F be a finite Galois extension and let C be a union of conjugacy classes of $\mathrm{Gal}(K/F)$, and W is a finitely generated subgroup of $F^\times$ of finite rank (≥ 1) modulo its torsion subgroup, and k is an integer (> 0). In [12], under G.R.H. it is shown that the density of the set $M(F, K, C, W, k)$ of prime ideals $\mathfrak{p}$ of F exists. Here $\mathfrak{p}$ is in $M(F, K, C, W, k)$ if and only if for a prime ideal $\mathfrak{P}$ of K lying above $\mathfrak{p}$, the Frobenius automorphism $\sigma_{K/F}(\mathfrak{P})$ is in C, $\mathrm{ord}_{\mathfrak{p}}(w) = 0$ for all $w \in W$ and the index $[(o_F/\mathfrak{p})^\times : \{w$

mod $\mathfrak{p} \mid w \in W\}]$ divides k. Hence the index is bounded. In contrast to this, our case allows that $[(o_F/\mathfrak{p})^\times : E(\mathfrak{p})]$ tends to infinity.

3. Case of rational primes

Hereafter, we study the distribution of units modulo (p), writing $E(p)$ for $E((p))$, where p is a rational prime, and we restrict ourselves to the case where $K = F$ is a Galois extension of $\mathbb{Q}$, and we let $\eta \in \mathrm{Gal}(F/\mathbb{Q})$. From Corollary 3.3 to the end of this section, we will assume $\eta \in Z(\mathrm{Gal}(F/\mathbb{Q}))$. Because, this yields that for $\epsilon \in o_F^\times$, we have

$$\epsilon^\eta \equiv \epsilon^p \bmod p$$

if a prime number p corresponds to η.

3.1. *Structure of $o_F^\times$ as an η-module*

Lemma 3.1. *Let d be a natural number. For a divisor m of d, we put*

$$\Theta_m(x) = \sum_{k=0}^{d-1} \Bigg(\sum_{\substack{a \bmod m \\ (a,m)=1}} \zeta_m^{ak} \Bigg) x^k \in \mathbb{Z}[x].$$

Then we have

$$\sum_{m|d} \Theta_m(x) = d, \quad x^d - 1 \,\Big|\, \Theta_m(x)\Phi_m(x).$$

Proof. Let ζ be a d-th root of unity; then there exist a divisor m of d and an integer a so that $\zeta = \zeta_m^a$, where m and $a \bmod m$ $((a,m) = 1)$ are uniquely determined, and hence we have

$$\sum_{m|d} \Bigg(\sum_{\substack{a \bmod m \\ (a,m)=1}} \zeta_m^{ak} \Bigg) = \sum_{j \bmod d} \zeta_d^{jk} = \begin{cases} 0 & \text{if } k \not\equiv 0 \bmod d, \\ d & \text{if } k \equiv 0 \bmod d, \end{cases}$$

which yields

$$\sum_{m|d} \Theta_m(x) = d.$$

Next, let us show that $(x^d - 1)/\Phi_m(x) \mid \Theta_m(x)$. Since a root of $(x^d - 1)/\Phi_m(x) = 0$ is not a primitive m-th root of unity, we have only to show that $\Theta_m(\zeta_d^b) = 0$ for every integer b so that ζ_d^b, being a d-th root of unity, is not a primitive m-th root of unity. Let b be an integer; then for an integer a satisfying $(a, m) = 1$, we have

$$\begin{aligned} \zeta_d^{da/m+b} = 1 &\Rightarrow ad/m + b \equiv 0 \bmod d \\ &\Rightarrow am_1 + b \equiv 0 \bmod mm_1 && (\text{ putting } d = mm_1) \\ &\Rightarrow b = m_1 m_2, m_2 \equiv -a \bmod m && (\text{ for } \exists m_2 \in \mathbb{Z}) \\ &\Rightarrow \zeta_d^b = \zeta_m^{m_2} \text{ is a primitive } m\text{-th root of unity.} \end{aligned}$$

This shows for $(a, m) = 1$ that if ζ_d^b is not a primitive m-th root of unity, then $\zeta_d^{ad/m+b} \neq 1$ and so we have

$$\Theta_m(\zeta_d^b) = \sum_{\substack{a \bmod m \\ (a,m)=1}} \sum_{k=0}^{d-1} (\zeta_d^{ad/m+b})^k = 0.$$

This completes the proof of $(x^d - 1)/\Phi_m(x) \mid \Theta_m(x)$. □

Lemma 3.2. *For a divisor m of d, decompose $x^d - 1$ as*

$$\Phi_m(x)\Psi_m(x) = x^d - 1.$$

Then there exist polynomials $u_m(x), v_m(x)$ in $\mathbb{Z}[x]$ satisfying

$$u_m(x)\Phi_m(x) + v_m(x)\Psi_m(x) = d. \tag{3.1}$$

Proof. We put directly as follows :

$$\begin{aligned} u_m(x) &= x\Psi_m'(x) - \deg \Psi_m(x) \cdot \Psi_m(x) \in \mathbb{Z}[x], \\ v_m(x) &= x\Phi_m'(x) - \deg \Phi_m(x) \cdot \Phi_m(x) \in \mathbb{Z}[x]. \end{aligned}$$

Then we have

$$\begin{aligned} &u_m(x)\Phi_m(x) + v_m(x)\Psi_m(x) \\ &= \big(x\Psi_m'(x) - \deg \Psi_m(x) \cdot \Psi_m(x)\big)\Phi_m(x) \\ &\qquad\qquad + \big(x\Phi_m'(x) - \deg \Phi_m(x) \cdot \Phi_m(x)\big)\Psi_m(x) \\ &= x\big(\Psi_m'(x)\Phi_m(x) + \Phi_m'(x)\Psi_m(x)\big) \\ &\qquad\qquad - \big(\deg \Psi_m(x) + \deg \Phi_m(x)\big)\Phi_m(x)\Psi_m(x) \\ &= x(x^d - 1)' - d(x^d - 1) \\ &= d. \end{aligned}$$

□

We fix $\eta \in \mathrm{Gal}(F/\mathbb{Q})$ and denote the order of η by d :

$$\#\langle \eta \rangle = d. \tag{3.2}$$

Proposition 3.1. *Let m be a positive divisor of d. Then*

$$U(m) = \left\{ u \in o_F^\times \,\middle|\, u^{\Phi_m(\eta)} \in W_F \right\} \tag{3.3}$$

is an η-stable subgroup of $o_F^\times$, and we have

$$o_F^{\times\,\Theta_m(\eta)},\ o_F^{\times\,\Psi_m(\eta)} \subset U(m), \quad o_F^{\times\,d} \subset \prod_{m|d} U(m) \subset o_F^\times.$$

If η is in the center of $\mathrm{Gal}(F/\mathbb{Q})$, *then $U(m)$ is* $\mathrm{Gal}(F/\mathbb{Q})$*-stable.*

Proof. It is easy to see that $U(m)$ is η-stable, and if we assume that η is in the center of $\mathrm{Gal}(F/\mathbb{Q})$, $u^{\rho\Phi_m(\eta)} = u^{\Phi_m(\eta)\rho}$ holds for $\rho \in \mathrm{Gal}(F/\mathbb{Q})$ and $u \in o_F^\times$, and so $U(m)$ is a $\mathrm{Gal}(F/\mathbb{Q})$-stable subgroup. By previous lemmas, we know that

$$\Theta_m(x)\Phi_m(x) \equiv \Psi_m(x)\Phi_m(x) \equiv 0 \bmod x^d - 1,$$

whence $\Theta_m(\eta)\Phi_m(\eta) = \Psi_m(\eta)\Phi_m(\eta) = 0$, which implies the left-hand side inclusion. Then Lemma 3.1 yields

$$u^d = u^{\sum_{m|d} \Theta_m(\eta)} = \prod_{m|d} u^{\Theta_m(\eta)},$$

which implies the right-hand side inclusion. □

Proposition 3.2.

(i) *Let u_m, v_m be elements in $U(m)$ and $\zeta \in W_F$. Then*

$$\prod_{m|d} u_m = \zeta \prod_{m|d} v_m$$

implies

$$u_m/v_m \in W_F, \quad \forall m \mid d.$$

(ii) *Let q be a natural number and $u_m \in U(m), \zeta \in W_F$. If*

$$\prod_{m|d} u_m \equiv \zeta \bmod q,$$

then there are roots κ_m's of unity in W_F such that

$$u_m^d \equiv \kappa_m \bmod q, \quad \forall m \mid d.$$

Proof. To prove the assertions, we may assume $\zeta = v_m = 1$, taking the quotient of both sides and absorbing ζ^{-1} into u_1. Let $u_m(x), v_m(x), \Psi_m(x)$ be those in Lemma 3.2. The assertion (i) is proved as follows. Noting that $u_m = \prod_{\substack{n|d \\ n\neq m}} u_n^{-1}$, we have by (3.1)

$$\begin{aligned} u_m^d &= \prod_{\substack{n|d \\ n\neq m}} u_n^{-u_m(\eta)\Phi_m(\eta)-v_m(\eta)\Psi_m(\eta)} \\ &= \left(\prod_{\substack{n|d \\ n\neq m}} u_n\right)^{-v_m(\eta)\Psi_m(\eta)} \cdot u_m^{u_m(\eta)\Phi_m(\eta)}, \end{aligned}$$

which we rewrite as

$$u_m^d = \left(\prod_{\substack{n|d \\ n\neq m}} (u_n^{\Phi_n(\eta)})^{-v_m(\eta)(\Psi_m/\Phi_n)(\eta)}\right) \cdot u_m^{\Phi_m(\eta)u_m(\eta)},$$

where we note that $\Phi_n(x)$ divides $\Psi_m(x)$ if $n \neq m$. Hence, recalling (3.3), $u_n^{\Phi_n(\eta)} \in W_F$ implies $u_m^d \in W_F$ and so $u_m \in W_F$.

Next, we assume $\prod u_m \equiv 1 \bmod q$; then similarly as above we have

$$u_m^d \equiv \prod_{\substack{n|d \\ n\neq m}} (u_n^{\Phi_n(\eta)})^{-v_m(\eta)(\Psi_m/\Phi_n)(\eta)} \cdot u_m^{\Phi_m(\eta)u_m(\eta)} \bmod q,$$

and the right-hand side is in W_F and denoted by κ_m. □

Remark 3.1. In *(ii)*, a stronger conclusion $u_m \equiv \kappa_m \bmod q$ does not hold in general.

Considering $U(m)/W_F$ as a $\mathbb{Z}$-lattice, we put

$$V(m) = U(m)/W_F \otimes_{\mathbb{Z}} \mathbb{Q}.$$

$f(x) \in \mathbb{Z}[x]$ acts on $U(m)$ by $u \mapsto u^{f(\eta)}$, and so $U(m)/W_F$ is a $\mathbb{Z}[x]$-module annihilated by $\Phi_m(x)$. Hence $\mathbb{Q}[x]/(\Phi_m(x))$ acts on $V(m)$. Thus $V(m)$ is a vector space over $\mathbb{Q}(\zeta_m)$, and thus the following Lemma 3.3 (Exercise 2 on p.282 in [3]) is clear. Note that $f(x) \in \mathbb{Q}[x]/(\Phi_m(x))$ acts on $\mathbb{Q}(\zeta_m)$ by $\alpha \mapsto f(\zeta_m)\alpha$ ($\alpha \in \mathbb{Q}(\zeta_m)$).

Lemma 3.3. *As η-modules, we have*

$$\mathbb{Q} \otimes_{\mathbb{Z}} (o_F^\times/W_F) \cong \bigoplus_i \mathbb{Q}[d_i],$$

where $\mathbb{Q}[d_i]$ is $\mathbb{Q}(\zeta_{d_i})$ viewed as a representation space of η, on which η acts by $\alpha^\eta = \zeta_{d_i}\alpha$ for $\alpha \in \mathbb{Q}(\zeta_{d_i})$, and d_i is a divisor of the order d of η.

Now let $1, \zeta_{d_i}, \cdots, \zeta_{d_i}^{\varphi(d_i)-1}$ be a basis of $\mathbb{Z}[\zeta_{d_i}]$ ($\subset \mathbb{Q}[d_i]$) and put

$$\Phi_{d_i}(x) = a_0 + a_1 x + \cdots + a_{\varphi(d_i)-1} x^{\varphi(d_i)-1} + x^{\varphi(d_i)}.$$

Then we have

$$1^\eta = \zeta_{d_i}, \quad \zeta_{d_i}^\eta = \zeta_{d_i}^2, \cdots,$$
$$\left\{\zeta_{d_i}^{\varphi(d_i)-1}\right\}^\eta = \zeta_{d_i}^{\varphi(d_i)} = -\left(a_0 + a_1\zeta_{d_i} + \cdots + a_{\varphi(d_i)-1}\zeta_{d_i}^{\varphi(d_i)-1}\right).$$

Hence denoting by U_i the subgroup of $o_F^\times$ corresponding to $n\mathbb{Z}[\zeta_{d_i}]$ for an appropriate natural number n in Lemma 3.3 where n is sufficiently large to kill the ambiguity of W_F, $\prod U_i$ is a subgroup of finite index of $o_F^\times$, and each $U = U_i$ has a basis $U = \langle \epsilon_0, \epsilon_1, \cdots, \epsilon_{\varphi(d_i)-1}\rangle$ ($\epsilon_j \leftrightarrow \zeta_{d_i}^j$) such that

$$\epsilon_i^\eta = \epsilon_{i+1}\ (i = 0, 1, \cdots, \varphi(d_i) - 2), \qquad \epsilon_0^{a_0}\epsilon_1^{a_1}\cdots\epsilon_{\varphi(d_i)-1}^{a_{\varphi(d_i)-1}}\epsilon_{\varphi(d_i)-1}^{\eta} = 1.$$

Lemma 3.4. *For the subgroup $U = U_i$ above, and $h(x) \in \mathbb{Z}[x]$, we have*

$$U^{h(\eta)} \subset W_F \Leftrightarrow \Phi_{d_i}(x) \,\big|\, h(x)$$

Proof. The assertion follows from

$$U^{h(\eta)} \subset W_F \Leftrightarrow h(\zeta_{d_i})\mathbb{Z}[\zeta_{d_i}] = 0 \Leftrightarrow h(\zeta_{d_i}) = 0 \Leftrightarrow \Phi_{d_i}(x) \,\big|\, h(x). \qquad \square$$

Corollary 3.1. *Let $g(x)$ be the polynomial defined in the previous section for $K = F$ and η, i.e. the monic polynomial in $\mathbb{Z}[x]$ of minimal degree such that*

$$\left\{\epsilon^{g(\eta)} \,\middle|\, \epsilon \in o_F^\times\right\}$$

is a finite group. Then $g(x)$ is equal to

$$\mathrm{lcm}_i\, \Phi_{d_i}(x) = \prod_{m|d,\, U(m) \neq W_F} \Phi_m(x).$$

Proof. By Lemma 3.4, the following equivalence holds for $h(x) \in \mathbb{Z}[x]$

$$\left\{\epsilon^{h(\eta)} \,\middle|\, \epsilon \in o_F^\times\right\} \subset W_F \Leftrightarrow U_i^{h(\eta)} \subset W_F \text{ for } \forall i \Leftrightarrow \Phi_{d_i} \mid h \text{ for } \forall i.$$

Thus we have $g(x) = \mathrm{lcm}_i\, \Phi_{d_i}(x)$. Similarly for $U(m)$ in (3.3), it is easy to see

$$\left\{\epsilon^{h(\eta)} \,\middle|\, \epsilon \in o_F^\times\right\} \subset W_F \Leftrightarrow U(m)^{h(\eta)} \subset W_F \text{ for } \forall m \Leftrightarrow \Phi_m \mid h \text{ for } \forall m,$$

where m should satisfy the condition $U(m) \neq W_F$. Because, the first equivalence is obvious, and the right-hand side divisibility implies the middle inclusion. Assume the middle inclusion. If $\Phi_m(x) \nmid h(x)$, then there are polynomials $f_1(x), f_2(x) \in \mathbb{Z}[x]$ such that $f_1(x)\Phi_m(x) + f_2(x)h(x) = e \in \mathbb{Z}\,(e \neq 0)$, and hence $U(m)^e \subset W_F$, which yields $U(m) \subset W_F$. This contradicts $U(m) \neq W_F$ and hence we obtain the right-hand side divisibility. Thus we have $g(x) = \prod_{m|d, U(m)\neq W_F} \Phi_m(x)$. □

Theorem 3.1. *For an integer p $(\neq \pm 1)$, we have*

$$[o_F^\times : W_F o_F^{\times\,\eta - p}] = \prod_i |\Phi_{d_i}(p)|,$$

where d_i's are those in Lemma 3.3.

To prove the theorem, we need some lemmas.

Lemma 3.5. *For $\epsilon \in o_F^\times$ and for an integer p $(\neq \pm 1)$, the inclusion $\epsilon^{\eta-p} \in W_F$ implies $\epsilon \in W_F$. Moreover, let $o_F^\times \supset U \supset V \supset W_F$ be η-groups; then the mapping $\phi : \epsilon \mapsto \epsilon^{\eta-p}$ from U to $W_F U^{\eta-p}$ induces an isomorphism*

$$U/V \cong W_F U^{\eta-p}/W_F V^{\eta-p}.$$

Proof. Suppose $\epsilon^{\eta-p} \in W_F$ for $\epsilon \in o_F^\times$. Inductively, it is easy to see that there is an element $\kappa_n \in W_F$ such that

$$\epsilon^{\eta^n} = \kappa_n \epsilon^{p^n}.$$

The assumption $\eta^d = \mathrm{id}$ yields $\epsilon = \kappa_d \epsilon^{p^d}$. Hence we have $\epsilon^{1-p^d} = \kappa_d \in W_F$, which implies $\epsilon \in W_F$ by $1 - p^d \neq 0$.

Next, suppose $\phi(\epsilon) \in W_F V^{\eta-p}$ for $\epsilon \in U$; then $\epsilon^{\eta-p} = \zeta v^{\eta-p}$ $(\zeta \in W_F, v \in V)$ holds. Thus $(\epsilon/v)^{\eta-p} = \zeta$ implies $\epsilon/v \in W_F$ and $\epsilon \in V$, which completes the proof. □

Lemma 3.6. *Let U_1, U_2, U be η-subgroups of $o_F^\times$ and suppose*

$$U_1U_2 \subset U, \quad U_1 \cap U_2 = W_F, \quad [U : U_1U_2] < \infty.$$

Then we have

$$[U : W_F U^{\eta-p}] = [U_1 : W_F U_1^{\eta-p}][U_2 : W_F U_2^{\eta-p}].$$

Proof. A canonical mapping $(u_1, u_2) \mapsto u_1u_2$ from $U_1 \times U_2$ to U_1U_2 induces a surjective homomorphism

$$f : U_1 \times U_2 \to U_1U_2/W_F(U_1U_2)^{\eta-p},$$

and it is easy to see $\ker f \supset W_F U_1^{\eta-p} \times W_F U_2^{\eta-p}$ and

$$\begin{aligned}(u_1, u_2) \in \ker f &\Rightarrow u_1 u_2 \in W_F (U_1 U_2)^{\eta-p} \\ &\Rightarrow u_1 u_2 = \zeta (v_1 v_2)^{\eta-p} \qquad (\zeta \in W_F,\ v_i \in U_i) \\ &\Rightarrow \zeta' := \frac{u_1}{v_1^{\eta-p}} = \zeta \frac{v_2^{\eta-p}}{u_2} \in U_1 \cap U_2 = W_F \\ &\Rightarrow u_1 = \zeta' v_1^{\eta-p},\ u_2 = (\zeta/\zeta') v_2^{\eta-p}\end{aligned}$$

Hence we have $\ker f \subset W_F U_1^{\eta-p} \times W_F U_2^{\eta-p}$ and so $\ker f = W_F U_1^{\eta-p} \times W_F U_2^{\eta-p}$, which implies

$$[U_1 U_2 : W_F (U_1 U_2)^{\eta-p}] = [U_1 : W_F U_1^{\eta-p}][U_2 : W_F U_2^{\eta-p}]. \tag{3.4}$$

Lemma 3.5 yields

$$[U : U_1 U_2] = [W_F U^{\eta-p} : W_F (U_1 U_2)^{\eta-p}]. \tag{3.5}$$

From (3.4) and (3.5) we have

$$\begin{aligned}[U : W_F U^{\eta-p}] &= \frac{[U : U_1 U_2][U_1 U_2 : W_F (U_1 U_2)^{\eta-p}]}{[W_F U^{\eta-p} : W_F (U_1 U_2)^{\eta-p}]} \\ &= [U_1 : W_F U_1^{\eta-p}][U_2 : W_F U_2^{\eta-p}].\end{aligned}$$

□

Lemma 3.7. *Suppose for an η-subgroup U we have*

$$U = W_F \left\{ \epsilon^{f(\eta)} \,\middle|\, f(x) \in \mathbb{Z}[x] \right\}$$

for some $\epsilon \in o_F^\times$. Let $h(x) \in \mathbb{Z}[x]$ be a primitive polynomial of minimal degree such that $U^{h(\eta)} \subset W_F$. Then $U/W_F U^{\eta-p}$ is a cyclic group generated by a coset $\epsilon W_F U^{\eta-p}$ and the following holds:

$$[U : W_F U^{\eta-p}] = |h(p)|,$$

$$W_F U^{\eta-p} = W_F \left\{ \epsilon^{A(\eta)} \,\middle|\, \begin{array}{l} A(x) \in \mathbb{Z}[x] \text{ with } \deg A(x) < \deg h(x) \text{ and} \\ A(p) \equiv 0 \bmod h(p) \end{array} \right\}.$$

Proof. By the assumption $\eta^d = \mathrm{id}$, we have $U^{\eta^d - 1} = \{1\} \subset W_F$ and so the polynomial $h(x)$ referred to in the lemma exists. Dividing $x^d - 1$ by $h(x)$, we write

$$x^d - 1 = q(x)h(x) + r(x) \quad (q(x), r(x) \in \mathbb{Q}[x],\ \deg r(x) < \deg h(x)),$$

and choose a non-zero integer a such that $aq(x), ar(x) \in \mathbb{Z}[x]$. Then by virtue of

$$\epsilon^{ar(\eta)} = \epsilon^{a(\eta^d - 1) - aq(\eta)h(\eta)} = (\epsilon^{h(\eta)})^{-aq(\eta)} \in W_F,$$

we have $r(x) = 0$ by the choice of $h(x)$. Hence $x^d - 1 = q(x)h(x)$, and we may assume that $h(x)$ is monic. We let its degree be n.

Because of $\epsilon \in U$, $\epsilon^{\eta-p} \in U^{\eta-p}$ is clear and hence we have $\epsilon^{\eta} W_F U^{\eta-p} = \epsilon^p W_F U^{\eta-p}$, and so $U/W_F U^{\eta-p}$ is a cyclic subgroup generated by $\epsilon W_F U^{\eta-p}$.

Since $h(x)$ is a monic polynomial of degree n, we have

$$U = W_F \langle \epsilon, \epsilon^{\eta}, \cdots, \epsilon^{\eta^{n-1}} \rangle.$$

We note that $\epsilon^{A(\eta)} \in W_F$ for a polynomial $A(x) \in \mathbb{Z}[x]$ with $\deg A(x) < \deg h(x)$ implies $A(x) = 0$. Because, $\epsilon^{A(\eta)} \in W_F$ yields $U^{A(\eta)} \subset W_F$ and therefore the definition of $h(x)$ implies $A(x) = 0$.

Now, let us show the second assertion; put

$$v = \epsilon^{A(\eta)} \in U, \quad A(x) = a_0 + a_1 x + \cdots + a_{n-1} x^{n-1} \in \mathbb{Z}[x].$$

We shall show that $v \in W_F U^{\eta-p}$ is equivalent to $A(p) \equiv 0 \bmod h(p)$. To this end, write

$$h(x) = \sum_{i=0}^{n} h_i x^i, \quad h_n = 1.$$

First, assume $v = \kappa u^{\eta-p} \in W_F U^{\eta-p}$ and put $u = \epsilon^{b_0 + b_1 \eta + \cdots + b_{n-1}\eta^{n-1}}$; then we have

$$\begin{aligned} v &= \kappa \epsilon^{(b_0 + b_1\eta + \cdots + b_{n-1}\eta^{n-1})(\eta - p)} \\ &= \kappa \epsilon^{b_0\eta + b_1\eta^2 + \cdots + b_{n-1}\eta^n - p(b_0 + b_1\eta + \cdots + b_{n-1}\eta^{n-1})} \\ &= \kappa \epsilon^{b_{n-1}h(\eta)} \\ &\quad \times \epsilon^{b_0\eta + b_1\eta^2 + \cdots + b_{n-2}\eta^{n-1} - b_{n-1}(h_0 + \cdots + h_{n-1}\eta^{n-1}) - p(b_0 + b_1\eta + \cdots + b_{n-1}\eta^{n-1})}. \end{aligned}$$

By the choice of $h(x)$, $\epsilon^{b_{n-1}h(\eta)} \in W_F$, whence comparing the exponent of ϵ, we obtain

$$a_0 = -b_{n-1}h_0 - pb_0, \quad a_k = b_{k-1} - b_{n-1}h_k - pb_k, \quad 1 \le k \le n-1.$$

Hence, putting $B(x) = \sum_{k=1}^{n} b_{k-1} x^{k-1}$, we get

$$B(x) = \sum_{k=1}^{n-1} (b_{n-1}h_k + pb_k + a_k) x^{k-1} + b_{n-1} x^{n-1},$$

which we may rewrite as

$$b_{n-1}(h(x) - h_0)/x + p(B(x) - b_0)/x + \sum_{k=1}^{n-1} a_k x^{k-1},$$

whence

$$(x-p)B(x) = b_{n-1}h(x) - b_{n-1}h_0 - pb_0 + \sum_{k=1}^{n-1} a_k x^k = b_{n-1}h(x) + A(x).$$

Substituting $x = p$, we have $A(p) = -b_{n-1}h(p)$, i.e., $h(p) \mid A(p)$.

If, conversely $h(p) \mid A(p)$ holds, then we define $b_{n-1} \in \mathbb{Z}$ by $A(p) = -b_{n-1}h(p)$. Then $x - p$ divides $b_{n-1}h(x) + A(x)$ and we may put

$$b_{n-1}h(x) + A(x) = (x-p)B(x), \quad B(x) \in \mathbb{Z}[x].$$

The leading coefficient of $B(x)$ is b_{n-1} and so we may put $B(x) = \sum_{k=0}^{n-1} b_k x^k$ for some integers $b_0, \cdots, b_{n-2}$. Then we have $v = \epsilon^{A(\eta)} = (\epsilon^{h(\eta)})^{-b_{n-1}}(\epsilon^{B(\eta)})^{\eta-p}$. Since by the choice of $h(x)$, $\epsilon^{h(\eta)} \in W_F$, we obtain $v \in W_F U^{\eta-p}$. Thus we have shown the equivalence and the last assertion in the lemma.

Since $U/W_F U^{\eta-p}$ is generated by $\epsilon W_F U^{\eta-p}$, the index $[U : W_F U^{\eta-p}]$ is equal to the order of $\epsilon W_F U^{\eta-p}$. Applying the last assertion to $A(x) = m \in \mathbb{Z}$, we conclude that the condition $\epsilon^m \in W_F U^{\eta-p}$ is equivalent to $m \equiv 0 \bmod h(p)$. Hence the second assertion $[U : W_F U^{\eta-p}] = |h(p)|$ follows. □

Proof of Theorem 3.1. By Lemma 3.3, there are η-subgroups U_i of $o_F^\times$ such that $U_i/W_F \cong \mathbb{Z}[\zeta_{d_i}]$ and U_i/W_F's form a direct product in $o_F^\times/W_F$. $[U_i : W_F U_i^{\eta-p}] = |\Phi_{d_i}(p)|$ follows from Lemma 3.7, and then Lemma 3.6 completes the proof of the theorem. □

Corollary 3.2. *Suppose that* $p(\neq \pm 1)$ *is an integer; then we have for* $U(m)$ *defined in* (3.3)

$$\left[U(m) : W_F U(m)^{\eta-p}\right] = |\Phi_m(p)|^r,$$
$$U(m)^{\Phi_m(p)} \subset W_F U(m)^{\eta-p},$$

where r *is defined by* $r\varphi(m) = \mathrm{rank}_{\mathbb{Z}} U(m)$.

Proof. Recalling that $\mathbb{Q} \otimes_{\mathbb{Z}} (U(m)/W_F)$ is a vector space over $\mathbb{Q}(\zeta_m)$, we denote its dimension by r. Therefore $U(m)$ contains a subgroup which is isomorphic to a direct product of r copies of $\mathbb{Z}[\zeta_m]$ as η-modules. Then the first equation follows from Lemmas 3.6 and 3.7.

Let $u \in U(m)$; then $u^{\Phi_m(\eta)-\Phi_m(p)} = u^{\Phi_m(\eta,p)(\eta-p)} \in U(m)^{\eta-p}$ (cf. Notation) and $U(m)^{\Phi_m(\eta)} \subset W_F$ (cf. (3.3)) together imply $u^{\Phi_m(p)} \in W_F U(m)^{\eta-p}$. □

From now on, we assume $\eta \in Z(\mathrm{Gal}(F/\mathbb{Q}))$.

Corollary 3.3. *Suppose that η be in the center of* $\mathrm{Gal}(F/\mathbb{Q})$*, and a prime number* p ($\nmid 2D_F$) *satisfies* $\eta = \sigma_{F/\mathbb{Q}}(p)$ *(cf. Notation). Then for a canonical surjective mapping*

$$\iota_p : o_F^\times \to E(p)/W_F$$

defined by

$$\epsilon \mapsto \epsilon \bmod p$$

we have

$$\#E(p) = w\frac{[o_F^\times : W_F {o_F^\times}^{\eta-p}]}{[\ker \iota_p : W_F {o_F^\times}^{\eta-p}]} = \frac{w\prod_i |\Phi_{d_i}(p)|}{[\ker \iota_p : W_F {o_F^\times}^{\eta-p}]}, \tag{3.6}$$

where $w = \#W_F$.

Proof. Since η is in the center of $\mathrm{Gal}(F/\mathbb{Q})$, it follows from $\eta = \sigma_{F/\mathbb{Q}}(p)$ that $\epsilon^{\eta-p} \equiv 1 \bmod p$ holds for $\epsilon \in o_F^\times$, and so we have $W_F(o_F^\times)^{\eta-p} \subset \ker \iota_p$, whence by the homomorphism theorem

$$\#(E(p)/W_F) = \#E(p)/w = \frac{[o_F^\times : W_F {o_F^\times}^{\eta-p}]}{[\ker \iota_p : W_F {o_F^\times}^{\eta-p}]} = \frac{\prod_i |\Phi_{d_i}(p)|}{[\ker \iota_p : W_F {o_F^\times}^{\eta-p}]}$$

by Theorem 3.1. □

In regard to this, under the assumption $\eta \in Z(\mathrm{Gal}(F/\mathbb{Q}))$, we put

$$\mathrm{Rel}_\eta = \gcd_p \left[\ker \iota_p : W_F {o_F^\times}^{\eta-p}\right], \tag{3.7}$$

where prime numbers p satisfy $\sigma_{F/\mathbb{Q}}(p) = \eta$ and $p \nmid 2D_{F_{\tilde{\Delta}}}$ for a constant $\tilde{\Delta}$ defined in the next subsection. Then we have

$$\#E(p) \,\Big|\, \frac{w\prod_i |\Phi_{d_i}(p)|}{\mathrm{Rel}_\eta}\,.$$

This upper bound seems to be the best one. Although we do not know how to evaluate Rel_η, there is a candidate $\kappa(\eta)$ for it, which is a divisor of Rel_η by definition. We will explain it in the next subsection, and in §4, we describe $\kappa(\eta)$ explicitly for several types of algebraic number fields.

3.2. Rel_η *and* $\kappa(\eta)$

In this subsection, we define a candidate $\kappa(\eta)$ for Rel_η and in §3.3, we rewrite it to evaluate easily it, and in §4, we write it down explicitly for several types of algebraic number fields. Computer experiments convince us of the truth of the conjecture. As before, let $\eta \in Z(\mathrm{Gal}(F/\mathbb{Q}))$ and let the polynomial $g(x)$ be as in Corollary 3.1. We put

$$\tilde{\Delta} = \max\left\{t \in \mathbb{N} \,\middle|\, \zeta_t^{g(\rho)} = 1 \text{ for } \forall\rho \in \mathrm{Gal}(F(\zeta_t)/\mathbb{Q}) \text{ with } \rho_{|F} = \eta\right\} \tag{3.8}$$

and for a divisor m of d (cf. (3.2)) and an extension $\tilde{\eta} \in \mathrm{Gal}(F_{\tilde{\Delta}}/\mathbb{Q})$ of η, put

$$\tau_m = \tau_m(\tilde{\eta}) = \max\left\{t \,\middle|\, \zeta_t^{\Phi_m(\tilde{\eta})} = 1,\ t|\tilde{\Delta}\right\},$$

where $F_{\tilde{\Delta}} = F\left(\sqrt[\tilde{\Delta}]{o_F^\times}\right)$. The existence of $\tilde{\Delta}$ is guaranteed by Proposition 5.1. By defining an integer a by

$$\zeta_{\tilde{\Delta}}^{\tilde{\eta}} = \zeta_{\tilde{\Delta}}^a,$$

it is easy to see that

$$\tau_m = (\Phi_m(a), \tilde{\Delta}), \tag{3.9}$$

and if $o_F^\times = U(m)$, then we have $g(x) = \Phi_m(x)$ and $\tau_m = \tilde{\Delta}$ by Corollary 3.1. Note that for a prime number p ($\nmid 2D_{F_{\tilde{\Delta}}}$) with $\tilde{\eta} \in \sigma_{F_{\tilde{\Delta}}/\mathbb{Q}}(p)$, τ_m divides $\Phi_m(p)$ because of $\zeta_{\tau_m}^{\Phi_m(p)} \equiv \zeta_{\tau_m}^{\Phi_m(\tilde{\eta})} = 1 \bmod \mathfrak{p}$ for a prime ideal $\mathfrak{p}$ lying above p, and we may put

$$\mathcal{R}(\tilde{\eta}) = \left\{ \prod_{m|d} v_m \,\middle|\, \begin{array}{l} \text{(i) } v_m \in U(m), \\ \text{(ii) } \prod_{m|d} v_m^{\Phi_m(p)/\tau_m} \equiv \zeta \bmod p \text{ for } \exists\zeta \in W_F \end{array} \right\}. \tag{3.10}$$

This is well-defined by Proposition 3.2 and forms a group. Moreover it is independent of the choice of a prime p, which follows from the following proposition.

Proposition 3.3. *Let $\tilde{\eta} \in \mathrm{Gal}(F_{\tilde{\Delta}}/\mathbb{Q})$ be an extension of η. Suppose that a prime number p ($\nmid 2D_{F_{\tilde{\Delta}}}$) satisfies $\sigma_{F_{\tilde{\Delta}}/\mathbb{Q}}(p) \ni \tilde{\eta}$, and $v_m \in U(m)$, $\zeta \in W_F$; then we have*

$$\prod_{m|d} v_m^{\Phi_m(p)/\tau_m} \equiv \zeta \bmod p$$
$$\Leftrightarrow \prod_{m|d} \sqrt[\tau_m]{v_m}^{\Phi_m(\rho\tilde{\eta}\rho^{-1})} = \zeta \textit{ for } \forall\rho \in \mathrm{Gal}(F_{\tilde{\Delta}}/\mathbb{Q}).$$

Proof. Suppose a prime ideal $\mathfrak{p}$ of $F_{\tilde{\Delta}}$ lying above p satisfy $\tilde{\eta} = \sigma_{F_{\tilde{\Delta}}/\mathbb{Q}}(\mathfrak{p})$; then we have, for $\rho \in \mathrm{Gal}(F_{\tilde{\Delta}}/\mathbb{Q})$

$$\prod_{m|d} v_m^{\Phi_m(p)/\tau_m} \equiv \zeta \bmod p$$

$$\Rightarrow \zeta \equiv \prod_{m|d} \sqrt[\tau_m]{v_m}^{\Phi_m(p)} \equiv \prod_{m|d} \sqrt[\tau_m]{v_m}^{\Phi_m(\rho\tilde{\eta}\rho^{-1})} \bmod \mathfrak{p}^{\rho^{-1}}$$

$$\Rightarrow \zeta = \prod_{m|d} \sqrt[\tau_m]{v_m}^{\Phi_m(\rho\tilde{\eta}\rho^{-1})},$$

on noting that the right-hand side is a root of unity in $F_{\tilde{\Delta}}$ by $\eta \in Z(\mathrm{Gal}(F/\mathbb{Q}))$ and $v_m^{\Phi_m(\eta)} \in W_F$.

We may trace the above argument in the reverse way to prove the converse. □

For a prime number $p\,(\nmid 2D_{F_{\tilde{\Delta}}})$ satisfying $\sigma_{F_{\tilde{\Delta}}/\mathbb{Q}}(p) \ni \tilde{\eta}$, we define a mapping

$$\phi_p : \prod_{m|d} U(m) \to o_F^\times / W_F$$

by

$$\phi_p\Big(\prod_{m|d} v_m\Big) = \prod_{m|d} v_m^{\Phi_m(p)/\tau_m}. \tag{3.11}$$

It is well-defined in view of $\Phi_m(p)/\tau_m \in \mathbb{Z}$ and Proposition 3.2 and we see that

$$\iota_p \circ \phi_p(\mathcal{R}(\tilde{\eta})) = \{1\}, \tag{3.12}$$

by the definition of $\mathcal{R}(\tilde{\eta})$ and ι_p in Corollary 3.3.

Proposition 3.4. *For $v_m \in U(m)$ and an integer $p\ (\neq \pm 1)$, we have*

$$\phi_p\Big(\prod v_m\Big) \in W_F {o_F^\times}^{\eta - p} \Leftrightarrow \prod v_m^{\Phi_m(\eta,p)\tilde{\Delta}/\tau_m} \in W_F {o_F^\times}^{\tilde{\Delta}},$$

where $\Phi_m(x,y) = (\Phi_m(x) - \Phi_m(y))/(x-y)$ as in the notation.

Proof. It is easy to see by (3.11)

$$\phi_p\Big(\prod v_m\Big) \in W_F {o_F^\times}^{\eta - p} \Leftrightarrow \prod v_m^{\Phi_m(p)/\tau_m} \cdot \epsilon^{-(\eta-p)} \in W_F, \quad \exists \epsilon \in o_F^\times,$$

which is equivalent to

$$\prod v_m^{\Phi_m(p)\tilde{\Delta}/\tau_m} \cdot \epsilon^{-(\eta-p)\tilde{\Delta}} \in W_F, \tag{3.13}$$

noting that for $u \in o_F^\times$, $u \in W_F$ if and only if $u^{\tilde{\Delta}} \in W_F$. Since $v_m^{\Phi_m(\eta)\tilde{\Delta}/\tau_m} \in W_F$ holds by (3.9) and (3.3), (3.13) is equivalent to

$$\begin{aligned} & \prod v_m^{(\Phi_m(\eta)-\Phi_m(p))\tilde{\Delta}/\tau_m} \cdot \epsilon^{(\eta-p)\tilde{\Delta}} \in W_F \\ \Leftrightarrow\ & \Big(\prod v_m^{\Phi_m(\eta,p)\tilde{\Delta}/\tau_m} \cdot \epsilon^{\tilde{\Delta}}\Big)^{\eta-p} \in W_F \\ \Leftrightarrow\ & \prod v_m^{\Phi_m(\eta,p)\tilde{\Delta}/\tau_m} \cdot \epsilon^{\tilde{\Delta}} \in W_F, \quad \text{(by Lemma 3.5)} \end{aligned}$$

which completes the proof. □

Proposition 3.5. *For an extension* $\tilde{\eta} \in \mathrm{Gal}(F_{\tilde{\Delta}}/\mathbb{Q})$ *of* η *and a prime number* $p\,(\nmid 2D_{F_{\tilde{\Delta}}})$ *satisfying* $\tilde{\eta} \in \sigma_{F_{\tilde{\Delta}}/\mathbb{Q}}(p)$, *we put*

$$\mathcal{R}(\tilde{\eta}, p) = \phi_p(\mathcal{R}(\tilde{\eta}))W_F o_F^{\times\,\eta-p}(\subset o_F^\times),$$

where the image of ϕ_p *is viewed in* $o_F^\times$ *and define an integer* a *by* $\zeta_{\tilde{\Delta}}^{\tilde{\eta}} = \zeta_{\tilde{\Delta}}^a$. *Then, we have*

$$\begin{aligned} & [\mathcal{R}(\tilde{\eta}, p) : W_F o_F^{\times\,\eta-p}] \\ & = \left[\mathcal{R}(\tilde{\eta}) : \left\{ \prod_{m|d} v_m \,\middle|\, v_m \in U(m),\ \prod_{m|d} v_m^{\Phi_m(\eta,a)\tilde{\Delta}/\tau_m} \in W_F o_F^{\times\,\tilde{\Delta}} \right\}\right], \end{aligned}$$

which is independent of the choice of p.

Proof. Let us see first $\phi_p^{-1}(W_F o_F^{\times\,\eta-p}) \subset \mathcal{R}(\tilde{\eta})$. Let $v_m \in U(m)$ and suppose $\phi_p(\prod v_m) \in W_F o_F^{\times\,\eta-p}$; we must show that $\prod v_m \in \mathcal{R}(\tilde{\eta})$. The supposition yields

$$\begin{aligned} & \phi_p\Big(\prod v_m\Big) \in W_F o_F^{\times\,\eta-p} \\ \Rightarrow\ & \prod v_m^{\Phi_m(\eta,p)\tilde{\Delta}/\tau_m} \in W_F o_F^{\times\,\tilde{\Delta}} && \text{(by Proposition 3.4)} \\ \Rightarrow\ & \prod v_m^{(\Phi_m(\eta)-\Phi_m(p))\tilde{\Delta}/\tau_m} \in W_F o_F^{\times\,\tilde{\Delta}(\eta-p)} \\ \Rightarrow\ & \prod v_m^{\Phi_m(p)\tilde{\Delta}/\tau_m} = \zeta\epsilon^{\tilde{\Delta}(\eta-p)} && (\zeta \in W_F, \epsilon \in o_F^\times) \\ \Rightarrow\ & \prod v_m^{\Phi_m(p)/\tau_m} = \zeta'\epsilon^{\eta-p}. \end{aligned}$$

Here ζ' is a root of unity and lies in W_F, because $\prod v_m^{\Phi_m(p)/\tau_m},\ \epsilon^{\eta-p} \in F$. Therefore, we have

$$\prod v_m^{\Phi_m(p)/\tau_m} = \zeta'\epsilon^{\eta-p} \equiv \zeta' \bmod p$$

and so $\prod v_m \in \mathcal{R}(\tilde{\eta})$. Now we have by the second homomorphism theorem

$$[\mathcal{R}(\tilde{\eta}, p) : W_F o_F^{\times\,\eta-p}] = [\phi_p(\mathcal{R}(\tilde{\eta})) : \phi_p(\mathcal{R}(\tilde{\eta})) \cap W_F o_F^{\times\,\eta-p}],$$

and it is equal to

$$\begin{aligned}&[\mathcal{R}(\tilde{\eta}):\phi_p^{-1}(W_F o_F^{\times\,\eta-p})] \qquad (\text{ by } \phi_p^{-1}(W_F o_F^{\times\,\eta-p})\subset\mathcal{R}(\tilde{\eta}))\\ &=\left[\mathcal{R}(\tilde{\eta}):\left\{\prod v_m\,\middle|\, v_m\in U(m),\ \prod v_m^{\Phi_m(\eta,p)\tilde{\Delta}/\tau_m}\in W_F o_F^{\times\,\tilde{\Delta}}\right\}\right],\end{aligned}$$

by Proposition 3.4. The definition of a implies $a\equiv p \bmod \tilde{\Delta}$ and therefore $v_m^{\Phi_m(\eta,p)-\Phi_m(\eta,a)}\in o_F^{\times\,\tilde{\Delta}}$, which completes the proof. □

Since by (3.12), $\mathcal{R}(\tilde{\eta},p)\subset\ker\iota_p$ holds, we have

$$[\mathcal{R}(\tilde{\eta},p):W_F o_F^{\times\,\eta-p}]\,\Big|\,[\ker\iota_p:W_F o_F^{\times\,\eta-p}],$$

and

$$[\mathcal{R}(\tilde{\eta},p):W_F o_F^{\times\,\eta-p}]\,\Big|\,\gcd_{\substack{\sigma_{F_{\tilde{\Delta}}/\mathbb{Q}}(p)\,\ni\,\tilde{\eta},\\ p\nmid 2D_{F_{\tilde{\Delta}}}}}[\ker\iota_p:W_F o_F^{\times\,\eta-p}],$$

since the left index is independent of the choice of $p(\nmid 2D_{F_{\tilde{\Delta}}})$ by Proposition 3.5. Therefore, putting

$$\kappa(\eta)=\gcd_{\tilde{\eta}|_F=\eta}[\mathcal{R}(\tilde{\eta},p):W_F o_F^{\times\,\eta-p}] \tag{3.14}$$

we have

$$\kappa(\eta)\,\Big|\,\gcd_{\substack{\sigma_{F/\mathbb{Q}}(p)\,=\,\eta,\\ p\nmid 2D_{F_{\tilde{\Delta}}}}}[\ker\iota_p:W_F o_F^{\times\,\eta-p}],$$

where the right-hand side is Rel_η by definition (cf.(3.7)). Hence we have

$$\kappa(\eta)\,\Big|\,\mathrm{Rel}_\eta\,\Big|\,[\ker\iota_p:W_F o_F^{\times\,\eta-p}]$$

for prime numbers p which satisfies $\eta\in\sigma_{F/\mathbb{Q}}(p)$ and $p\nmid 2D_{F_{\tilde{\Delta}}}$.

Thus, we have shown with (3.6),

Theorem 3.2.

$$\#E(p)\,\Big|\,\frac{w\prod_i|\Phi_{d_i}(p)|}{\kappa(\eta)}. \tag{3.15}$$

We expect $\kappa(\eta)=\mathrm{Rel}_\eta$, and we conjecture that for infinitely many primes p, "|" is replaced by "=" in (3.15).

Note that if there is at least one prime p such that

$$\kappa(\eta)=w\prod|\Phi_{d_i}(p)|\Big/\#E(p)\ (=[\ker\iota_p:W_F o_F^{\times\,\eta-p}]),$$

then $\kappa(\eta) = \mathrm{Rel}_\eta$ holds. The computer experiment supports this in all the examples in §4.

To proceed to the next step, we need to know in terms of Frobenius automorphisms the condition on p for which $[\ker \iota_p : W_F o_F^{\times\,\eta-p}]/\kappa(\eta)$ is a multiple of a natural number m (cf. Proposition 2.2). Successful cases are some number fields with $\mathrm{rank}\, o_F^\times = 1$ [8], and cubic abelian fields [9].

Remark 3.2. It is desirable to generalize the prime number case to more general situation in case of "modulo prime ideal" in the previous section.

3.3. *Evaluation of $\kappa(\eta)$*

In this subsection, we give another description of the index $[\mathcal{R}(\tilde{\eta}, p) : W_F o_F^{\times\,\eta-p}]$ and $\kappa(\eta)$ convenient for evaluation.

3.3.1. *Action of automorphisms*

First, we study the explicit action of $\tilde{\eta}$ on $\sqrt[\Delta]{\epsilon}$ as a preparation. Let Δ be a natural number and U a $\mathrm{Gal}(F/\mathbb{Q})$-stable subgroup of $o_F^\times$ such that

$$U = \langle \zeta_w, u_1, \cdots, u_s \rangle, \tag{3.16}$$

where ζ_w is a primitive w-th root of unity and u_i's are multiplicatively independent. Therefore $\{u_1, \cdots, u_s\}$ is a basis of U/W_F as a $\mathbb{Z}$-module. For a polynomial with integral coefficients

$$h(x) = h_n x^n + h_{n-1} x^{n-1} + \cdots + h_0,$$

we assume

$$U^{h(\eta)} \subset W_F \tag{3.17}$$

and write, as in the introduction

$$h(x, y) = (h(x) - h(y))/(x - y) = \sum_{t=1}^{n} h_t \sum_{k=0}^{t-1} x^{t-k-1} y^k.$$

We suppose $\eta \in Z(\mathrm{Gal}(F/\mathbb{Q}))$ as before, and let $\tilde{\eta}$ be an extension of η to $\mathrm{Gal}(F_\Delta/\mathbb{Q})$, and fix a Δ-th root $\sqrt[\Delta]{u_j} \in F_\Delta$ once and for all. Write

$$\zeta_{\Delta w}^{\tilde{\eta}} = \zeta_{\Delta w}^{a}, \quad \sqrt[\Delta]{u_i}^{\tilde{\eta}} = \zeta_{\Delta w}^{a_i} \prod_j \sqrt[\Delta]{u_j}^{a_{ij}}, \tag{3.18}$$

and similarly for $\rho \in \mathrm{Gal}(F_\Delta/\mathbb{Q})$

$$\zeta_{\Delta w}^{\rho} = \zeta_{\Delta w}^{b}, \quad \sqrt[\Delta]{u_i}^{\rho} = \zeta_{\Delta w}^{b_i} \prod_j \sqrt[\Delta]{u_j}^{b_{ij}}, \tag{3.19}$$

and

$$\boldsymbol{a} = \begin{pmatrix} a_1 \\ \vdots \\ a_s \end{pmatrix}, \quad \boldsymbol{b} = \begin{pmatrix} b_1 \\ \vdots \\ b_s \end{pmatrix}, \quad A = (a_{ij}),\ B = (b_{ij}).$$

Lemma 3.8. *We have*

$$\rho\eta = \eta\rho \text{ on } U \Leftrightarrow b\boldsymbol{a} + A\boldsymbol{b} \equiv a\boldsymbol{b} + B\boldsymbol{a} \bmod w, \text{ and } AB = BA.$$

Proof. The assertion follows, comparing

$$u_i^{\eta\rho} = \left(\zeta_w^{a_i} \prod u_j^{a_{ij}}\right)^{\rho} = \zeta_w^{ba_i} \prod_j \left\{\zeta_w^{b_j} \prod_k u_k^{b_{jk}}\right\}^{a_{ij}} = \zeta_w^{ba_i + \sum_j a_{ij} b_j} \prod_k u_k^{\sum_j a_{ij} b_{jk}}$$

and

$$u_i^{\rho\eta} = \left(\zeta_w^{b_i} \prod u_j^{b_{ij}}\right)^{\eta} = \zeta_w^{ab_i} \prod_j \left\{\zeta_w^{a_j} \prod_k u_k^{a_{jk}}\right\}^{b_{ij}} = \zeta_w^{ab_i + \sum_j b_{ij} a_j} \prod_k u_k^{\sum_j b_{ij} a_{jk}}.$$

□

Lemma 3.9. *Putting $A^k = (a_{ij}(k))$ for each non-negative integer k, we have*

$$\sqrt[\Delta]{u_i}^{\tilde{\eta}^k} = \zeta_{\Delta w}^{\alpha_i(k)} \prod_j \sqrt[\Delta]{u_j}^{a_{ij}(k)}, \tag{3.20}$$

where $\alpha_i(k)$ is defined by

$$^t(\alpha_1(k), \cdots, \alpha_s(k)) = \begin{cases} 0 & \text{if } k = 0, \\ ^t(a_1, a_2, \cdots, a_s) & \text{if } k = 1, \\ (a^{k-1} + a^{k-2}A + \cdots + A^{k-1})\boldsymbol{a} & \text{if } k > 1. \end{cases} \tag{3.21}$$

Proof. The case of $k = 0, 1$ is clear. Inductively, we see the assertion, using

$$\begin{aligned} \sqrt[\Delta]{u_i}^{\tilde{\eta}^{k+1}} &= \zeta_{\Delta w}^{\alpha_i(k)a} \prod_j \left\{\zeta_{\Delta w}^{a_j} \prod_k \sqrt[\Delta]{u_k}^{a_{jk}}\right\}^{a_{ij}(k)} \\ &= \zeta_{\Delta w}^{\alpha_i(k)a + \sum_j a_{ij}(k) a_j} \prod_k \sqrt[\Delta]{u_k}^{\sum_j a_{ij}(k) a_{jk}}. \end{aligned}$$

□

Lemma 3.10. *We have*

$$u_i^{h(\eta,a)} \in W_F \prod_j u_j^{h(a,A)_{(i,j)}},$$

where $h(a,A)_{(i,j)}$ is the (i,j)-entry of $h(a,A)$.

Proof. The assertion follows from

$$\begin{aligned} u_i^{h(\eta,a)} &= u_i^{\sum_{t=1}^{n}\sum_{k=0}^{t-1} h_t a^{t-k-1}\eta^k} \\ &\in W_F \prod_{j=1}^{s} u_j^{\sum_{t=1}^{n}\sum_{k=0}^{t-1} h_t a^{t-k-1} a_{ij}(k)} \qquad \text{(by Lemma 3.9)} \\ &= W_F \prod_j u_j^{h(a,A)_{(i,j)}}. \end{aligned}$$

□

Lemma 3.11. *For an integral vector* $\boldsymbol{x} = (x_1, \ldots, x_s)$*, we put*

$$\sqrt[\Delta]{\epsilon} = \prod \sqrt[\Delta]{u_i}^{x_i}.$$

Then we have

$$\sqrt[\Delta]{\epsilon}^{h(\tilde{\eta})} = \prod_i \sqrt[\Delta]{u_i}^{h(\tilde{\eta})x_i} = \zeta_{\Delta w}^{\boldsymbol{x}h(a,A)\boldsymbol{a}}$$

and

$$h(A) = 0.$$

Proof. We have by (3.20)

$$\sqrt[\Delta]{u_i}^{h(\tilde{\eta})} = \prod_{k=0}^{n} \sqrt[\Delta]{u_i}^{h_k\tilde{\eta}^k} = \zeta_{\Delta w}^{\sum_k h_k \alpha_i(k)} \prod_j \sqrt[\Delta]{u_j}^{\sum_k a_{ij}(k)h_k}.$$

The assumption $u_i^{h(\eta)} \in W_F$ (cf. (3.17)) yields $\sum_k a_{ij}(k)h_k = 0$, i.e., $h(A) = 0$, and $\sum_k h_k\alpha_i(k)$ is the i-th component of $h(a, A)\boldsymbol{a}$ by (3.21), from which the first assertion follows. □

Lemma 3.12. *For* ρ^{-1}*, we put*

$$\zeta_{\Delta w}^{\rho^{-1}} = \zeta_{\Delta w}^{b'}, \quad \sqrt[\Delta]{u_i}^{\rho^{-1}} = \zeta_{\Delta w}^{b'_i} \prod_j \sqrt[\Delta]{u_j}^{b'_{ij}}, \quad \boldsymbol{b}' = {}^t(b'_1, \cdots, b'_s). \tag{3.22}$$

Then we have

$$bb' \equiv 1 \bmod \Delta w, \quad b'\boldsymbol{b} \equiv -B\boldsymbol{b}' \bmod \Delta w, \quad (b'_{ij}) = B^{-1}. \tag{3.23}$$

Proof. By (3.19), we have

$$\zeta_{\Delta w} = \zeta_{\Delta w}^{\rho\rho^{-1}} = \zeta_{\Delta w}^{bb'},$$

and the equation

$$\sqrt[\Delta]{u_i} = \sqrt[\Delta]{u_i}^{\rho\rho^{-1}} = \zeta_{\Delta w}^{b_i b' + \sum_j b_{ij}b'_j} \prod_k \sqrt[\Delta]{u_k}^{\sum_j b_{ij}b'_{jk}}.$$

The first equation implies the first congruence in (3.23) and together with multiplicative independence of u_i's the second equation implies both $(b'_{ij}) = B^{-1}$ and the second congruence $b'\boldsymbol{b} + B\boldsymbol{b}' \equiv 0 \bmod \Delta w$. □

Lemma 3.13. *Putting*

$$\sqrt[\Delta]{u_i}^{\rho\tilde{\eta}\rho^{-1}} = \zeta_{\Delta w}^{q_i} \prod \sqrt[\Delta]{u_j}^{q_{ij}}, \quad \boldsymbol{q} = {}^t(q_1, \cdots, q_s), \tag{3.24}$$

we have

$$q_{ij} = a_{ij}, \quad \boldsymbol{q} \equiv b'\{(a-A)\boldsymbol{b} + B\boldsymbol{a}\} \bmod \Delta w.$$

Proof. By $u_i^{\rho\eta\rho^{-1}} = u_i^{\eta}$, we have $q_{ij} = a_{ij}$. Since we have

$$\begin{aligned}
&\sqrt[\Delta]{u_i}^{\rho\tilde{\eta}\rho^{-1}} \\
&= \Big\{\zeta_{\Delta w}^{b_i} \prod_j \sqrt[\Delta]{u_j}^{b_{ij}}\Big\}^{\tilde{\eta}\rho^{-1}} && \text{(by (3.19))} \\
&= \Big\{\zeta_{\Delta w}^{b_i a + \sum_j b_{ij} a_j} \prod_j \Big\{\prod_k \sqrt[\Delta]{u_k}^{a_{jk}}\Big\}^{b_{ij}}\Big\}^{\rho^{-1}} && \text{(by (3.18))} \\
&= \zeta_{\Delta w}^{b'(b_i a + \sum_j b_{ij} a_j)} \prod_{j,k} \Big\{\zeta_{\Delta w}^{b'_k} \prod_\ell \sqrt[\Delta]{u_\ell}^{b'_{kl}}\Big\}^{b_{ij} a_{jk}} && \text{(by (3.22))} \\
&= \zeta_{\Delta w}^{b'(b_i a + \sum_j b_{ij} a_j) + \sum_{j,k} b'_k b_{ij} a_{jk}} \prod_\ell \sqrt[\Delta]{u_\ell}^{a_{i\ell}}, && \text{(by Lemma 3.8)}
\end{aligned}$$

it is easy to see by comparing this with (3.24)

$$\begin{aligned}
\boldsymbol{q} &\equiv b'(a\boldsymbol{b} + B\boldsymbol{a}) + BA\boldsymbol{b}' \\
&\equiv b'(a\boldsymbol{b} + B\boldsymbol{a}) + AB\boldsymbol{b}' && \text{(by Lemma 3.8)} \\
&\equiv b'(a\boldsymbol{b} + B\boldsymbol{a}) - Ab'\boldsymbol{b} && \text{(by Lemma 3.12)} \\
&\equiv b'\{(a-A)\boldsymbol{b} + B\boldsymbol{a}\} \bmod \Delta w.
\end{aligned}$$

□

Corollary 3.4. *For*

$$\sqrt[\Delta]{\epsilon} = \prod \sqrt[\Delta]{u_i}^{x_i},$$

we have

$$\left(\sqrt[\Delta]{\epsilon}^{h(\rho\tilde{\eta}\rho^{-1})}\right)^b = \zeta_{\Delta w}^{\boldsymbol{x}\{h(a)\boldsymbol{b} + h(a,A)B\boldsymbol{a}\}}.$$

Proof. Using Lemmas 3.11 and 3.13, we have

$$\sqrt[\Delta]{\epsilon}^{h(\rho\tilde{\eta}\rho^{-1})} = \prod_i \sqrt[\Delta]{u_i}^{h(\rho\tilde{\eta}\rho^{-1})x_i} = \zeta_{\Delta w}^{\boldsymbol{x}h(a,A)\boldsymbol{q}} = \zeta_{\Delta w}^{\boldsymbol{x}h(a,A)b'\{(a-A)\boldsymbol{b} + B\boldsymbol{a}\}}.$$

Noting that $h(a,A)(a-A) = h(a) - h(A) = h(a)$, we have

$$h(a,A)\{(a-A)\boldsymbol{b} + B\boldsymbol{a}\} \equiv h(a)\boldsymbol{b} + h(a,A)B\boldsymbol{a} \mod \Delta w,$$

which completes the proof. □

3.3.2. *Evaluation of* $\kappa(\eta)$

With the preparation in the previous subsubsection, we may now give a formula for the index $[\mathcal{R}(\tilde{\eta},p) : W_F o_F^{\times\,\eta-p}]$, which is easier to evaluate. For $\tilde{\eta} \in \mathrm{Gal}(F_{\tilde{\Delta}}/\mathbb{Q})$ with $\tilde{\eta}_{|F} = \eta$, we recall (3.10) on the form

$$\mathcal{R}(\tilde{\eta}) = \left\{ \prod_{m|d} v_m \;\middle|\; \begin{array}{l} \text{(i) } v_m \in U(m), \\ \text{(ii) } \prod_{m|d} \sqrt[\tau_m]{v_m}^{\Phi_m(\rho\tilde{\eta}\rho^{-1})} = \prod_{m|d} \sqrt[\tau_m]{v_m}^{\Phi_m(\tilde{\eta})} \in W_F \text{ for } \forall\rho \end{array} \right\},$$

by Proposition 3.3.

Lemma 3.14. *Recalling* (3.16),(3.18),(3.19), *we put*

$$\begin{aligned}
U(m) &= \langle \zeta_w, u_{m,1}, \cdots, u_{m,s_m} \rangle, \\
\sqrt[\tilde{\Delta}]{u_{m,j}}^{\tilde{\eta}} &= \zeta_{\tilde{\Delta}w}^{a_{m,j}} \prod_k \sqrt[\tilde{\Delta}]{u_{m,k}}^{a_{jk}^{(m)}}, & \zeta_{\tilde{\Delta}w}^{\tilde{\eta}} &= \zeta_{\tilde{\Delta}w}^{a}, \\
\sqrt[\tilde{\Delta}]{u_{m,i}}^{\rho} &= \zeta_{\tilde{\Delta}w}^{b_{m,i}} \prod_j \sqrt[\tilde{\Delta}]{u_{m,j}}^{b_{ij}^{(m)}}, & \zeta_{\tilde{\Delta}w}^{\rho} &= \zeta_{\tilde{\Delta}w}^{b}, \\
{}^t\boldsymbol{a}_m &= (a_{m,1}, \cdots, a_{m,s_m}), & A^{(m)} &= (a_{ij}^{(m)}), \\
{}^t\boldsymbol{b}_m &= (b_{m,1}, \cdots, b_{m,s_m}), & B^{(m)} &= (b_{ij}^{(m)}).
\end{aligned}$$

Then for

$$v_m = \prod_j u_{m,j}^{y_{m,j}}, \quad \boldsymbol{y}_m = (y_{m,1}, \cdots, y_{m,s_m}), \tag{3.25}$$

the condition $\prod_m v_m \in \mathcal{R}(\tilde{\eta})$ *is equivalent to*

$$(\#)\begin{cases} \sum_m (\tilde{\Delta}/\tau_m)\boldsymbol{y}_m \Phi_m(a, A^{(m)})\boldsymbol{a}_m \equiv 0 \mod \tilde{\Delta}, \\ \sum_m (\tilde{\Delta}/\tau_m)\boldsymbol{y}_m \{\Phi_m(a)\boldsymbol{b}_m + \Phi_m(a, A^{(m)})(B^{(m)} - b)\boldsymbol{a}_m\} \equiv 0 \mod \tilde{\Delta}w \end{cases}$$

for every $\rho \in \mathrm{Gal}(F_{\tilde{\Delta}}/\mathbb{Q})$.

Proof. By putting

$$x_{m,j} = y_{m,j}\tilde{\Delta}/\tau_m, \quad \boldsymbol{x}_m = (x_{m,1}, \cdots, x_{m,s_m}),$$

the condition $\prod_m v_m \in \mathcal{R}(\tilde{\eta})$ amounts to clearly

$$\prod_m \Big(\prod_j \sqrt[\tilde{\Delta}]{u_{m,j}}^{x_{m,j}}\Big)^{\Phi_m(\rho\tilde{\eta}\rho^{-1})} = \prod_m \Big(\prod_j \sqrt[\tilde{\Delta}]{u_{m,j}}^{x_{m,j}}\Big)^{\Phi_m(\tilde{\eta})} \in W_F.$$

By Lemma 3.11, we have

$$\Big(\prod_j \sqrt[\tilde{\Delta}]{u_{m,j}}^{x_{m,j}}\Big)^{\Phi_m(\tilde{\eta})} = \zeta_{\tilde{\Delta}w}^{\boldsymbol{x}_m \Phi_m(a,A^{(m)})\boldsymbol{a}_m},$$

while by Corollary 3.4

$$\Big(\prod_j \sqrt[\tilde{\Delta}]{u_{m,j}}^{x_{m,j}}\Big)^{\Phi_m(\rho\tilde{\eta}\rho^{-1})} = \zeta_{\tilde{\Delta}w}^{b'\boldsymbol{x}_m\{\Phi_m(a)\boldsymbol{b}_m + \Phi_m(a,A^{(m)})B^{(m)}\boldsymbol{a}_m\}},$$

on putting $\zeta_{\tilde{\Delta}w}^{\rho^{-1}} = \zeta_{\tilde{\Delta}w}^{b'}$. Hence the condition $\prod_m v_m \in \mathcal{R}(\tilde{\eta})$ is equivalent to

$$\prod_m \zeta_{\tilde{\Delta}w}^{b'\boldsymbol{x}_m\{\Phi_m(a)\boldsymbol{b}_m + \Phi_m(a,A^{(m)})B^{(m)}\boldsymbol{a}_m\}} = \prod_m \zeta_{\tilde{\Delta}w}^{\boldsymbol{x}_m \Phi_m(a,A^{(m)})\boldsymbol{a}_m} \in W_F. \quad (3.26)$$

It is easy to see that

$$\begin{aligned} &\textstyle\prod_m \zeta_{\tilde{\Delta}w}^{\boldsymbol{x}_m \Phi_m(a,A^{(m)})\boldsymbol{a}_m} \in W_F \\ \Leftrightarrow &\textstyle\prod_m \zeta_{\tilde{\Delta}}^{\boldsymbol{x}_m \Phi_m(a,A^{(m)})\boldsymbol{a}_m} = 1 \\ \Leftrightarrow &\textstyle\sum_m \boldsymbol{x}_m \Phi_m(a, A^{(m)})\boldsymbol{a}_m \equiv 0 \bmod \tilde{\Delta} \\ \Leftrightarrow &\textstyle\sum_m (\tilde{\Delta}/\tau_m)\boldsymbol{y}_m \Phi_m(a, A^{(m)})\boldsymbol{a}_m \equiv 0 \bmod \tilde{\Delta}, \end{aligned}$$

and the equality in (3.26) is equivalent to

$$\sum_m b'\boldsymbol{x}_m\{\Phi_m(a)\boldsymbol{b}_m + \Phi_m(a, A^{(m)})B^{(m)}\boldsymbol{a}_m\} \equiv \sum_m \boldsymbol{x}_m \Phi_m(a, A^{(m)})\boldsymbol{a}_m \bmod \tilde{\Delta}w,$$

which completes the proof. □

Remark 3.3. The condition (#) depends only on

(i) $\boldsymbol{y}_m \bmod \tau_m$,
(ii) $b, \boldsymbol{b}_m \bmod w$, i.e., on $\rho_{|F}$,
(iii) $\boldsymbol{a}_m \bmod w\tau_m$.

Further,

(iv) $\boldsymbol{a}_m \bmod w$ is uniquely determined by η.

With respect to the assertion (i), we have only to note that

$$\begin{aligned}&\Phi_m(a)\boldsymbol{b}_m + \Phi_m(a, A^{(m)})(B^{(m)} - b)\boldsymbol{a}_m \\ &= \Phi_m(a, A^{(m)})\{(a - A^{(m)})\boldsymbol{b}_m + (B^{(m)} - b)\boldsymbol{a}_m\} \\ &\equiv 0 \bmod w\end{aligned}$$

by Lemma 3.8.

(3.9) implies $\Phi_m(a) \equiv 0 \bmod \tau_m$, which implies the assertion on $\boldsymbol{b}_m$. The assertion on b follows from the first equation of (#). The statements (iii),(iv) are obvious.

Lemma 3.15. *Let $v_m \in U(m)$ and $\boldsymbol{y}_m$ be as in the previous lemma; then the condition*

$$\prod_{m|d} v_m^{\Phi_m(\eta,a)\tilde{\Delta}/\tau_m} \in W_F o_F^{\times\tilde{\Delta}}$$

is equivalent to

$$(\natural) \qquad \prod_{m|d}\prod_j u_{m,j}^{\sum_i y_{m,i}\Phi_m(a,A^{(m)})_{(i,j)}\tilde{\Delta}/\tau_m} \in W_F o_F^{\times\tilde{\Delta}}.$$

If a stronger condition $o_F^\times = \prod U(m)$ holds, then it is equivalent to

$$\boldsymbol{y}_m \Phi_m(a, A^{(m)}) \equiv 0 \bmod \tau_m.$$

Proof. By Lemma 3.10, we know that $u_{m,i}^{\Phi_m(\eta,a)} \in W_F \prod_j u_{m,j}^{\Phi_m(a,A^{(m)})_{(i,j)}}$, which yields

$$\begin{aligned}&\textstyle\prod_{m|d} v_m^{\Phi_m(\eta,a)\tilde{\Delta}/\tau_m} \\ &= \textstyle\prod_{m|d}\prod_i u_{m,i}^{y_{m,i}\Phi_m(\eta,a)\tilde{\Delta}/\tau_m} && (\text{ by } (3.25)\ ) \\ &\in W_F \textstyle\prod_{m|d}\prod_i\prod_j u_{m,j}^{y_{m,i}\Phi_m(a,A^{(m)})_{(i,j)}\tilde{\Delta}/\tau_m} \\ &= W_F \textstyle\prod_{m|d}\prod_j u_{m,j}^{\sum_i y_{m,i}\Phi_m(a,A^{(m)})_{(i,j)}\tilde{\Delta}/\tau_m},\end{aligned}$$

whence follows the first assertion. If $o_F^\times = \prod U(m)$ holds, then $\{u_{m,j}\}$ is a basis of $o_F^\times/W_F$, and so the first equivalence implies the second one. □

Proposition 3.6. *Let $u_{m,i}, y_{m,i}, \boldsymbol{y}_m$ be as in Lemma 3.14, and let $\tilde{\eta}$ be an extension of η, and let $p(\nmid 2D_{F_{\tilde{\Delta}}})$ be a prime number satisfying $\tilde{\eta} \in \sigma_{F_{\tilde{\Delta}}/\mathbb{Q}}(p)$; then we have*

$$\left[\mathcal{R}(\tilde{\eta}, p) : W_F o_F^{\times\,\eta-p}\right] = \left[V_1(\tilde{\eta}) : V_2(\tilde{\eta})\right],$$

where

$$V_1(\tilde{\eta}) = W_F\left\{ \prod_m \prod_j u_{m,j}^{y_{m,j}} \,\middle|\, \boldsymbol{y}_m \textit{ satisfies } (\#) \textit{ in Lemma 3.14} \right\},$$

$$V_2(\tilde{\eta}) = W_F\left\{ \prod_m \prod_j u_{m,j}^{y_{m,j}} \,\middle|\, \boldsymbol{y}_m \textit{ satisfies } (\natural) \textit{ in Lemma 3.15} \right\}.$$

If $o_F^\times = \prod U(m)$ *holds, then we have*

$$V_2(\tilde{\eta}) = W_F\left\{ \prod_m \prod_j u_{m,j}^{y_{m,j}} \,\middle|\, \boldsymbol{y}_m \Phi_m(a, A^{(m)}) \equiv 0 \bmod \tau_m \textit{ for } \forall m|d \right\}.$$

Proof. The assertion follows easily from Proposition 3.5 and the previous two lemmas. □

Suppose $o_F^\times = \prod U(m)$; then $\{u_{m,i}\}$ is a basis of $o_F^\times / W_F$, whence comparing exponents, we may assume

$$V_1(\tilde{\eta}) = \left\{ \{\boldsymbol{y}_m \bmod \tau_m\} \,\middle|\, \boldsymbol{y}_m\text{'s satisfy } (\#) \text{ in Lemma 3.14} \right\}, \tag{3.27}$$

$$V_2(\tilde{\eta}) = \left\{ \{\boldsymbol{y}_m \bmod \tau_m\} \,\middle|\, \boldsymbol{y}_m \Phi_m(a, A^{(m)}) \equiv 0 \bmod \tau_m \text{ for } \forall m|d \right\}. \tag{3.28}$$

The inclusion $V_2(\tilde{\eta}) \subset V_1(\tilde{\eta})$ follows from their original definitions, but we can check it directly when $o_F^\times = \prod U(m)$ holds, as follows: Suppose $\{\boldsymbol{y}_m\} \in V_2(\tilde{\eta})$; then the first equality of $(\#)$ is obvious. Noting that

$$\Phi_m(a) = \Phi_m(a) - \Phi_m(A^{(m)}) = \Phi_m(a, A^{(m)})(a - A^{(m)}),$$

we have

$$\begin{aligned} &\Phi_m(a)\boldsymbol{b}_m + \Phi_m(a, A^{(m)})(B^{(m)} - b)\boldsymbol{a}_m \\ &= \Phi_m(a, A^{(m)})(a\boldsymbol{b}_m - A^{(m)}\boldsymbol{b}_m + B^{(m)}\boldsymbol{a}_m - b\boldsymbol{a}_m) \end{aligned}$$

and then Lemma 3.8 yields the second equality of $(\#)$.

4. Examples

4.1. *Case of* $\eta = \mathrm{id}$

We assume $\eta = \mathrm{id}$ throughout this subsection. Then obviously, we have d (= the order of η) $= 1, o_F^\times = U(1)$, and the polynomial $g(x)$ defined in Corollary 3.1 is equal to

$$g(x) = x - 1.$$

Corollary 5.1 in the appendix yields

$$\tilde{\Delta} = \tau_1 = w.$$

Let $\tilde{\eta}$ be an extension of η. Since $g(x, y) = 1$ yields $V_2(\tilde{\eta}) = 0$, we have for $r = \mathrm{rank}_{\mathbb{Z}} o_F^{\times}$, (3.27), (3.28) read

$$V_1(\tilde{\eta})/V_2(\tilde{\eta}) = \left\{ \boldsymbol{y} \in (\mathbb{Z}/w\mathbb{Z})^r \,\middle|\, \begin{cases} \boldsymbol{y}\boldsymbol{a} \equiv 0 \bmod w, \\ \boldsymbol{y}\{(a-1)\boldsymbol{b} + (B-b)\boldsymbol{a})\} \equiv 0 \bmod w^2 \end{cases} \right\}$$

where we put, as in §3.3.1 with (ii) in Remark 3.3

$$\begin{aligned} &o_F^{\times} = \langle \zeta_w, u_1, \cdots, u_r \rangle, \\ &\zeta_{w^2}^{\tilde{\eta}} = \zeta_{w^2}^{a}, \quad \sqrt[w]{u_i}^{\tilde{\eta}} = \zeta_{w^2}^{a_i} \sqrt[w]{u_i}, \\ &{}^t\boldsymbol{a} = (a_1, \cdots, a_r), \qquad\qquad A = 1_r, \\ &\zeta_w^{\rho} = \zeta_w^{b}, u_i^{\rho} = \zeta_w^{b_i} \prod_j u_j^{b_{ij}}, \\ &{}^t\boldsymbol{b} = (b_1, \cdots, b_r), \qquad\qquad B = (b_{ij}). \end{aligned}$$

Now, we note that the assumption $\eta = \mathrm{id}$ yields in the above

$$a \equiv 1 \bmod w, \quad \boldsymbol{a} \equiv 0 \bmod w$$

and so

$$a = 1 + w\overline{a}, \quad \boldsymbol{a} = w\overline{\boldsymbol{a}},$$

say. Then we have

$$V_1(\tilde{\eta})/V_2(\tilde{\eta}) = \left\{ \boldsymbol{y} \bmod w \,\middle|\, \boldsymbol{y}(\overline{a}\boldsymbol{b} + (B-b)\overline{\boldsymbol{a}}) \equiv 0 \bmod w \right\}.$$

Then, putting

$$R(\rho) = (B - b1_r, \boldsymbol{b}), \tag{4.1}$$

and replacing $\overline{a}, \overline{\boldsymbol{a}}$ by $a, \boldsymbol{a}$ we have

$$\begin{aligned} V(\tilde{\eta}) &:= V_1(\tilde{\eta})/V_2(\tilde{\eta}) \\ &= \left\{ \boldsymbol{y} \bmod w \,\middle|\, \boldsymbol{y}R(\rho) \begin{pmatrix} \boldsymbol{a} \\ a \end{pmatrix} \equiv 0 \bmod w \text{ for } \forall \rho \in \mathrm{Gal}(F/\mathbb{Q}) \right\}, \end{aligned}$$

where redefining $a, \boldsymbol{a}$ as above

$$\zeta_{w^2}^{\tilde{\eta}} = \zeta_{w^2} \cdot \zeta_w^{a}, \quad \sqrt[w]{u_i}^{\tilde{\eta}} = \zeta_w^{a_i} \sqrt[w]{u_i}, \quad {}^t\boldsymbol{a} = (a_1, \cdots, a_r) \tag{4.2}$$

$$\zeta_w^{\rho} = \zeta_w^{b}, \quad u_i{}^{\rho} = \zeta_w^{b_i} \prod_j u_j{}^{b_{ij}}, \quad {}^t\boldsymbol{b} = (b_1, \cdots, b_r), \quad B = (b_{ij}). \tag{4.3}$$

We note that if $\rho_{|F} = \mathrm{id}$, then (4.3) implies $b \equiv 1 \bmod w$, $\boldsymbol{b} \equiv 0 \bmod w$ and B is the identity matrix, and hence $R(\rho) \equiv 0 \bmod w$. By denoting B, b corresponding to ρ by $B(\rho)$, $b(\rho)$, respectively, it is easy to see that

$$R(\rho_1\rho_2) = B(\rho_1)R(\rho_2) + b(\rho_2)R(\rho_1). \tag{4.4}$$

In the following, we evaluate $\kappa(\eta) = \gcd \#V(\tilde{\eta})$ (cf. (3.14)) for several types of algebraic number fields, and furthermore we show that for $\tilde{\eta} \in \mathrm{Gal}(F_w/F)$, there is $\tilde{\eta}_0$ such that $\mathcal{R}(\tilde{\eta}) \supset \mathcal{R}(\tilde{\eta}_0)$ and $\kappa(\eta) = \#V(\tilde{\eta}_0)$. This is not necessarily true if $\eta \neq \mathrm{id}$. Once we find a prime number p such that $\sigma_{F/\mathbb{Q}}(p) = \eta$ and $w(p-1)^r/\#E(p) = \kappa(\eta)$, we have $\kappa(\eta) = \mathrm{Rel}_\eta$.

4.1.1. *Case of real quadratic fields*

Let F be a real quadratic field and let ϵ (> 1) be the fundamental unit with $N(\epsilon) = (-1)^s$; then clearly $\tilde{\Delta} = w = 2$. We take ϵ as u_1; then for $\rho(\neq \mathrm{id}) \in \mathrm{Gal}(F/\mathbb{Q})$,

$$(-1)^\rho = -1, \quad \epsilon^\rho = (-1)^s\epsilon^{-1}$$

imply $B = (-1), b = 1, \boldsymbol{b} = (s)$ and so

$$R(\rho) = (-2, s) \equiv (0, s) \bmod 2.$$

Therefore

$$V(\tilde{\eta}) = \left\{ x \bmod 2 \,\middle|\, x(0,s)\begin{pmatrix} a \\ a \end{pmatrix} = sax \equiv 0 \bmod 2 \right\}.$$

Here a is defined by $\zeta_4^{\tilde{\eta}} = (-1)^a\zeta_4$ as in (4.2). Let us see that

$$\kappa(\eta) = \gcd \#V(\tilde{\eta}) = \begin{cases} 2 & \text{if } N(\epsilon) = 1, \\ 1 & \text{if } N(\epsilon) = -1. \end{cases}$$

The first is obvious because of $s = 0$, and for the second, we have only to take $\tilde{\eta}_0$ so that $a = 1$. This is compatible with [8], where we have shown that the expected density of the set (cf. (3.15))

$$\{p \,|\, \#E(p) = 2(p-1)/\kappa(\eta),\ \sigma_{F/\mathbb{Q}}(p) = \mathrm{id}\}$$

is positive.

4.1.2. *Case of real cubic abelian fields*

Let F be a real cubic abelian field and σ a generator of $\mathrm{Gal}(F/\mathbb{Q})$; then we have $\tilde{\Delta} = w = 2$ and as a set of fundamental units, we can take u_1, u_2 so that

$$(-1)^\sigma = (-1)^1, \quad u_1^\sigma = u_2, \quad u_2^\sigma = (u_1 u_2)^{-1}$$

and $N_{F/\mathbb{Q}}(u_1) = N_{F/\mathbb{Q}}(u_2) = 1$ [9]. Thus we have (cf. (4.1))

$$R(\sigma) = \left(\begin{pmatrix} 0 & 1 \\ -1 & -1 \end{pmatrix} - 1_2, \begin{pmatrix} 0 \\ 0 \end{pmatrix} \right) \equiv \begin{pmatrix} 1\,1\,0 \\ 1\,0\,0 \end{pmatrix} \bmod 2,$$

$$R(\sigma^2) \equiv \begin{pmatrix} 0\,1\,0 \\ 1\,1\,0 \end{pmatrix} \bmod 2,$$

which yield

$$V(\tilde{\eta})$$
$$= \left\{ (x_1, x_2) \bmod 2 \,\middle|\, x_1(a_1 + a_2) + x_2 a_1 \equiv x_1 a_2 + x_2(a_1 + a_2) \equiv 0 \bmod 2 \right\}.$$

Here a_1, a_2 are defined by $\sqrt{u_i}^{\tilde{\eta}} = (-1)^{a_i}\sqrt{u_i}$ in (4.2). We can choose $\tilde{\eta}_0$ which corresponds to $a_1 = a_2 = 1$ and then $\#V(\tilde{\eta}_0) = 1$, i.e. $\kappa(\eta) = \gcd V(\tilde{\eta}) = 1$. This is compatible with [9], where the expected density of the set

$$\{p \,|\, \#E(p) = 2(p-1)^2, \quad \sigma_{F/\mathbb{Q}}(p) = \mathrm{id}\}$$

is explicitly given.

4.1.3. *Case of non-cyclic abelian fields of degree* 4

Let $F = \mathbb{Q}(\sqrt{d_1}, \sqrt{d_2})$, where d_1, d_2 (> 1) are natural numbers and let F_1, F_2, F_3 be three real quadratic subfields of F. Let ϵ_i (> 1) be the fundamental unit of F_i, $N_{F_i/\mathbb{Q}}\epsilon_i = (-1)^{s_i}$, $s_i = 0, 1$. Put

$$Q = \left[o_F^\times : \langle -1, \epsilon_1, \epsilon_2, \epsilon_3 \rangle \right];$$

then the type of a set $\{u_1, u_2, u_3\}$ of fundamental units of F is given as follows [10]:

$$\begin{array}{llllll}
\text{(i)} & u_1 = \epsilon_1, & u_2 = \epsilon_2, & u_3 = \epsilon_3 & (Q=1) \\
\text{(ii)} & u_1 = \sqrt{\epsilon_1}, & u_2 = \epsilon_2, & u_3 = \epsilon_3 & (Q=2) \\
\text{(iii)} & u_1 = \sqrt{\epsilon_1}, & u_2 = \sqrt{\epsilon_2}, & u_3 = \epsilon_3 & (Q=4) \\
\text{(iv)} & u_1 = \sqrt{\epsilon_1\epsilon_2}, & u_2 = \epsilon_2, & u_3 = \epsilon_3 & (Q=2) \\
\text{(v)} & u_1 = \sqrt{\epsilon_1\epsilon_2}, & u_2 = \sqrt{\epsilon_3}, & u_3 = \epsilon_2 & (Q=4) \\
\text{(vi)} & u_1 = \sqrt{\epsilon_1\epsilon_2}, & u_2 = \sqrt{\epsilon_2\epsilon_3}, & u_3 = \sqrt{\epsilon_3\epsilon_1} & (Q=4) \\
\text{(vii)} & u_1 = \sqrt{\epsilon_1\epsilon_2\epsilon_3}, & u_2 = \epsilon_2, & u_3 = \epsilon_3 & (Q=2).
\end{array}$$

In the case (ii) – (vi), $s_i = 0$ is supposed if ϵ_i appears in the symbol $\sqrt{}$, and $s_1 = s_2 = s_3$ is supposed for the case (vii). We denote by σ_i the non-trivial automorphism fixing ϵ_i and so $\mathrm{Gal}(F/\mathbb{Q}) = \{\sigma_1, \sigma_2, \sigma_3 = \sigma_1\sigma_2, \mathrm{id}\}$. $b = b(\rho) = 1$ in (4.3) is equal to 1 because of $w = 2$ and hence we have by (4.4)

$$R(\sigma_3) = R(\sigma_1\sigma_2) = B(\sigma_1)R(\sigma_2) + R(\sigma_1).$$

Proposition 4.1. *Let η be the identity; then the value of $\kappa(\eta) = \gcd \#V(\tilde{\eta})$ is given as follows:*
Case (i)

$$\kappa(\eta) = \begin{cases} 8 & \textit{if } s_1 + s_2 + s_3 = 0, \\ 4 & \textit{if } s_1 + s_2 + s_3 = 1, \\ 2 & \textit{otherwise}, \end{cases}$$

Case (ii)

$$\kappa(\eta) = \begin{cases} 4 & \textit{if } \begin{cases} s_2 = s_3 = 0 \ \textit{or} \\ s_2 = 0, \quad s_3 = 1, \quad u_1 u_1^{\sigma_2} = -1 \ \textit{or} \\ s_2 = 1, \quad s_3 = 0, \quad u_1 u_1^{\sigma_3} = -1, \end{cases} \\ 2 & \textit{otherwise}, \end{cases}$$

Case (iii)

$$\kappa(\eta) = \begin{cases} 4 & \textit{if } u_1 u_1^{\sigma_2} = u_2 u_2^{\sigma_1} = -1, \\ 2 & \textit{otherwise}, \end{cases}$$

Case (iv)

$$\kappa(\eta) = \begin{cases} 2 & \textit{if } \ u_1^{\sigma_1} u_1^{\sigma_2} = -1 \ \textit{and } s_3 = 1, \\ 4 & \textit{otherwise}, \end{cases}$$

Case (v), (vi)

$$\kappa(\eta) = 2,$$

Case (vii)

$$\kappa(\eta) = \begin{cases} 4 & \text{if } s_1 = 0, \\ 1 & \text{if } s_1 = 1. \end{cases}$$

Proof. We prove the case (vii). Proofs of the other cases are similar. For $u_1 = \sqrt{\epsilon_1\epsilon_2\epsilon_3}$, $u_2 = \epsilon_2$, $u_3 = \epsilon_3$, we have

$$(-1)^{\sigma_1} = (-1)^{\sigma_2} = -1,$$

and

$$\begin{array}{lll} u_1^{\sigma_1} = (-1)^{\kappa_1} u_1 u_2^{-1} u_3^{-1}, & u_2^{\sigma_1} = (-1)^{s_2} u_2^{-1}, & u_3^{\sigma_1} = (-1)^{s_3} u_3^{-1}, \\ u_1^{\sigma_2} = (-1)^{\kappa_2} u_1^{-1} u_2, & u_2^{\sigma_2} = u_2, & u_3^{\sigma_2} = (-1)^{s_3} u_3^{-1}, \end{array}$$

for some $\kappa_1, \kappa_2 = 0, 1$, and put $s_1 = s_2 = s_3 = s$. Then it is easy to see that

$$R(\sigma_1) \equiv \begin{pmatrix} 0 & 1 & 1 & \kappa_1 \\ 0 & 0 & 0 & s \\ 0 & 0 & 0 & s \end{pmatrix} \bmod 2,$$

$$R(\sigma_2) \equiv \begin{pmatrix} 0 & 1 & 0 & \kappa_2 \\ 0 & 0 & 0 & 0 \\ 0 & 0 & 0 & s \end{pmatrix} \bmod 2,$$

$$R(\sigma_1\sigma_2) \equiv \begin{pmatrix} 0 & 0 & 1 & \kappa_1 + \kappa_2 + s \\ 0 & 0 & 0 & s \\ 0 & 0 & 0 & 0 \end{pmatrix} \bmod 2,$$

and hence, putting ${}^t\boldsymbol{a} = (a_1, a_2, a_3, a)$, we have in due order

$$\begin{aligned} &(x_1, x_2, x_3) \in V(\tilde{\eta}) \\ \Leftrightarrow &\begin{cases} (0, x_1, x_1, x_1\kappa_1 + x_2 s + x_3 s)\boldsymbol{a} \equiv 0 \bmod 2 \\ (0, x_1, 0, x_1\kappa_2 + x_3 s)\boldsymbol{a} \equiv 0 \bmod 2 \\ (0, 0, x_1, x_1(\kappa_1 + \kappa_2 + s) + x_2 s)\boldsymbol{a} \equiv 0 \bmod 2 \end{cases} \\ \Leftrightarrow &\begin{cases} x_1(a_2 + a_3 + \kappa_1 a) + x_2 sa + x_3 sa \equiv 0 \bmod 2 \\ x_1(a_2 + \kappa_2 a) + x_3 sa \equiv 0 \bmod 2 \\ x_1(a_3 + \kappa_1 a + \kappa_2 a + sa) + x_2 sa \equiv 0 \bmod 2. \end{cases} \end{aligned}$$

We divide the proof into two cases. For $s = 0$ we get

$$\begin{aligned} &(x_1, x_2, x_3) \in V(\tilde{\eta}) \\ \Leftrightarrow &\begin{cases} x_1(a_2 + a_3 + \kappa_1 a) \equiv 0 \bmod 2 \\ x_1(a_2 + \kappa_2 a) \equiv 0 \bmod 2 \\ x_1(a_3 + \kappa_1 a + \kappa_2 a) \equiv 0 \bmod 2 \end{cases} \\ \Rightarrow &\{(0, x_2, x_3) \mid x_2, x_3 \bmod 2\} \subset V(\tilde{\eta}), \end{aligned}$$

where the inclusion becomes the equality for $a = 0, a_2 = 1$. Therefore $\kappa(\eta) = 4$ holds.
For $s = 1$, we have

$$(x_1, x_2, x_3) \in V(\tilde{\eta}) \Leftrightarrow \begin{cases} x_1(a_2 + a_3 + \kappa_1 a) + x_2 a + x_3 a \equiv 0 \bmod 2, \\ x_1(a_2 + \kappa_2 a) + x_3 a \equiv 0 \bmod 2, \\ x_1(a_3 + \kappa_1 a + \kappa_2 a + a) + x_2 a \equiv 0 \bmod 2 \end{cases}$$

and since the coefficient matrix is regular for $a = 1$, we have

$$V(\tilde{\eta}) = \{(0,0,0)\}$$

and so $\kappa(\eta) = 1$. □

Remark 4.1. Suppose $F = \mathbb{Q}(\sqrt{d_1}, \sqrt{d_2})$ $(2 \le d_1, d_2 \le 500)$; then $\kappa(\eta) = \mathrm{Rel}_\eta$ is confirmed by finding a prime number p satisfying $\#E(p) = 2(p-1)^3/\kappa(\eta)$ by computer.

4.1.4. *Case of imaginary abelian fields of degree* 4

Let F be an imaginary abelian field of degree 4 and let F_0 be the real quadratic subfield in F, and ϵ_0 (> 1) the fundamental unit of F_0. Put

$$Q = \left[o_F^\times : W_F \, o_{F_0}^\times\right].$$

We define a fundamental unit ϵ of F as follows: In case of $Q = 1$, we put $\epsilon = \epsilon_0$.
Next, we assume $Q = 2$; let us see then that we can choose a fundamental unit ϵ of F so that

$$\epsilon_0 = \zeta_w \epsilon^2, \quad \epsilon^J = \zeta_w \epsilon,$$

where J means the complex conjugation. We agree as follows. We may suppose $\epsilon_0 = \zeta_w^a \epsilon^2$ with $a = 0, 1$ without loss of generality. Assume $a = 1$; $\epsilon\epsilon^J \in F_0$ implies $\epsilon\epsilon^J = \epsilon_0^n$, and so $\epsilon\epsilon^J = (\zeta_w \epsilon^2)^n$. Thus we have $\epsilon^J = \zeta_w^n \epsilon^{2n-1}$ and comparing the absolute values, $n = 1$, which yields $\epsilon^J = \zeta_w \epsilon$, and $\epsilon_0 = \epsilon\epsilon^J = \zeta_w \epsilon^2$. If $a = 0$, then $\epsilon_0 = \epsilon^2$ follows, which implies $\epsilon^J = \epsilon^{2n-1}$ by $\epsilon\epsilon^J = \epsilon_0^n$ for an integer n, whence yields $n = 1$ and $\epsilon^J = \epsilon$, contradicting $Q = 2$.

Proposition 4.2. *We have*

$$\kappa(\eta) = \begin{cases} 2 & \text{if} \quad \begin{array}{l} \text{either } Q = 1 \text{ and } N_{F_0/\mathbb{Q}}(\epsilon_0) = 1 \\ \quad \text{or } \quad Q = 1 \text{ and } \sqrt{-1} \in F, \end{array} \\ 1 & \text{otherwise.} \end{cases}$$

Proof. Putting

$$\zeta_{w^2}^{\tilde{\eta}} = \zeta_{w^2}\zeta_w^a, \quad \sqrt[w]{\epsilon}^{\tilde{\eta}} = \zeta_w^{a_1}\sqrt[w]{\epsilon}, \quad \zeta_w^\rho = \zeta_w^b, \quad \epsilon^\rho = \zeta_w^{b_1}\epsilon^{b_{11}},$$

we have

$$V(\tilde{\eta}) = \{x \bmod w \mid x((b_{11}-b)a_1 + b_1 a) \equiv 0 \bmod w \text{ for } \forall\rho \in \mathrm{Gal}(F/\mathbb{Q})\}.$$

• Case of $Q = 1,\ N_{F_0/\mathbb{Q}}(\epsilon_0) = 1$:
In this case, we show that

$$V(\tilde{\eta}) \supset V(\tilde{\eta}_0) = \{x \bmod w \mid x \equiv 0 \bmod w/2\} \quad (\tilde{\eta}_0 \leftrightarrow a_1 = 1),$$

from which we have $\kappa(\eta) = 2$. We note that $\epsilon^\rho = \epsilon$ or ϵ^{-1} by virtue of $\epsilon = \epsilon_0$, and so $b_1 = 0$ for every ρ and so

$$V(\tilde{\eta}) = \{x \bmod w \mid x(b_{11}-b)a_1 \equiv 0 \bmod w \text{ for } \forall\rho \in \mathrm{Gal}(F/\mathbb{Q})\}.$$

If $w = 2, 4, 6$, then the possibilities for b, b_{11} are

$$\begin{aligned} w = 2 &\Rightarrow b = 1, \quad b_{11} = \pm 1 \\ w = 4, 6 &\Rightarrow b = \pm 1, \ b_{11} = \pm 1 \end{aligned}$$

and hence the assertion above is true.
If $w = 8$, then we have $F = \mathbb{Q}(\zeta_8)$ and $\epsilon_0 = \sqrt{2}+1, N_{F_0/\mathbb{Q}}(\epsilon_0) = -1$, which contradicts the assumption.
If $w = 12$, then $F = \mathbb{Q}(\sqrt{-1}, \sqrt{3})$ holds, and the possibilities are

$$\text{either } b = \pm 1,\ b_{11} = 1, \text{ or } b = \pm 5,\ b_{11} = -1.$$

The automorphism ρ corresponding to $b = -1, b_{11} = 1$ implies $2x \equiv 0 \bmod 12$, i.e. $x \equiv 0 \bmod 6$ and so $V(\tilde{\eta}_0) = \{x \bmod 12 \mid x \equiv 0 \bmod 6\}$. Therefore $\#V(\tilde{\eta}_0) = 2$ holds.

• Case of $Q = 1, N_{F_0/\mathbb{Q}}(\epsilon_0) = -1, \sqrt{-1} \in F$ and $\zeta_8 \in F$:
In this case, we have $F = \mathbb{Q}(\zeta_8)$ and $\epsilon_0 = \sqrt{2} - 1$, and the possibilities are

b	1	3	5	7
b_{11}	1	-1	-1	1
b_1	0	4	4	0
$(b_{11}-b)a_1 + b_1 a$	0	$-4a_1 + 4a$	$-6a_1 + 4a$	$-6a_1$

which also implies $V(\tilde{\eta}) \supset V(\tilde{\eta}_0) = \{x \bmod w \mid x \equiv 0 \bmod w/2\}$ ($\tilde{\eta}_0 \leftrightarrow a = a_1 = 1$). This means $\kappa(\eta) = 2$.

• Case of $Q=1, N_{F_0/\mathbb{Q}}(\epsilon_0)=-1, \sqrt{-1}\in F$ and $\zeta_8\notin F$:
We note that $w=4$ and F is the composite of $\mathbb{Q}(\sqrt{-1})$ and F_0. Then we have the table:

$\begin{cases}\rho\\ b\end{cases}$	$\begin{cases}\rho_{\mid F_{F_0}}=\mathrm{id}\\ 1\end{cases}$	$\begin{cases}\rho_{\mid F_{F_0}}=\mathrm{id}\\ 3\end{cases}$	$\begin{cases}\rho_{\mid F_{F_0}}\neq\mathrm{id}\\ 1\end{cases}$	$\begin{cases}\rho_{\mid F_{F_0}}\neq\mathrm{id}\\ 3\end{cases}$
b_{11}	1	1	-1	-1
b_1	0	0	2	2
$(b_{11}-b)a_1+b_1a$	0	$-2a_1$	$-2a_1+2a$	$-4a_1+2a$

Therefore we have $V(\tilde{\eta})\supset V(\tilde{\eta}_0)=\{x \bmod w \mid x\equiv 0 \bmod w/2\}$ ($\tilde{\eta}_0\leftrightarrow a_1=1, a=0$) and $\kappa(\eta)=2$.

• Case of $Q=1, N_{F_0/\mathbb{Q}}(\epsilon_0)=-1, \sqrt{-1}\notin F$:
In case of $w=2$, we take $\tilde{\eta}_0$ corresponding to $a_1=0, a=1$, so that $b_1=1$ for $\rho_{\mid F_0}\neq\mathrm{id}$ implies $V(\tilde{\eta})\supset V(\tilde{\eta}_0)=\{0\}$, i.e. $\kappa(\eta)=1$. In case of $w=6$, $F=F_0(\zeta_3)$ holds, and we take $\tilde{\eta}_0$ corresponding to $a=a_1=1$. Then we have $V(\tilde{\eta})\supset V(\tilde{\eta}_0)=\{0\}$ and so $\kappa(\eta)=1$, considering ρ corresponding to $\zeta_w^\rho=\zeta_w, \epsilon_0^\rho=-\epsilon_0^{-1}$, for which we get $b=1, b_1=3, b_{11}=-1$.

• Case of $Q=2$:
Let ρ be the complex conjugation; then $b=-1$, $b_1=1$, $b_{11}=1$ hold. Therefore $\tilde{\eta}_0$ corresponding to $a_1=0$, $a=1$ gives $V(\tilde{\eta})\supset V(\tilde{\eta}_0)=\{0\}$ and $\kappa(\eta)=1$. □

The proposition explains the theoretical background of the constant Δ in [8], which is our $\kappa(\eta)$ where the positivity of the expected density of $\{p \mid \#E(p)=w(p-1)/\kappa(\eta), \sigma_{F/\mathbb{Q}}(p)=\mathrm{id}\}$ is shown.

4.1.5. *Case where F is the Galois closure of a real cubic field F_0 with negative discriminant*

We note that F is an S_3-extension of $\mathbb{Q}$, and so $w=2,4,6$. Let $\epsilon\,(>1)$ be the fundamental unit of F_0, and σ an automorphism of order 3 in $\mathrm{Gal}(F/\mathbb{Q})$; then $\zeta_w^\sigma=\zeta_w$ holds, and putting $\epsilon'=\epsilon^\sigma$, we see

$$\epsilon'^J=\epsilon^{\sigma^2}=\epsilon'^\sigma=\epsilon^{-1-\sigma},$$

where J denotes the complex conjugation. That $[o_F^\times : \langle\zeta_w,\epsilon,\epsilon'\rangle]=1$ or 3 is known [4]. We still suppose $\eta\in\mathrm{Gal}(F/\mathbb{Q})$ is the identity. Then we have

Proposition 4.3.

$$\kappa(\eta)=\begin{cases}3 & \textit{if } F_0 \textit{ is pure cubic and } [o_F^\times : \langle\zeta_w,\epsilon,\epsilon'\rangle]=1,\\ 1 & \textit{otherwise.}\end{cases}$$

Proof. For an extension $\tilde{\eta}$ of η, we write as (4.2), (4.3),

$$\zeta_{w^2}^{\tilde{\eta}} = \zeta_{w^2}^{aw+1}, \quad \sqrt[w]{u_i}^{\tilde{\eta}} = \zeta_w^{a_i} \sqrt[w]{u_i},$$
$$\zeta_w^{\rho} = \zeta_w^{b}, \quad u_1{}^{\rho} = \zeta_w^{b_1} u_1{}^{b_{11}} u_2{}^{b_{12}}, \quad u_2{}^{\rho} = \zeta_w^{b_2} u_1{}^{b_{21}} u_2{}^{b_{22}},$$

where $o_F^\times = \langle \zeta_w, u_1, u_2 \rangle$.

Suppose $[o_F^\times : \langle \zeta_w, \epsilon, \epsilon' \rangle] = 1$; then we have, on putting $u_1 = \epsilon, u_2 = \epsilon'$,

$$\begin{array}{lll} \zeta_w^{\sigma} = \zeta_w, & u_1^{\sigma} = u_2, & u_2^{\sigma} = u_1^{-1} u_2^{-1}, \\ \zeta_w^{\sigma^2} = \zeta_w, & u_1^{\sigma^2} = u_1^{-1} u_2^{-1}, & u_2^{\sigma^2} = u_1, \\ \zeta_w^{J} = \zeta_w^{-1}, & u_1^{J} = u_1, & u_2^{J} = u_1^{-1} u_2^{-1}, \\ \zeta_w^{\sigma J} = \zeta_w^{-1}, & u_1^{\sigma J} = u_1^{-1} u_2^{-1}, & u_2^{\sigma J} = u_2, \\ \zeta_w^{\sigma^2 J} = \zeta_w^{-1}, & u_1^{\sigma^2 J} = u_2, & u_2^{\sigma^2 J} = u_1^{-1} u_2^{-1}. \end{array}$$

Therefore we have $b_1 = b_2 = 0$ for $\forall \rho \in \mathrm{Gal}(F/\mathbb{Q})$, and

$$(x_1, x_2) \bmod w \in V(\tilde{\eta})$$

$$\Leftrightarrow$$

$$(x_1, x_2) \begin{pmatrix} b_{11} - b & b_{12} \\ b_{21} & b_{22} - b \end{pmatrix} \begin{pmatrix} a_1 \\ a_2 \end{pmatrix} \equiv 0 \bmod w \text{ for } \forall \rho$$

$$\Leftrightarrow$$

$$\begin{cases} x_1(-a_1 + a_2) + x_2(-a_1 - 2a_2) \equiv 0 \bmod w & (\rho = \sigma) \\ x_1(-2a_1 - a_2) + x_2(a_1 - a_2) \equiv 0 \bmod w & (\rho = \sigma^2) \\ 2x_1 a_1 - x_2 a_1 \equiv 0 \bmod w & (\rho = J) \\ -x_1 a_2 + 2x_2 a_2 \equiv 0 \bmod w & (\rho = \sigma J) \\ x_1(a_1 + a_2) + x_2(a_1 + a_2) \equiv 0 \bmod w & (\rho = \sigma^2 J). \end{cases} \tag{4.5}$$

If F_0 is pure cubic, then we have $w = 6$, and taking $\tilde{\eta}_0$ corresponding to $a_1 = a_2 = 1$, we get by (4.5)

$$(x_1, x_2) \bmod 6 \in V(\tilde{\eta}_0) \Leftrightarrow \begin{cases} -3x_2 \equiv 0 \bmod 6 \\ -3x_1 \equiv 0 \bmod 6 \\ 2x_1 - x_2 \equiv 0 \bmod 6 \\ -x_1 + 2x_2 \equiv 0 \bmod 6 \\ 2x_1 + 2x_2 \equiv 0 \bmod 6 \end{cases}$$
$$\Leftrightarrow \begin{cases} x_j \equiv 0 \bmod 2, \\ x_1 + x_2 \equiv 0 \bmod 3. \end{cases}$$

It is easy to see $V(\tilde{\eta}) \supset V(\tilde{\eta}_0)$ for any extension $\tilde{\eta}$, and therefore $\kappa(\eta) = 3$. Next, we suppose that F_0 is not pure cubic, which implies $\mathbb{Q}(\sqrt{D_{F_0}}) \neq \mathbb{Q}(\sqrt{-3})$ and $w = 2$ or 4. For $\tilde{\eta}_0$ corresponding to $a_1 = a_2 = 1$, we have (cf. (4.5))

$$(x_1, x_2) \bmod w \in V(\tilde{\eta}_0) \Leftrightarrow x_1 \equiv x_2 \equiv 0 \bmod w.$$

Therefore we have $\kappa(\eta) = 1$.

Now suppose $[o_F^\times : \langle \zeta_w, \epsilon, \epsilon' \rangle] = 3$. First we show that there exist $\epsilon_0 \in o_F^\times, e \in \mathbb{Z}$ so that

$$\begin{aligned} o_F^\times &= \langle \zeta_w, \epsilon_0, \epsilon_0^\sigma \rangle, \quad \epsilon_0^3 = \zeta_w^e \epsilon/\epsilon', \\ \epsilon_0^{\sigma} &= \zeta_w^e \epsilon_0^{-1-\sigma}, \qquad \epsilon_0^J = \zeta_w^{-e} \epsilon_0^{1+\sigma}, \quad \epsilon_0^{\sigma J} = \epsilon_0^{-\sigma}. \end{aligned} \tag{4.6}$$

Take a unit $\epsilon_0 \in o_F^\times \setminus \langle \zeta_w, \epsilon, \epsilon' \rangle$ and let $\epsilon_0^3 = \zeta_w^e \epsilon^b \epsilon'^c$ for $e, b, c \in \mathbb{Z}$. If $b \equiv c \equiv 0 \bmod 3$, then we may assume $b = c = 0$, i.e. that ϵ_0 is a root of unity. This contradicts $\epsilon_0 \notin \langle \zeta_w, \epsilon, \epsilon' \rangle$.
If $b \equiv 0 \bmod 3$, then $c \not\equiv 0 \bmod 3$ and $(\epsilon_0^\sigma)^3 = \zeta_w^e \epsilon'^b (\epsilon\epsilon')^{-c} = \zeta_w^e \epsilon^{-c} \epsilon'^{b-c}$ allows us to assume $b \not\equiv 0 \bmod 3$, taking ϵ_0^σ instead of ϵ_0. Thus we may assume $\epsilon_0^3 = \zeta_w^e \epsilon \epsilon'^c$, taking ϵ_0^{-1} instead, if necessary. If $c \equiv 0 \bmod 3$ holds, then $\epsilon = (\zeta_{3w}^{-e} \epsilon_0 \epsilon'^{-c/3})^3 \in F(\zeta_9)^3$, which contradicts Lemma 1.2 in [6]. If $c = 1$ holds, then $\epsilon_0^3 = \zeta_w^e \epsilon^{-\sigma^2}$ and ${\epsilon_0}^{3\sigma} = \zeta_w^e \epsilon^{-1}$, which yields the contradiction $\epsilon = (\zeta_{3w}^e \epsilon_0^{-\sigma})^3 \in F(\zeta_9)^3$ as above. Therefore we may assume

$$\epsilon_0^3 = \zeta_w^e \epsilon/\epsilon'.$$

Then we have $\epsilon_0^{3\sigma} = \zeta_w^e \epsilon'(\epsilon\epsilon') = \zeta_w^e \epsilon\epsilon'^2 = (\epsilon_0\epsilon')^3$. Hence there is a third root ω of unity so that

$$\epsilon_0^\sigma = \omega \epsilon_0 \epsilon', \tag{4.7}$$

which implies

$$\epsilon' = \omega^{-1} \epsilon_0^\sigma / \epsilon_0, \quad \epsilon = \zeta_w^{-e} \epsilon_0^3 \epsilon' = \zeta_w^{-e} \omega^{-1} \epsilon_0^2 \epsilon_0^\sigma.$$

This yields

$$o_F^\times = \langle \zeta_w, \epsilon_0, \epsilon_0^\sigma \rangle$$

and $\epsilon_0^{\sigma^2} = \zeta_w^e \epsilon_0^{-1-\sigma}$ in (4.6) follows from $\epsilon_0^{\sigma^2} = (\omega\epsilon_0\epsilon')^\sigma = \omega(\omega\epsilon_0\epsilon')(\epsilon^{-1}\epsilon'^{-1}) = \omega^3 \zeta_w^e \epsilon_0^{-1-\sigma} = \zeta_w^e \epsilon_0^{-1-\sigma}$. $\epsilon_0^{\sigma+\sigma J} = 1$ in (4.6) follows from $\epsilon_0^{\sigma+\sigma J} = (\epsilon_0^\sigma)^{1+J} > 0$ and the fact that $\epsilon_0^{\sigma+\sigma J}$ is a root of unity, since $(\epsilon_0^{\sigma+\sigma J})^3 = (\zeta_w^e \epsilon/\epsilon')^{\sigma+\sigma J} = \epsilon^{\sigma+\sigma J - \sigma^2 - \sigma^2 J} = 1$ by $\epsilon^{\sigma^2} = \epsilon^{\sigma J}$, $\epsilon^{\sigma^2 J} = \{\epsilon'^J\}^J = \epsilon^\sigma$. By (4.7), we have

$$\begin{aligned} \epsilon_0^J &= (\omega^{-1}\epsilon'^{-1}\epsilon_0^\sigma)^J = \omega \epsilon^{1+\sigma} \epsilon_0^{-\sigma} = \omega(\zeta_w^{-e}\omega^{-1}\epsilon_0^2\epsilon_0^\sigma)(\zeta_w^{-e}\omega^{-1}\epsilon_0^{2\sigma}\epsilon_0^{\sigma^2})\epsilon_0^{-\sigma} \\ &= \omega^{-1}\zeta_w^{-2e}\epsilon_0^{2+2\sigma+\sigma^2} = \omega^{-1}\zeta_w^{-e}\epsilon_0^{1+\sigma} \end{aligned}$$

by $\epsilon_0^{1+\sigma+\sigma^2} = \zeta_w^e$. We have only to show $\omega = 1$ to complete the proof of (4.6). It follows from

$$\epsilon_0^\sigma = \epsilon_0^{J\sigma^2 J} = (\omega^{-1}\zeta_w^{-e}\epsilon_0^{1+\sigma})^{\sigma^2 J} = \omega\zeta_w^e \epsilon_0^{\sigma^2 J + J} = \omega\zeta_w^e(\zeta_w^e\epsilon_0^{-1-\sigma})^J \epsilon_0^J = \omega\epsilon_0^\sigma.$$

Now, putting $u_1 = \epsilon_0, u_2 = \epsilon_0^\sigma$, we have by (4.6)

$$\begin{array}{lll}
\zeta_w^\sigma = \zeta_w, & u_1^\sigma = u_2, & u_2^\sigma = \zeta_w^e u_1^{-1} u_2^{-1}, \\
\zeta_w^{\sigma^2} = \zeta_w, & u_1^{\sigma^2} = \zeta_w^e u_1^{-1} u_2^{-1}, & u_2^{\sigma^2} = u_1, \\
\zeta_w^J = \zeta_w^{-1}, & u_1^J = \zeta_w^{-e} u_1 u_2, & u_2^J = u_2^{-1}, \\
\zeta_w^{\sigma J} = \zeta_w^{-1}, & u_1^{\sigma J} = u_2^{-1}, & u_2^{\sigma J} = u_1^{-1}, \\
\zeta_w^{\sigma^2 J} = \zeta_w^{-1}, & u_1^{\sigma^2 J} = u_1^{-1}, & u_2^{\sigma^2 J} = \zeta_w^{-e} u_1 u_2,
\end{array}$$

which yield

$(x_1, x_2) \bmod w \in V(\tilde{\eta})$

$\Leftrightarrow$

$$x_1((b_{11} - b)a_1 + b_{12}a_2 + b_1 a) + x_2(b_{21}a_1 + (b_{22} - b)a_2 + b_2 a) \equiv 0 \bmod w,$$

$\Leftrightarrow$

$$\begin{cases}
x_1(-a_1 + a_2) + x_2(-a_1 - 2a_2 + ea) \equiv 0 \bmod w & (\rho = \sigma) \\
x_1(-2a_1 - a_2 + ea) + x_2(a_1 - a_2) \equiv 0 \bmod w & (\rho = \sigma^2) \\
x_1(2a_1 + a_2 - ea) \equiv 0 \bmod w & (\rho = J) \\
x_1(a_1 - a_2) + x_2(-a_1 + a_2) \equiv 0 \bmod w & (\rho = \sigma J) \\
x_2(a_1 + 2a_2 - ea) \equiv 0 \bmod w & (\rho = \sigma^2 J)
\end{cases}$$

$$\Rightarrow \begin{cases}
x_1(a_1 - a_2) \equiv 0 \bmod w, \\
x_2(a_1 - a_2) \equiv 0 \bmod w,
\end{cases}$$

$\Rightarrow x_1 \equiv x_2 \equiv 0 \bmod w$ for $\tilde{\eta}_0$ ($\leftrightarrow a_1 - a_2 = 1$).

Hence we have $\kappa(\eta) = 1$ under the assumption $[o_F^\times : \langle \zeta_w, \epsilon, \epsilon' \rangle] = 3$. This completes the proof. □

When F_0 is defined by $x^3 + a_1 x + a_0 = 0$ with $0 \leq a_1, |a_0| \leq 100$, we have checked $\kappa(\eta) = \mathrm{Rel}_\eta$, by finding a prime number p satisfying $\#E(p) = w(p-1)^2/\kappa(\eta)$ by computer.

4.2. *Case of complex conjugation*

In this subsection, let F be an imaginary abelian extension of the rational number field $\mathbb{Q}$ with $[F : \mathbb{Q}] = 2n (\geq 4)$, and we assume that η is the complex conjugation J. Denote the maximal real subfield by F_0; then $d = 2$ and $g(x) = x - 1$ are obvious, and it is known that

$$Q = \left[o_F^\times : o_{F_0}^\times\right]$$

is 1 or 2.

Therefore, we have $o_F^\times = U(1)$ and so $\tilde{\Delta} = \tau_1 = 2$ by Corollary 5.1. Now Proposition 3.6 reads as follows:

For an extension $\tilde{\eta} \in \mathrm{Gal}(F_2/\mathbb{Q})$ of η, put

$$
\begin{aligned}
o_F^\times &= \langle \zeta_w, u_1, \cdots, u_r \rangle, \\
\zeta_{2w}^{\tilde{\eta}} &= \zeta_{2w}^{aw-1}, \\
\sqrt{u_i}^{\tilde{\eta}} &= \zeta_{2w}^{a_i} \prod_j \sqrt{u_j}^{a_{ij}}, \\
{}^t\boldsymbol{a} &= (a_1, \cdots, a_r), \quad A = (a_{ij}), \\
\zeta_w^\rho &= \zeta_w^b, u_i{}^\rho = \zeta_w^{b_i} \prod_j u_j{}^{b_{ij}}, \\
{}^t\boldsymbol{b} &= (b_1, \cdots, b_r), \quad B = (b_{ij}).
\end{aligned}
$$

Then we have (cf. (3.27),(3.28))

$$
V_1(\tilde{\eta}) = \left\{ \boldsymbol{x} \in \mathbb{Z}^r/2\mathbb{Z}^r \;\middle|\; \begin{array}{l} \boldsymbol{xa} \equiv 0 \bmod 2, \text{and for } \forall \rho \in \mathrm{Gal}(F/\mathbb{Q}) \\ \boldsymbol{x}((aw-2)\boldsymbol{b} + (B-b)\boldsymbol{a}) \equiv 0 \bmod 2w \end{array} \right\}, \tag{4.8}
$$

$$
V_2(\tilde{\eta}) = \{0\},
$$

and we note that $u_i^J = \zeta_w^{a_i} \prod_j u_j^{a_{ij}}$ determines $a_i \bmod w$ uniquely.

4.2.1. *Case of* $[F : \mathbb{Q}] = 4$

Let ϵ_0 (> 1) be the fundamental unit of F_0.

Proposition 4.4. *We have the following:*

$$
\kappa(\eta) = \begin{cases} 2 & \textit{if } N_{F_0/\mathbb{Q}}(\epsilon_0) = 1 \textit{ and } Q = 1, \\ 1 & \textit{otherwise.} \end{cases}
$$

Proof. Since $r = \operatorname{rank} o_F^\times = 1$, (4.8) amounts to

$$
V_1(\tilde{\eta}) = \left\{ x_1 \in \mathbb{Z}/2\mathbb{Z} \;\middle|\; \begin{array}{l} x_1 a_1 \equiv 0 \bmod 2, \text{ for } \forall \rho \in \mathrm{Gal}(F/\mathbb{Q}) \\ x_1\{(aw-2)b_1 + (b_{11} - b)a_1\} \equiv 0 \bmod 2w \end{array} \right\}. \tag{4.9}
$$

First, we assume $Q = 1$; then ϵ_0 being a fundamental unit of F, we can take ϵ_0 as the fundamental unit u_1 of F and $a_1 \equiv 0 \bmod w$ is clear and so the first congruence $x_1 a_1 \equiv 0 \bmod 2$ is satisfied for any $x_1 \in \mathbb{Z}$. If $\rho = \mathrm{id}$ on F_0, then we have $\rho = \mathrm{id}$ or J, and hence $b = \pm 1$, $b_1 \equiv 0 \bmod w$, and $b_{11} = 1$. Therefore, the above equation for this ρ is satisfied for all x_1 in this case. Suppose $\rho \neq \mathrm{id}$ on F_0 and put $N_{F_0/\mathbb{Q}}(\epsilon_0) = (-1)^s$; then noting $b_1 \equiv sw/2 \bmod w$, $b_{11} = -1$, $b \equiv 1 \bmod 2$, the second equation in (4.9) becomes

$$
x_1(aw-2)b_1 \equiv x_1(aw-2)sw/2 \equiv 0 \bmod 2w.
$$

Hence, if $s = 0$, this is satisfied for all x_1, and so $\kappa(\eta) = 2$. If $s = 1$, then it is equivalent to $x_1(aw-2)w/2 \equiv 0 \bmod 2w$. Taking $a = 0$, we have $x_1 \equiv 0 \bmod 2$, and so $\kappa(\eta) = 1$.

Next, we assume $Q = 2$; then we can choose a fundamental unit ϵ of F as in §4.1.4 so that

$$\epsilon_0 = \zeta_w \epsilon^2, \quad \epsilon^J = \zeta_w \epsilon.$$

Let $u_1 = \epsilon$; then a_1 is odd, and then the first congruence implies $x_1 \equiv 0 \bmod 2$ and so $\kappa(\eta) = 1$. □

We remark that this is also compatible with [8] and explains the theoretical background of the constant Δ there, which is $\kappa(\eta)$ here.

4.2.2. *Case of* $[F:\mathbb{Q}] = 6$

Proposition 4.5. *In this case, we have* $\kappa(\eta) = 1$.

Proof. Let σ be an element in $\mathrm{Gal}(F/\mathbb{Q})$ of order 3. Let us show that there is a system $\{u_1, u_2\}$ of fundamental units so that

$$u_1^\sigma = u_2, \quad u_2^\sigma = (u_1u_2)^{-1}, \quad u_i^J = u_i. \tag{4.10}$$

Because of $\operatorname{rank} o_F^\times = 2$, there is a system u_1, u_2 of fundamental units

$$u_1^\sigma = u_2, \quad u_2^\sigma = \zeta_w^c (u_1u_2)^{-1},$$

using the theory of integral representation of the cyclic group of prime order (the theorem on p. 508 [3]). Put $u_i^J = \zeta_w^{d_i} u_i$; then defining an integer e by $\zeta_w^\sigma = \zeta_w^e$, we have

$$u_1^{\sigma J} = u_2^J = \zeta_w^{d_2} u_2, \quad u_1^{J\sigma} = (\zeta_w^{d_1} u_1)^\sigma = \zeta_w^{ed_1} u_2,$$

and so $\mathrm{Gal}(F/\mathbb{Q})$ being abelian, we may assume $d_2 = ed_1$. The equality $u_1^{1+\sigma+\sigma^2} = \zeta_w^c$ yields $u_1^{(1+\sigma+\sigma^2)J} = \zeta_w^{-c}$ and $u_1^{J(1+\sigma+\sigma^2)} = (\zeta_w^{d_1} u_1)^{1+\sigma+\sigma^2} = \zeta_w^{(1+e+e^2)d_1+c}$, and hence $\zeta_w^{(1+e+e^2)d_1+2c} = 1$. Therefore we have $(1+e+e^2)d_1 + 2c \equiv 0 \bmod w$, which implies d_1 is even. Thus we have $(\zeta_w^{d_1/2} u_1)^J = \zeta_w^{d_1/2} u_1$, and so we may assume $d_1 = 0$. This necessitates $c \equiv 0 \bmod w/2$, i.e. $\zeta_w^c = \pm 1$. If $\zeta_w^c = -1$, taking $-u_i$ as u_i, we can assume $\zeta_w^c = 1$, which completes the proof of the above assertion (4.10).

Hence we have

$$\begin{aligned} \zeta_w^{J\sigma} &= \zeta_w^{-e}, & u_1^{J\sigma} &= u_2, & u_2^{J\sigma} &= u_1^{-1}u_2^{-1}, \\ \zeta_w^{J\sigma^2} &= \zeta_w^{-e^2}, & u_1^{J\sigma^2} &= u_1^{-1}u_2^{-1}, & u_2^{J\sigma^2} &= u_1, \end{aligned}$$

whence

$$\rho = J\sigma \Rightarrow b \equiv 1 \bmod 2, \quad B = \begin{pmatrix} 0 & 1 \\ -1 & -1 \end{pmatrix}, \quad \boldsymbol{b} \equiv 0 \bmod w,$$
$$\rho = J\sigma^2 \Rightarrow b \equiv 1 \bmod 2, \quad B = \begin{pmatrix} -1 & -1 \\ 1 & 0 \end{pmatrix}, \quad \boldsymbol{b} \equiv 0 \bmod w$$

and this implies, for an extension $\tilde{\eta}_0$ corresponding to $a_1 = a_2 = 1$

$$\begin{aligned}
&(x_1, x_2) \in V_1(\tilde{\eta}_0) \\
&\Leftrightarrow \begin{cases} \boldsymbol{xa} \equiv 0 \bmod 2, \\ \boldsymbol{x}((aw-2)\boldsymbol{b} + (B-b)\boldsymbol{a}) \equiv 0 \bmod 2w \end{cases} \\
&\Rightarrow (x_1, x_2)(B + 1_2)\begin{pmatrix} 1 \\ 1 \end{pmatrix} \equiv 0 \bmod 2 \quad (\rho = J\sigma, J\sigma^2) \\
&\Rightarrow \begin{cases} (x_1, x_2)\left(\begin{pmatrix} 0 & 1 \\ -1 & -1 \end{pmatrix} + 1_2\right)\begin{pmatrix} 1 \\ 1 \end{pmatrix} \equiv 0 \bmod 2, & (\rho = J\sigma) \\ (x_1, x_2)\left(\begin{pmatrix} -1 & -1 \\ 1 & 0 \end{pmatrix} + 1_2\right)\begin{pmatrix} 1 \\ 1 \end{pmatrix} \equiv 0 \bmod 2 & (\rho = J\sigma^2) \end{cases} \\
&\Rightarrow x_1 \equiv x_2 \equiv 0 \bmod 2.
\end{aligned}$$

This yields $V_1(\tilde{\eta}_0) = V_2(\tilde{\eta}_0) = 2\mathbb{Z}$, and so $\kappa(\eta) = 1$. □

Remark 4.2. When equations $y^3 - ay + b = 0$ and $x^2 + c = 0$ ($0 < a, b < 1000, 0 < c < 100, a, b, c \in \mathbb{Z}$), define a real cubic abelian subfield and an imaginary quadratic subfield of F, respectively, the equality $\kappa(\eta) = \mathrm{Rel}_\eta$ is confirmed by finding a prime number p so that $\#E(p) = w(p-1)^2$ with the aid of computer.

4.3. *Case where F is an imaginary abelian field with $[F:\mathbb{Q}] = 6$ and the order of $\eta \in \mathrm{Gal}(F/\mathbb{Q})$ is 3*

Proposition 4.6. *In this case, we have $\kappa(\eta) = 1$.*

Proof. As in §4.2.2 (cf. (4.10)), we may assume

$$o_F^\times = \langle \zeta_w, u_1, u_2 \rangle, \quad u_1^\eta = u_2, \quad u_2^\eta = (u_1 u_2)^{-1} \ (u_1, u_2 \in F_0), \tag{4.11}$$

where F_0 is the maximal real subfield of F. It is easy to see that $o_F^\times = U(3)$, i.e. $o_F^\times = \{\epsilon \in o_F^\times \mid \epsilon^{g(\eta)} \in W_F\}$, on putting $g(x) = \Phi_3(x) = x^2 + x + 1$. Hence we have $\tilde{\Delta} = \tau_3$ for every extension $\tilde{\eta}$ of η (cf. the remark just after (3.9)).

Lemma 4.1. *We have*

$$\tilde{\Delta} = \begin{cases} 7 & \textit{if } F = \mathbb{Q}(\zeta_7), \\ 3 & \textit{if } \zeta_3 \in F, \\ 1 & \textit{otherwise.} \end{cases}$$

Proof. Let p be a prime divisor of $\tilde{\Delta}$ and let $p^n \,||\, \tilde{\Delta}$. Then $\zeta_{p^n}^{g(\rho)} = 1$ for every extension $\rho \in \mathrm{Gal}(F(\zeta_{p^n})/\mathbb{Q})$ of η by definition (cf. (3.8)). Suppose that $\sigma_a \in \mathrm{Gal}(\mathbb{Q}(\zeta_{p^n})/\mathbb{Q})$ with $\zeta_{p^n}^{\sigma_a} = \zeta_{p^n}^a$ coincides with η on $F \cap \mathbb{Q}(\zeta_{p^n})$; then it is extended to an element $\tilde{\sigma_a} \in \mathrm{Gal}(F(\zeta_{p^n})/\mathbb{Q})$ with $\tilde{\sigma_a}_{|F} = \eta$ and so we have $a^2 + a + 1 \equiv 0 \bmod p^n$ by $\zeta_{p^n}^{g(\tilde{\sigma}_a)} = 1$. In particular, p is an odd prime ($\neq 5$), and

$$[\mathbb{Q}(\zeta_{p^n}) : F \cap \mathbb{Q}(\zeta_{p^n})] \leq \# \left\{ a \bmod p^n \,\middle|\, a^2 + a + 1 \equiv 0 \bmod p^n \right\}.$$

Let us check that the right-hand side is ≤ 2. In case of $p = 3$, this is obvious, since there is no solution of $x^2 + x + 1 \equiv 0 \bmod 9$. Suppose $p \neq 3$; then Lemma 5.2 with $f(x) = x^2 + x + 1, A(x) = 4, B(x) = -(2x+1), n = 3$ yields $\#\{a \bmod p^n \mid a^2 + a + 1 \equiv 0 \bmod p^n\} \leq 2$. Now the inequality $[\mathbb{Q}(\zeta_{p^n}) : F \cap \mathbb{Q}(\zeta_{p^n})] \leq 2$ yields

$$p^{n-1}(p-1) = [\mathbb{Q}(\zeta_{p^n}) : F \cap \mathbb{Q}(\zeta_{p^n})][F \cap \mathbb{Q}(\zeta_{p^n}) : \mathbb{Q}] \mid 12. \qquad (4.12)$$

Therefore the possibilities of p, n are $p = 13$, $n = 1$; $p = 7$, $n = 1$; $p = 3$, $n \leq 2$.

• Case of $p = 13, n = 1$

(4.12) implies $[F \cap \mathbb{Q}(\zeta_{13}) : \mathbb{Q}] = 6$, and so $F \subset \mathbb{Q}(\zeta_{13})$. Since the subgroup corresponding to F in $\mathrm{Gal}(\mathbb{Q}(\zeta_{13})/\mathbb{Q})$ is of order 2, it is generated by the complex conjugation. Hence F is a real subfield, which is a contradiction.

• Case of $p = 7, n = 1$

(4.12) implies $6 \leq 2[F \cap \mathbb{Q}(\zeta_7) : \mathbb{Q}]$, and then $[F \cap \mathbb{Q}(\zeta_7) : \mathbb{Q}] = 3, 6$.
Suppose $[F \cap \mathbb{Q}(\zeta_7) : \mathbb{Q}] = 6$; then $F = \mathbb{Q}(\zeta_7)$. Therefore we have $\eta = \sigma_2$ or $= \sigma_4$ and they satisfy $\zeta_7^{g(\eta)} = 1$. Thus we conclude that $7 \mid \tilde{\Delta}$ in case of $F = \mathbb{Q}(\zeta_7)$.
Suppose $[F \cap \mathbb{Q}(\zeta_7) : \mathbb{Q}] = 3$; then F_0 and $F \cap \mathbb{Q}(\zeta_7)$ coincides with the maximal real subfield of $\mathbb{Q}(\zeta_7)$. σ_a ($a = 3, 5$) induces an automorphism of order 3 in $\mathrm{Gal}(F \cap \mathbb{Q}(\zeta_7)/\mathbb{Q})$, and one of them coincides with η on $F \cap \mathbb{Q}(\zeta_7) = F_0$ but neither of them satisfies $a^2 + a + 1 \equiv 0 \bmod 7$. Thus this case does not occur.

• Case of $p = 3, n = 2$

In this case, $p^n = 9$, and there is no integer a which satisfies $a^2 + a + 1 \equiv 0 \bmod 9$. Hence this case does not happen.

• Case of $p=3, n=1$

In case of $F \ni \zeta_3$, an automorphism $\rho \in \mathrm{Gal}(F(\zeta_3)/\mathbb{Q}) = \mathrm{Gal}(F/\mathbb{Q})$, which is an extension of η fixes ζ_3 ($\in F \cap \mathbb{Q}(\zeta_3)$) because the order of η is three, and so $\zeta_3^{g(\rho)} = 1$, i.e. $3 \mid \tilde{\Delta}$. In case of $\zeta_3 \notin F$, $F \cap \mathbb{Q}(\zeta_3) = \mathbb{Q}$ holds, and so extending η to ρ by $\zeta_3^\rho = \zeta_3^{-1}$, we have $\zeta_3^{g(\rho)} \neq 1$, i.e. $3 \nmid \tilde{\Delta}$. □

Now we distinguish three cases. Suppose $\tilde{\Delta} = 1$; then $\tau_3 = 1$ and Proposition 3.5 implies $[\mathcal{R}(\tilde{\eta}, p) : W_F o_F^{\times\,\eta - p}] = 1$. Hence we get $\kappa(\eta) = 1$.

Suppose $\tilde{\Delta} = 3$; then $\zeta_3 \in F$ implies $3 \mid w$. As in Lemma 3.14, we put

$$\begin{aligned} \zeta_{3w}^{\tilde{\eta}} &= \zeta_{3w}^{a}, & \sqrt[3]{u_i}^{\tilde{\eta}} &= \zeta_{3w}^{a_i} \textstyle\prod_j \sqrt[3]{u_j}^{a_{ij}} \\ \zeta_w^{\rho} &= \zeta_w^{b}, & u_i{}^{\rho} &= \zeta_w^{b_i} \textstyle\prod_j u_j{}^{b_{ij}}, \end{aligned}$$

which yields $\boldsymbol{a} \equiv \boldsymbol{b} \equiv 0 \bmod w$ by (4.11), and since the quadratic subfield of F is $\mathbb{Q}(\zeta_3)$, we have $a \equiv 1 \bmod 3$ and so $\Phi_3(a) \equiv 0 \bmod 3$. Hence by (3.9), we have $\tau_3 = 3 = \tilde{\Delta}$ and then (3.27) means $V_1(\tilde{\eta}) = \{\boldsymbol{y} \bmod 3 \mid \boldsymbol{y}\Phi_3(a, A)(B - b)\boldsymbol{a} \equiv 0 \bmod 3w \text{ for } \forall \rho\}$. It is easy to see that

$$B - b = \begin{cases} \begin{pmatrix} 2 & 0 \\ 0 & 2 \end{pmatrix} & \text{if } \rho = J, \\ \begin{pmatrix} -a & 1 \\ -1 & -a-1 \end{pmatrix} & \text{if } \rho = \eta, \end{cases}$$

whence choosing $\tilde{\eta}$ with $\boldsymbol{a} = \begin{pmatrix} w \\ 2w \end{pmatrix}$, we have, by $a \equiv 1 \bmod 3$

$$(B - b)\boldsymbol{a} \equiv \begin{pmatrix} 2w \\ w \end{pmatrix}, \quad \begin{pmatrix} w \\ w \end{pmatrix} \bmod 3w.$$

Since $2x + y \equiv x + y \equiv 0 \bmod 3$ has only a trivial solution, $\boldsymbol{y} \in V_1(\tilde{\eta})$ satisfies $\boldsymbol{y}\Phi_3(a, A) \equiv 0 \bmod 3$. Thus $V_1(\tilde{\eta}) \subset V_2(\tilde{\eta})$ (cf. (3.27),(3.28)) holds, which means $\kappa(\eta) = 1$.

Suppose $\tilde{\Delta} = 7$, i.e. $F = \mathbb{Q}(\zeta_7)$ as above. As in Lemma 3.14, we put

$$\begin{aligned} \zeta_{7w}^{\tilde{\eta}} &= \zeta_{7w}^{a}, & \sqrt[7]{u_i}^{\tilde{\eta}} &= \zeta_{7w}^{a_i} \textstyle\prod_j \sqrt[7]{u_j}^{a_{ij}} \\ \zeta_w^{\rho} &= \zeta_w^{b}, & u_i{}^{\rho} &= \zeta_w^{b_i} \textstyle\prod_j u_j{}^{b_{ij}}. \end{aligned}$$

Since the order of η is 3, we have $a \equiv 2, 4 \bmod 7$, and $\boldsymbol{a} \equiv \boldsymbol{b} \equiv 0 \bmod w$ (cf. (4.11)). Therefore we have $\Phi_3(a) \equiv 0 \bmod 7$, from which $\tau_3 = 7 = \tilde{\Delta}$ follows by (3.9) and hence $\boldsymbol{y} \in V_1(\tilde{\eta})$ (cf. (3.27)) yields

$$\boldsymbol{y}\Phi_3(a, A)(B - b)\boldsymbol{a} \equiv 0 \bmod \tilde{\Delta}w.$$

Here, since we have

$$B - b = \begin{cases} \begin{pmatrix} 2 & 0 \\ 0 & 2 \end{pmatrix} & \text{if } \rho = J, \\ \begin{pmatrix} -a & 1 \\ -1 & -a-1 \end{pmatrix} & \text{if } \rho = \eta, \end{cases}$$

taking $\boldsymbol{a} = \begin{pmatrix} w \\ w \end{pmatrix}$, we obtain

$$(B-b)\boldsymbol{a} = \begin{pmatrix} 2w \\ 2w \end{pmatrix}, \quad \begin{pmatrix} (-a+1)w \\ (-a-2)w \end{pmatrix}.$$

The determinant of $\begin{pmatrix} 2 & -a+1 \\ 2 & -a-2 \end{pmatrix}$ is not 0 mod 7, and so $\boldsymbol{y} \in V_1(\tilde{\eta})$ yields $\boldsymbol{y}\Phi_3(a, A) \equiv 0 \bmod 7$ and so $\boldsymbol{y} \in V_2(\tilde{\eta})$. Therefore we conclude that $\kappa(\eta) = 1$ as above. □

Remark 4.3. In all the examples given here, we have $o_F^\times = U(m)$ for a single m. If F is a real cyclic extension of degree 4, then we see that there is a system of fundamental units $\{u_1, u_2, u_3\}$ such that

$$u_1^\sigma = \zeta u_1^{-1}, \quad u_2^\sigma = u_3^{-1}, \quad u_3^\sigma = \zeta' u_2 \qquad (\zeta, \zeta' = \pm 1), \tag{4.13}$$

$$u_1^\sigma = \zeta u_1^{-1}, \quad u_2^\sigma = u_3^{-1}, \quad u_3^\sigma = u_1^{-1} u_2 \qquad (\zeta = \pm 1), \tag{4.14}$$

where σ is a generator of $\mathrm{Gal}(F/\mathbb{Q})$. Let $\eta = \sigma^2$; then we have

$$\begin{aligned} &U(1) = \langle -1, u_1 \rangle, \quad U(2) = \langle -1, u_2, u_3 \rangle && \text{for (4.13)}, \\ &U(1) = \langle -1, u_1 \rangle, \quad U(2) = \langle -1, u_1 u_2^{-2}, u_1^{-1} u_2 u_3 \rangle && \text{for (4.14)}, \end{aligned}$$

and so $o_F^\times \neq U(1)U(2)$ in case of *(4.14)*. We can see that

$$\kappa(\eta) = \begin{cases} 2 & \text{if } \zeta' = -1 \\ 4 & \text{if } \zeta' = 1 \end{cases} \text{ for (4.13)}, = \begin{cases} 1 & \text{if } \zeta = -1 \\ 2 & \text{if } \zeta = 1 \end{cases} \text{ for (4.14)},$$

and as far as we have checked, $\kappa(\eta) = \mathrm{Rel}_\eta$ is true.

5. Appendix

5.1. *Divisors of* $f(p)$

Proposition 5.1. *Let L be a Galois extension of $\mathbb{Q}$ and $\eta \in \mathrm{Gal}(L/\mathbb{Q})$, and let $f(x) \in \mathbb{Z}[x]$ be a polynomial in $\mathbb{Q}[x]$ with $(f(x), f'(x)) = 1$. Then there exists the maximum δ of natural numbers m such that*

$$\textit{if } \tilde{\eta} \in \mathrm{Gal}(L(\zeta_m)/\mathbb{Q}) \textit{ and } \tilde{\eta}_{|L} = \eta, \textit{ then } \zeta_m^{f(\tilde{\eta})} = 1 \textit{ holds.} \tag{5.1}$$

We have the following expression for δ:

$$\delta = \gcd_{\eta\in\sigma_{L/\mathbb{Q}}(p), p\nmid 2D_L\delta} f(p) = \lim_{x\to\infty} \gcd_{\eta\in\sigma_{L/\mathbb{Q}}(p), p>x} f(p).$$

We need a few lemmas to prove Proposition 5.1.

Lemma 5.1 (Newton Approximation). *Let q be a prime number and $f(x) \in \mathbb{Z}_q[x]$. If $a \in \mathbb{Z}_q$ satisfies $|f(a)|_q < |f'(a)|_q^2$, then there is a solution $\alpha \in \mathbb{Z}_q$ of $f(x) = 0$ such that $|\alpha - a|_q \leq |f(a)/f'(a)|_q$.*

Proof. See p.83 in [1]. □

Lemma 5.2. *Let q be a prime number and let $f(x) \in \mathbb{Z}[x]$ be the polynomial in the proposition. Taking integral polynomials $A(x), B(x) \in \mathbb{Z}[x]$ and a natural number n such that*

$$A(x)f(x) + B(x)f'(x) = n,$$

we define the integer s by $q^s || n$.

Then for any natural number $t \geq 2s+1$

$$\#\{a \bmod q^t \mid f(a) \equiv 0 \bmod q^t\} \leq \deg f(x) \cdot q^s$$

holds.

Proof. If $t \geq 2s+1$ and $f(a) \equiv 0 \bmod q^t$, then we have $q^{s+1} \nmid f'(a)$ and hence $|f(a)|_q < |f'(a)|_q^2$. Hence, by Lemma 5.1, there is an element $\alpha \in \mathbb{Z}_q$ such that $f(\alpha) = 0$ and $|\alpha - a|_q \leq |f(a)/f'(a)|_q$. $q^{t-s} \mid f(a)/f'(a)$ implies $a \equiv \alpha \bmod q^{t-s}$. Since the number of roots α is less than or equal to $\deg f(x)$, we obtain the assertion. □

Lemma 5.3. *The maximal integer δ in Proposition 5.1 exists.*

Proof. For relatively prime natural numbers m_1 and m_2, the condition (5.1) holds for $m = m_1m_2$ if and only if it holds for $m = m_1, m_2$. Hence, assuming that m is a power q^t of a prime q, we have only to show that it holds for a finitely many such integers m. Suppose that $\eta' \in \mathrm{Gal}(\mathbb{Q}(\zeta_{q^t})/\mathbb{Q})$ coincides η on $L \cap \mathbb{Q}(\zeta_{q^t})$; the number of such η''s is equal to $[\mathbb{Q}(\zeta_{q^t}) : L \cap \mathbb{Q}(\zeta_{q^t})]$. Since η' is extended to an element of $\mathrm{Gal}(L(\zeta_{q^t})/\mathbb{Q})$ whose restriction on L is η, the condition (5.1) yields $\zeta_{q^t}^{f(a)} = 1$, where a is defined by $\zeta_{q^t}^{\eta'} = \zeta_{q^t}^a$. Hence $a \bmod q^t$ corresponding to η' satisfies $f(a) \equiv 0 \bmod q^t$. Let n, s be those in Lemma 5.2. If, then $t \geq 2s+1$ holds, then Lemma 5.2

implies $[\mathbb{Q}(\zeta_{q^t}) : L \cap \mathbb{Q}(\zeta_{q^t})] \leq \deg f(x) \cdot q^s$ and so $\varphi(q^t) \leq \deg f(x) \cdot q^s [L \cap \mathbb{Q}(\zeta_{q^t}) : \mathbb{Q}] \leq \deg f(x) \cdot n[L : \mathbb{Q}]$. Hence q^t is bounded. □

Proof of Proposition 5.1. Put

$$G = \gcd_{\eta \in \sigma_{L/\mathbb{Q}}(p), p > x} f(p)$$

for a large number x $(> 2D_{I,}\delta)$. Take any extension $\tilde{\eta}$ of η in $\mathrm{Gal}(L(\zeta_G)/\mathbb{Q})$; then we have $\zeta_G^{f(\tilde{\eta})} \equiv \zeta_G^{f(q)} \equiv 1 \bmod \mathfrak{q}$ if q $(> xG)$ is a prime number such that $\sigma_{L(\zeta_G)/\mathbb{Q}}(q) \ni \tilde{\eta}$. If $\zeta_G^{f(\tilde{\eta})} \neq 1$, then for a prime divisor ℓ of the order of $\zeta_G^{f(\tilde{\eta})}$, we have $\ell \mid G$ and $\zeta_\ell - 1 \in \mathfrak{q}$. Hence we get $\ell = q$ and so $q \mid G$. This contradicts $q > G$. Thus we obtain $\zeta_G^{f(\tilde{\eta})} = 1$ and so $G \mid \delta$.
Conversely, take a prime p so that $\eta \in \sigma_{L/\mathbb{Q}}(p)$ and $p \nmid 2D_L\delta$. For $\tilde{\eta} \in \sigma_{L_\delta/\mathbb{Q}}(p)$, $\zeta_\delta^{f(\tilde{\eta})} = 1$ holds, whence $\zeta_\delta^{f(p)} \equiv 1 \bmod \mathfrak{p}$ follows. Then $\zeta_\delta^{f(p)} = 1$ should occur and δ divides $f(p)$. Thus we have $\delta \mid G$. □

In general, $\delta = \gcd_{\eta \in \sigma_{L/\mathbb{Q}}(p), p \nmid 2D_L} f(p)$ does not hold.

Corollary 5.1. *Suppose $f(x) = x - 1$ in the proposition. If η is the identity (resp. the complex conjugation), then we have $\delta = w$ (resp. $\delta = 2$).*

Proof. Suppose δ satisfies the condition (5.1). By the Galois theory, there are $[\mathbb{Q}(\zeta_\delta) : L \cap \mathbb{Q}(\zeta_\delta)]$ extensions $\tilde{\eta}$ of η to $\mathrm{Gal}(L(\zeta_\delta)/\mathbb{Q})$, and then the supposition $f(x) = x - 1$ implies $\zeta_\delta^{\tilde{\eta}-1} = 1$, i.e. $\zeta_\delta^{\tilde{\eta}} = \zeta_\delta$. Thus the extension to $\mathrm{Gal}(L(\zeta_\delta)/\mathbb{Q})$ is uniquely determined as the identity. This means $[\mathbb{Q}(\zeta_\delta) : L \cap \mathbb{Q}(\zeta_\delta)] = 1$, that is $\mathbb{Q}(\zeta_\delta) \subset L$ and hence $\delta | w$. Hence, in case that η is the identity, w divides δ clearly and so $\delta = w$, and if η is the complex conjugation, then $\zeta_\delta^{-1} = \zeta_\delta^{\tilde{\eta}} = \zeta_\delta$ yields $\delta = 2$. □

5.2. *Structure of Galois group extended by roots of units*

The aim of this subsection is to study the structure of

$$\mathrm{Gal}\left(L\left(\sqrt[p^n]{o_L^\times}\right) \Big/ L\left(\sqrt[p^n]{W_L}\right)\right)$$

for an algebraic number field L. Let us recall [11]

Theorem 5.1. *Let L be a field of characteristic 0 and let p be a prime and $a \in L \setminus L^p$. Then we have*

(i) *If $p \neq 2$, then $x^{p^n} - a$ is irreducible over L for every natural number n,*

(ii) *If $p=2$, then x^2-a is irreducible over L, and $a \notin -4L^4$ if and only if $x^{2^n}-a$ is irreducible over L for any integer $n\,(\geq 2)$.*

Proposition 5.2. *Let L be an algebraic number field. Then the following hold for any natural number n:*

(i) *For $\epsilon \in o_L^\times$ with $\epsilon \notin W_L L^2$, we have $-4\epsilon \notin L(\zeta_{2^n})^4$.*
(ii) *For $\epsilon \in o_L^\times$ with $\epsilon \notin W_L L^2$, $x^{2^n}-\epsilon$ is reducible over $L(\zeta_{2^n})$ if and only if $\sqrt{\epsilon} \in L(\zeta_{2^n})$.*
(iii) *If, for $\epsilon_i \in o_L^\times$ $(i=1,2)$ $\sqrt{\epsilon_i} \in L(\zeta_{2^n})$ and $\epsilon_i \notin W_L L^2$ hold for $i=1$ and 2, then $\epsilon_1\epsilon_2 \in W_L L^2$.*

Proof. Proof of (i): Suppose $-4\epsilon \in L(\zeta_{2^n})^4$; then there is an element $\alpha \in L(\zeta_{2^n})$ such that $\alpha^4=-4\epsilon$. Putting $M=L(\alpha)$, we see that $L \subset M \subset L(\zeta_{2^n})$. Since $L(\zeta_{2^n})$ is abelian over L, M is a Galois extension of L. The assumption on ϵ and (ii) in Theorem 5.1 imply that $x^4+4\epsilon$ is irreducible over L and so $[M:L]=4$ is valid. Since M is a Galois extension of L, M contains a conjugate $\sqrt{-1}\alpha$ of α and so $\sqrt{-1}$.

In case of $\sqrt{-1} \notin L(\sqrt{-\epsilon})$, we have $L(\sqrt{-\epsilon})(\sqrt{-1}) = M = L(\sqrt{-\epsilon})(\sqrt{\alpha^2})$ $(\alpha^2 = \pm 2\sqrt{-\epsilon} \in L(\sqrt{-\epsilon}))$, which yields

$$\alpha^2/(-1) \in L(\sqrt{-\epsilon})^2.$$

Then $-\alpha^2 = \mp 2\sqrt{-\epsilon} = (c+d\sqrt{-\epsilon})^2$ holds for $^\exists c, {}^\exists d \in L$, and this implies $\mp 2\sqrt{-\epsilon} = c^2-\epsilon d^2+2cd\sqrt{-\epsilon}$, which yields a contradiction $\epsilon=(c/d)^2$. Thus $\sqrt{-1} \in L(\sqrt{-\epsilon})$ holds.

If $\sqrt{-1} \notin L$, then $L(\sqrt{-1}) = L(\sqrt{-\epsilon})$ and so $\sqrt{-1}/\sqrt{-\epsilon} \in L$ follows, which implies the contradiction $\epsilon \in L^2$.

Suppose $\sqrt{-1} \in L$. We note that $\mathbb{Q}(\zeta_{2^n}) = \mathbb{Q}(\sqrt{-1})\mathbb{Q}(\zeta_{2^n}+\zeta_{2^n}^{-1})$ and $\mathbb{Q}(\zeta_{2^n}+\zeta_{2^n}^{-1})/\mathbb{Q}$ is cyclic, and so a subfield containing $\sqrt{-1}$ of $\mathbb{Q}(\zeta_{2^n})$ coincides with $\mathbb{Q}(\zeta_{2^m})$ for some integer m. Hence by $\sqrt{-1} \in L \subset M \subset L(\zeta_{2^n})$ there is an integer such that $L\cap\mathbb{Q}(\zeta_{2^n}) = \mathbb{Q}(\zeta_{2^m})$, $M\cap\mathbb{Q}(\zeta_{2^n}) = \mathbb{Q}(\zeta_{2^{m+2}})$. Hence $x^4-\zeta_{2^m}$ is irreducible over L by (ii) in Theorem 5.1. Thus $M = L(\sqrt[4]{-4\epsilon}) = L(\sqrt[4]{\zeta_{2^m}})$ holds. By Kummer's theory, we have $\sqrt[4]{-4\epsilon}/\sqrt[4]{\zeta_{2^m}}^{\,a} \in L$ for $^\exists a \in \mathbb{Z}$, which implies a contradiction $\epsilon \in -4\zeta_{2^m}^a L^4 \subset W_L(o_L^\times)^2$. Thus we have completed the proof of (i).

Proof of (ii): We know

$$\begin{aligned} & x^{2^n}-\epsilon \text{ is irreducible over } L(\zeta_{2^n}) \\ \Leftrightarrow\ & x^2-\epsilon \text{ is irreducible over } L(\zeta_{2^n}) \text{ and } -4\epsilon \notin L(\zeta_{2^n})^4 \text{ if } n \geq 2 \\ \Leftrightarrow\ & x^2-\epsilon \text{ is irreducible over } L(\zeta_{2^n}). \end{aligned}$$

The first equivalence follows from Theorem 5.1. The second follows from (i). This completes the proof of the case (ii).

Proof of (iii): If $L(\sqrt{\epsilon_1}) = L(\sqrt{\epsilon_2})$, then $\epsilon_1/\epsilon_2 \in L^2$ holds and the assertion (iii) is clear. We assume $L(\sqrt{\epsilon_1}) \neq L(\sqrt{\epsilon_2})$ hereafter.

We need the following:

For a natural number n, the subfields of $\mathbb{Q}(\zeta_{2^n})$ are

$$\begin{aligned} N_a &= \quad \mathbb{Q}(\zeta_{2^a}), & [N_a : \mathbb{Q}] &= 2^{a-1} \ (a = 1, 2, \cdots, n), \\ N_{a,+} &= \mathbb{Q}(\zeta_{2^a} + \zeta_{2^a}^{-1}), & [N_{a,+} : \mathbb{Q}] &= 2^{a-2} \ (a = 3, 4, \cdots, n), \\ N_{a,-} &= \mathbb{Q}(\zeta_{2^a} - \zeta_{2^a}^{-1}), & [N_{a,-} : \mathbb{Q}] &= 2^{a-2} \ (a = 3, 4, \cdots, n). \end{aligned}$$

To show this, we have only to verify that subfields of $N_n = \mathbb{Q}(\zeta_{2^n})$ not contained in N_{n-1} are $N_{n,+}$ and $N_{n,-}$ for $n \geq 3$. Since $\mathrm{Gal}(N_n/\mathbb{Q}) \cong \mathbb{Z}/2\mathbb{Z} \oplus \mathbb{Z}/2^{n-2}\mathbb{Z}$ and the subgroup corresponding to N_{n-1} is of order 2 and generated by $g := (0 \mod 2) \oplus (2^{n-3} \mod 2^{n-2}) \in \mathbb{Z}/2\mathbb{Z} \oplus \mathbb{Z}/2^{n-2}\mathbb{Z}$, the assertion above follows from the fact that subgroups of $\mathrm{Gal}(N_n/\mathbb{Q})$ which do not contain g is $(0 \mod 2) \oplus (0 \mod 2^{n-2}), \langle(1 \mod 2) \oplus (0 \mod 2^{n-2})\rangle, \langle(1 \mod 2) \oplus (2^{n-3} \mod 2^{n-2})\rangle$. They correspond to $N_n, N_{n,+}, N_{n,-}$ respectively, since $(1 \mod 2) \oplus (0 \mod 2^{n-2})$ corresponds to the complex conjugation and $(0 \mod 2) \oplus (1 \mod 2^{n-2})$ corresponds to $\zeta_{2^n} \to \zeta_{2^n}^5$, and $5^{2^{n-3}} \equiv 1 + 2^{n-1} \mod 2^n$.

By virtue of $\mathrm{Gal}(L(\zeta_{2^n})/L) \cong \mathrm{Gal}(\mathbb{Q}(\zeta_{2^n})/L \cap \mathbb{Q}(\zeta_{2^n}))$, we have

$$\begin{aligned} &[L(\sqrt{\epsilon_i}) \cap \mathbb{Q}(\zeta_{2^n}) : L \cap \mathbb{Q}(\zeta_{2^n})] \\ &= [\mathbb{Q}(\zeta_{2^n}) : L \cap \mathbb{Q}(\zeta_{2^n})]/[\mathbb{Q}(\zeta_{2^n}) : L(\sqrt{\epsilon_i}) \cap \mathbb{Q}(\zeta_{2^n})] \\ &= [L(\zeta_{2^n}) : L]/[L(\sqrt{\epsilon_i}, \zeta_{2^n}) : L(\sqrt{\epsilon_i})] \\ &= [L(\zeta_{2^n}) : L]/[L(\zeta_{2^n}) : L(\sqrt{\epsilon_i})] = [L(\sqrt{\epsilon_i}) : L] = 2. \end{aligned}$$

Since $L \subset L(\sqrt{\epsilon_i}) \subset L(\zeta_{2^n})$ and $L(\sqrt{\epsilon_1}) \neq L(\sqrt{\epsilon_2})$, fields $L(\sqrt{\epsilon_1}) \cap \mathbb{Q}(\zeta_{2^n})$ and $L(\sqrt{\epsilon_2}) \cap \mathbb{Q}(\zeta_{2^n})$ are different quadratic extensions of $L \cap \mathbb{Q}(\zeta_{2^n})$. Using the classification above, we get

$$L \cap \mathbb{Q}(\zeta_{2^n}) = \mathbb{Q} \text{ or } N_{a,+} \ \ (a = 3, 4, \cdots, n-1),$$

on noting that

the quadratic extensions of N_a $(2 \leq a \leq n-1)$ in N_n are only N_{a+1},
the quadratic extensions of $N_{a,+}$ $(3 \leq a \leq n-1)$ in N_n are N_a, $N_{a+1,+}$, $N_{a+1,-}$,
the quadratic extensions of $N_{a,-}$ $(3 \leq a \leq n)$ in N_n are only N_a.

(iii.1) Suppose $L \cap \mathbb{Q}(\zeta_{2^n}) = \mathbb{Q}$.

We note that $L(\sqrt{\epsilon_1}), L(\sqrt{\epsilon_2}), L(\sqrt{\epsilon_1\epsilon_2})$ are quadratic extensions of L contained in $L(\zeta_{2^n})$ and they are equal to $L(\sqrt{-1}), L(\sqrt{2}), L(\sqrt{-2})$. If $L(\sqrt{\epsilon_j}) = L(\sqrt{-1})$ ($j = 1$ or 2), then $\sqrt{\epsilon_j}/\sqrt{-1} \in L$ and $-\epsilon_j \in L^2$ follows, which is a contradiction. Hence we have $L(\sqrt{\epsilon_1\epsilon_2}) = L(\sqrt{-1})$, whence $-\epsilon_1\epsilon_2 \in L^2$ follows.

For the remaining cases, we may assume $L(\sqrt{2}) = L(\sqrt{\epsilon_1})$, $L(\sqrt{-2}) = L(\sqrt{\epsilon_2})$; then $\epsilon_1 \in 2L^2$, $\epsilon_2 \in -2L^2$ yield $\epsilon_1\epsilon_2 \in -L^2$.

(iii.2) Suppose $L \cap \mathbb{Q}(\zeta_{2^n}) = N_{a,+}$ for $a = 3, 4, \cdots, n-1$.

Put $\kappa = \zeta_{2^a} + \zeta_{2^a}^{-1}$; then $N_{a,+} = \mathbb{Q}(\kappa)$, $N_{a+1,+} = \mathbb{Q}(\sqrt{\kappa+2})$ and $N_{a+1,-} = \mathbb{Q}(\sqrt{\kappa-2})$ hold, and $\mathbb{Q}(\sqrt{\kappa+2})$, $\mathbb{Q}(\sqrt{\kappa-2})$ and $\mathbb{Q}(\zeta_{2^a})$ are quadratic extensions of $N_{a,+}$ in $\mathbb{Q}(\zeta_{2^n})$. Also $L(\sqrt{\epsilon_1})$, $L(\sqrt{\epsilon_2})$, $L(\sqrt{\epsilon_1\epsilon_2})$ are quadratic extensions of L in $L(\zeta_{2^n})$ and should be equal to $L(\sqrt{\kappa+2})$, $L(\sqrt{\kappa-2})$, $L(\zeta_2^n)$. Since $\sqrt{-1}(\zeta_{2^a} - \zeta_{2^a}^{-1})$ is real, it is in $N_{a,+}$ and it follows that $(\kappa+2)(\kappa-2) = \kappa^2 - 4 = (\zeta_{2^a} - \zeta_{2^a}^{-1})^2 = -(\sqrt{-1}(\zeta_{2^a} - \zeta_{2^a}^{-1}))^2 \in -N_{a,+}^2 \subset -L^2$. Hence we have $\epsilon_1\epsilon_2, \epsilon_1(\epsilon_1\epsilon_2)$ or $\epsilon_2(\epsilon_1\epsilon_2) \in -L^2$. The assumption $\epsilon_i \notin W_L L^2$ now implies $-\epsilon_1\epsilon_2 \in L^2$. □

Corollary 5.2. *Let L be an algebraic number field and suppose $\epsilon \in o_L^\times$ satisfies $\epsilon \notin W_L L^2$. If $\sqrt{\epsilon} \in L(\zeta_{2^m})$, then $x^{2^m} - \sqrt{\epsilon}$ is irreducible over $L(\zeta_{2^m})$.*

Proof. The assumption $\sqrt{\epsilon} \in L(\zeta_{2^m})$ implies $m \geq 2$. By Theorem 5.1, $x^{2^m} - \sqrt{\epsilon}$ is irreducible over $L(\zeta_{2^m})$ if and only if $\sqrt[4]{\epsilon} \notin L(\zeta_{2^m})$ and $-4\sqrt{\epsilon} \notin L(\zeta_{2^m})^4$. Suppose that $x^{2^m} - \sqrt{\epsilon}$ is reducible over $L(\zeta_{2^m})$; first, assume $m \geq 3$. If $f := \sqrt[4]{\epsilon} \in L(\zeta_{2^m})$, then $-4\epsilon = (\zeta_8\sqrt{2}f)^4 \in L(\zeta_{2^m})^4$, which contradicts the assertion (i) of Proposition 5.2. If $-4\sqrt{\epsilon} \in L(\zeta_{2^m})^4$, then putting $-4\sqrt{\epsilon} = f_1^4$ ($f_1 \in L(\zeta_{2^m})$), we have $-4\epsilon = (f_1^2/(\zeta_8\sqrt{2}))^4 \in L(\zeta_{2^m})^4$, which is also a contradiction. Therefore $x^{2^m} - \sqrt{\epsilon}$ is irreducible over $L(\zeta_{2^m})$ in case of $m \geq 3$.

Now, applying the above to $m = 3$, $x^8 - \sqrt{\epsilon} = (x^2)^4 - \sqrt{\epsilon}$ is irreducible over $L(\zeta_8)$. Hence $x^4 - \sqrt{\epsilon}$ is irreducible over $L(\zeta_4)$, which completes the proof of the case $m = 2$. □

Corollary 5.3. *Let L be an algebraic number field and let $\epsilon_1, \epsilon_2 \in o_L^\times$ satisfy $\epsilon_1 \notin W_L L^2$. If $\sqrt{\epsilon_1} \in L(\zeta_{2^m})$ for a natural number m, then $\sqrt{\epsilon_1}\epsilon_2 \notin L(\zeta_{2^m})^2$.*

Proof. By applying Corollary 5.2 to $\epsilon = \epsilon_1\epsilon_2^2$, $x^{2^m} - \sqrt{\epsilon_1}\epsilon_2$ is irreducible over $L(\zeta_{2^m})$ and so is $x^2 - \sqrt{\epsilon_1}\epsilon_2$. Therefore $\sqrt{\epsilon_1}\epsilon_2 \notin L(\zeta_{2^m})^2$. □

The following is the main result of this subsection.

Theorem 5.2. *Let L be an algebraic number field. Denote by r the rank of $o_L^\times$. Let p be a prime number and n a natural number. Then we have the following:*

(i) *Suppose either $p \neq 2$ or that $p = 2$ and $o_L^\times \cap L(\sqrt[2^n]{W_L})^2 \subset W_L \cdot (o_L^\times)^2$. Then we have*

$$\operatorname{Gal}\Big(L\Big(\sqrt[p^n]{o_L^\times}\Big)\Big/L\Big(\sqrt[p^n]{W_L}\Big)\Big) \cong (\mathbb{Z}/p^n\mathbb{Z})^r .$$

(ii) *Suppose that $p = 2$ and $o_L^\times \cap L(\sqrt[2^n]{W_L})^2 \not\subset W_L \cdot (o_L^\times)^2$. Then we have $n \geq 2$ and*

$$\operatorname{Gal}\Big(L\Big(\sqrt[p^n]{o_L^\times}\Big)\Big/L\Big(\sqrt[p^n]{W_L}\Big)\Big) \cong \mathbb{Z}/2^{n-1}\mathbb{Z} \times (\mathbb{Z}/2^n\mathbb{Z})^{r-1}.$$

We need more lemmas.

Lemma 5.4. *Suppose that p is an odd prime number and $\epsilon \in o_L^\times$ is not in $W_L(o_L^\times)^p$. Then $\epsilon \notin L(\zeta_{p^n})^p$ for $\forall n \geq 1$.*

Proof. Suppose $\epsilon \in L(\zeta_{p^n})^p$; then $L \subset L(\sqrt[p]{\epsilon}) \subset L(\zeta_{p^n})$ clearly and

$$\operatorname{Gal}\big(L(\zeta_{p^n})/L\big) \cong \operatorname{Gal}\big(\mathbb{Q}(\zeta_{p^n})/L \cap \mathbb{Q}(\zeta_{p^n})\big).$$

Since $L(\zeta_{p^n})/L$ is an abelian extension, $L(\sqrt[p]{\epsilon})/L$ is a Galois extension. Hence in view of $\sqrt[p]{\epsilon} \notin L$ and $p \neq 2$, $x^p - \epsilon$ is irreducible over L. Thus we have $[L(\sqrt[p]{\epsilon}) : L] = p$ and so the conjugate $\zeta_p \sqrt[p]{\epsilon}$ is in $L(\sqrt[p]{\epsilon})$. Therefore $\zeta_p \in L(\sqrt[p]{\epsilon})$ and $L \subset L(\zeta_p) \subset L(\sqrt[p]{\epsilon})$ follows. $[L(\sqrt[p]{\epsilon}) : L] = p$ and $[L(\zeta_p) : L] \mid p-1$ imply $L(\zeta_p) = L$. Therefore we have $L \cap \mathbb{Q}(\zeta_{p^n}) \supset \mathbb{Q}(\zeta_p)$ and hence $L \cap \mathbb{Q}(\zeta_{p^n}) = \mathbb{Q}(\zeta_{p^m})$ and $L(\sqrt[p]{\epsilon}) \cap \mathbb{Q}(\zeta_{p^n}) = \mathbb{Q}(\zeta_{p^{m+1}})$ hold for $1 \leq {}^\exists m < n$ by $p \neq 2$. Since $L(\sqrt[p]{\epsilon})$ $(\subset L(\zeta_{p^n}))$ is the composite of L and $L(\sqrt[p]{\epsilon}) \cap \mathbb{Q}(\zeta_{p^n}) = \mathbb{Q}(\zeta_{p^{m+1}})$, we have $L(\sqrt[p]{\epsilon}) = L(\zeta_{p^{m+1}})$ and hence $f := \sqrt[p]{\epsilon}/\sqrt[p]{\zeta_{p^m}}^{\,a} \in L$ for ${}^\exists a \in \mathbb{Z}$ by virtue of $\zeta_{p^m} \in L$. Therefore $\epsilon = \zeta_{p^m}^a f^p \in W_L(o_L^\times)^p$ and this contradicts the assumption on ϵ. $\square$

Lemma 5.5. *Suppose $\epsilon_1 \in o_L^\times$ is not in $W_L(o_L^\times)^p$ and let n (≥ 2) be a natural number. Under the further assumption of $\epsilon_1 \notin L(\sqrt[2^n]{W_L})^2$ in case of $p = 2$, we have*

$$\operatorname{Gal}\Big(L\big(\sqrt[p^n]{W_L}, \sqrt[p^n]{\epsilon_1}\big)\Big/L\big(\sqrt[p^n]{W_L}\big)\Big) \cong \mathbb{Z}/p^n\mathbb{Z}.$$

Proof. We note that $L(\sqrt[p^n]{W_L}) = L(\zeta_{p^m})$ for some integer m $(\geq n)$ and by Kummer's theory, it suffices to prove that $x^{p^n} - \epsilon_1$ is irreducible over $L(\sqrt[p^n]{W_L})$.
In the case of $p \neq 2$, $x^{p^n} - \epsilon_1$ is irreducible over $L(\sqrt[p^n]{W_L})$ if and only if $x^p - \epsilon_1$ is so over $L(\sqrt[p^n]{W_L})$, which is true by virtue of Lemma 5.4. Suppose that $p = 2$ and $x^{2^n} - \epsilon_1$ is reducible over $L(\sqrt[2^n]{W_L})$; then either $\sqrt{\epsilon_1} \in L(\sqrt[2^n]{W_L})$ or $-4\epsilon_1 \in L(\sqrt[2^n]{W_L})^4$ occurs. However, neither of them can not occur by the assumption or (i) in Proposition 5.2. Thus $x^{2^n} - \epsilon_1$ is irreducible over $L(\sqrt[2^n]{W_L})$. □

Lemma 5.6. *Under the assumption in* (ii) *in Theorem 5.2, $n \geq 2$ holds and there is an element $\epsilon_1 \in o_L^\times$ such that $\epsilon_1 \notin W_L(o_L^\times)^\ell$ for $\forall \ell \geq 2$, $\epsilon_1 \in L(\sqrt[2^n]{W_L})^2$ and*

$$\mathrm{Gal}\Big(L(\sqrt[2^n]{W_L}, \sqrt[2^n]{\epsilon_1}) \Big/ L(\sqrt[2^n]{W_L})\Big) \cong \mathbb{Z}/2^{n-1}\mathbb{Z}.$$

Proof. Suppose $\epsilon \in o_L^\times \cap L(\zeta_{2m})^2$ for $m := \#W_L$; then we have $\epsilon = (a + b\zeta_{2m})^2$ for $\exists a, \exists b \in L$, which implies $\epsilon = a^2 + b^2\zeta_m, ab = 0$ and hence $\epsilon \in W_L(o_L^\times)^2$. Thus we have $o_L^\times \cap L(\sqrt{W_L})^2 \subset W_L(o_L^\times)^2$, and hence $n \geq 2$.

By the assumption in (ii), there is an element $\epsilon \in o_L^\times$ such that $\epsilon = \alpha^2$ $(\alpha \in L(\sqrt[2^n]{W_L}))$ and $\epsilon \notin W_L(o_L^\times)^2$. Write $\epsilon = \zeta\epsilon_1^k$ where $\zeta \in W_L$ and $\epsilon_1 \in o_L^\times$ with $\epsilon_1 \notin W_L(o_L^\times)^\ell$ for $\forall \ell \geq 2$. The condition $\epsilon \notin W_L(o_L^\times)^2$ implies $2 \nmid k$. Then we have $\alpha = \sqrt{\epsilon} = \sqrt{\zeta}\sqrt{\epsilon_1}{\epsilon_1}^{(k-1)/2} \in L(\sqrt[2^n]{W_L})$. Therefore $\sqrt{\epsilon_1} \in L(\sqrt[2^n]{W_L})$ holds. Defining the integer a by $\#W_L = 2^a b$ $(2 \nmid b)$, Corollary 5.2 implies that $x^{2^{n+a}} - \sqrt{\epsilon_1} = (x^{2^{a+1}})^{2^{n-1}} - \sqrt{\epsilon_1}$ is irreducible over $L(\zeta_{2^{n+a}}) = L(\sqrt[2^n]{W_L})$. Therefore $x^{2^{n-1}} - \sqrt{\epsilon_1}$ is also irreducible over $L(\sqrt[2^n]{W_L})$. □

Lemma 5.7. *Let $\epsilon_1, \cdots, \epsilon_r$ be a system of fundamental units of L. Suppose the assumptions of* (i), (ii) *in Theorem* 5.2 *in each case, with one extra-condition $\sqrt{\epsilon_1} \in L(\sqrt[2^n]{W_L})$ in case of* (ii). *Then we have*

$$\sqrt[p^{a+1}]{\epsilon_{s+1}} \notin L\left(\sqrt[p^n]{W_L}, \sqrt[p^n]{\epsilon_1}, \cdots, \sqrt[p^n]{\epsilon_s}, \sqrt[p^a]{\epsilon_{s+1}}\right)$$

for $0 \leq \forall a < n$ and $1 \leq \forall s < r$.

Proof. Suppose $\sqrt[p^{a+1}]{\epsilon_{s+1}} \in L(\sqrt[p^n]{W_L}, \sqrt[p^n]{\epsilon_1}, \cdots, \sqrt[p^n]{\epsilon_s}, \sqrt[p^a]{\epsilon_{s+1}})$; then we have, for some integers $a_1, a_2, \cdots, a_{s+1}$

$$f := \sqrt[p^n]{\epsilon_1}^{\,a_1} \cdots \sqrt[p^n]{\epsilon_s}^{\,a_s} \sqrt[p^a]{\epsilon_{s+1}}^{\,a_{s+1}} / \sqrt[p^{a+1}]{\epsilon_{s+1}} \in L(\sqrt[p^n]{W_L}),$$

since $L(\sqrt[p^n]{W_L}, \sqrt[p^n]{\epsilon_1} \cdots, \sqrt[p^a]{\epsilon_{s+1}})$ is a Kummer extension of $L(\sqrt[p^n]{W_L})$.

In case of (i):

Define the integer b by

$$p^b \| (a_1, \cdots, a_s, (a_{s+1}p - 1)p^{n-a-1}), \quad 0 \leq b \leq n - a - 1.$$

Then we have

$$\epsilon := \epsilon_1^{a_1/p^b} \cdots \epsilon_s^{a_s/p^b} \epsilon_{s+1}^{(a_{s+1}p-1)p^{n-a-1-b}} = f^{p^{n-b}} = (f^{p^{n-b-1}})^p \in L(\sqrt[p^n]{W_L})^p.$$

By noting that ϵ is not in $W_L(o_L^\times)^p$, the equation above contradicts Lemma 5.4 if $p \neq 2$. Suppose $p = 2$; the assumption of (i) implies $\epsilon \in W_L(o_L^\times)^2$, which contradicts the choice of the integer b.

In case of (ii):

Define the integer b by

$$2^b \| (2a_1, a_2, \cdots, a_s, (2a_{s+1} - 1)2^{n-a-1}), \quad 0 \leq b \leq n - a - 1.$$

Then we have

$$\epsilon := \epsilon_1^{2a_1/2^b} \epsilon_2^{a_2/2^b} \cdots \epsilon_s^{a_s/2^b} \epsilon_{s+1}^{(2a_{s+1}-1)2^{n-a-1-b}} = (\sqrt{\epsilon_1}^{\,2a_1/2^b} f^{2^{n-b-1}})^2$$
$$\in L(\sqrt[2^n]{W_L})^2.$$

Suppose $2^b \| 2a_1$; then put

$$\eta_1 = \epsilon_1 \text{ and } \eta_2 = \epsilon_1^{(2a_1/2^b-1)/2} \epsilon_2^{a_2/2^b} \cdots \epsilon_s^{a_s/2^b} \epsilon_{s+1}^{(2a_{s+1}-1)2^{n-a-1-b}}.$$

We have $\eta_1, \eta_2 \in o_L^\times$, $\sqrt{\eta_1} \in L(\sqrt[2^n]{W_L})$ and $\sqrt{\eta_1}\,\eta_2 = \epsilon \in L(\sqrt[2^n]{W_L})^2$. This contradicts Corollary 5.3, since $L(\sqrt[2^n]{W_L}) = L(\zeta_{2^m})$ for some m.

Suppose $2^b \mid a_1$; then $\epsilon \in L(\sqrt[2^n]{W_L})^2$ as above and one of $A_2 := a_2/2^b, \cdots, A_s := a_s/2^b, A_{s+1} := (2a_{s+1} - 1)2^{n-a-1-b}$ is odd, and ${\epsilon_1}^{2a_1/2^b} = ({\epsilon_1}^{a_1/2^b})^2 \in L(\sqrt[2^n]{W_L})^2$. Hence we have

$$\epsilon\epsilon_1^{-2a_1/2^b} = \epsilon_2^{A_2} \cdots \epsilon_{s+1}^{A_{s+1}} \in L\left(\sqrt[2^n]{W_L}\right)^2.$$

Applying (iii) of Proposition 5.2 to ϵ_1 and $\epsilon_2^{A_2} \cdots \epsilon_{s+1}^{A_{s+1}}$, we have the inclusion $\epsilon_1\epsilon_2^{A_2} \cdots \epsilon_{s+1}^{A_{s+1}} \in W_L L^2$, which is a contradiction. □

Proof of Theorem 5.2. Let $\epsilon_1, \cdots, \epsilon_r$ be a system of fundamental units of L, and we may suppose that ϵ_1 is a unit given in Lemma 5.6 in the case (ii). Then Lemmas 5.5, 5.6, 5.7 imply

$$\left[L\left(\sqrt[p^n]{o_L^\times}\right) : L\left(\sqrt[p^n]{W_L}\right)\right] = \begin{cases} p^{rn} & \text{in the case (i),} \\ 2^{n-1+(r-1)n} & \text{in the case (ii).} \end{cases}$$

Since $L\left(\sqrt[p^n]{o_L^\times}\right) \Big/ L(\sqrt[p^n]{W_L})$ is a Kummer extension, this completes the proof. □

Acknowledgments

I thank Professor S. Kanemitsu for many helpful suggestions, and this work was partially supported by Grant-in-Aid for Scientific Research (C), The Ministry of Education, Culture, Sports, Science and Technology of Japan.

References

1. J.W.S. Cassels and A. Fröhlich, *Algebraic Number Theory,* Academic Press, 1967.
2. Y-M. J. Chen, Y. Kitaoka and J. Yu, Distribution of units of real quadratic number fields, *Nagoya Math. J.*, **158** (2000), 167–184.
3. C.W. Curtis and I. Reiner, *Representation theory of finite groups and associative algebras,* Interscience, 1962.
4. T. Honda, Pure cubic fields whose class numbers are multiple of three, *J. Number Theory,* **3** (1971), 7–12.
5. M. Ishikawa and Y. Kitaoka, On the distribution of units modulo prime ideals in real quadratic fields, *J. reine angew. Math.*, **494** (1998), 65–72.
6. Y. Kitaoka, Distribution of units of a cubic field with negative discriminant, *J. Number Theory,* **91** (2001), 318–355.
7. Y. Kitaoka, Distribution of units of an algebraic number field, *in Galois Theory and Modular Forms,* (2003), 287–303. Developments in Mathematics, Kluwer Academic Publishers.
8. Y. Kitaoka, Distribution of units of an algebraic number fields with only one fundamental unit, *Proc. Japan Acad.*, **80A** (2004), 86–89.
9. Y. Kitaoka, Distribution of units of a cubic abelian field modulo prime numbers, *J. Math. Soc. Japan,* (2)**58** (2006), 563–584.
10. T. Kubota, Über den bizyklischen biquadratischen Zahlkörper, *Nagoya Math. J.*, **10** (1955), 65–85.
11. S. Lang, *Algebra,* Springer-Verlag, 2002.
12. H.W. Lenstra, Jr., On Artin's conjecture and Euclid' algorithm in global fields, *Inventiones math.*, **42** (1977), 201–224.
13. K. Masima, On the distribution of units in the residue class field of real quadratic fields and Artin's conjecture *(in Japanese)*, RIMS *Kokyuroku,* **1026** (1998), 156–166.
14. H. Roskam, A quadratic analogue of Artin's conjecture on primitive roots, *J. Number Theory,* **81** (2000), 93–109.

SIGN CHANGES OF FOURIER COEFFICIENTS AND EIGENVALUES OF CUSP FORMS

WINFRIED KOHNEN

Universität Heidelberg, Mathematisches Institut,
INF 288, D-69120 Heidelberg, Germany
E-mail: winfried@mathi.uni-heidelberg.de

We give a survey about recent results on sign changes of Fourier coefficients and eigenvalues of cusp forms, both in the elliptic case and in the case of Siegel modular forms.

1. Introduction

Fourier coefficients of elliptic cusp forms are mysterious objects and in general no simple arithmetical formulas are known for them. If one checks tables, one finds e.g. that quite often sign changes of those coefficients occur and it seems a natural assignment to try to understand them. For example, one may ask if there are infinitely many sign changes or when the first sign change occurs, or one may study sign changes in short intervals.

This might be particularly interesting when the cusp form is a normalized Hecke eigenform and so the Fourier coefficients are equal to the Hecke eigenvalues.

In this article we would like to give a survey on recent results obtained in this direction. In the last section we will also address the case of Siegel modular forms of genus two, where the situation gets more involved.

2. The starting point

The result in the following Theorem seems to be well-known. However, we are not able to give a precise reference where it appeared first. As a substitute, we refer to the joint paper with M. Knopp and W. Pribitkin [10] for an extension to quite general subgroups of $SL_2(\mathbb{R})$ and a discussion of related topics.

As usual, we define

$$\Gamma_0(N) := \left\{ \begin{pmatrix} a & b \\ c & d \end{pmatrix} \in \Gamma_1 \;\middle|\; c \equiv 0 \pmod{N} \right\},$$

where of course $\Gamma_1 := SL_2(\mathbb{Z})$ denotes the full modular group.

Theorem 2.1. *Let f be a non-zero cusp form of even integral weight k on $\Gamma_0(N)$ and suppose that its Fourier coefficients $a(n)$ are real for all $n \geq 1$. Then the sequence $(a(n))_{n\in\mathbb{N}}$ has infinitely many sign changes, i.e. there are infinitely many n such that $a(n) > 0$ and there are infinitely many n such that $a(n) < 0$.*

Proof. It is sufficient to assume that $a(n) \geq 0$ for all but finitely many n and to derive a contradiction. Let

$$L_f(s) = \sum_{n\geq 1} a(n) n^{-s}, \qquad \Re(s) \gg 1$$

be the Hecke L-series of f. Our assumption and a very classical result of Landau (published in 1909) then imply that either $L_f(s)$ converges everywhere or $L_f(s)$ has a singularity at the real point of its line of convergence.

However, according to a classical result of Hecke, $L_f(s)$ extends to an entire function, and the former case occurs. Hence we conclude in particular that $a(n) \ll n^c$ and hence also $a(n)^2 \ll n^c$ for all real c. As a consequence, the Rankin-Selberg zeta function

$$R_{f,f}(s) = \sum_{n\geq 1} a(n)^2 n^{-s}, \qquad \Re(s) \gg 1$$

must converge everywhere, hence is entire. This contradicts a classical result of Rankin-Selberg according to which the latter has a pole at $s = k$ of residue proportional (up to a non-zero constant depending only on k and N) to the square of the Petersson norm of f which is non-zero, since f is non-zero. This completes the proof. □

According to the above Theorem, a reasonable question to ask is if it is possible to obtain a bound on the first sign change, say in terms of k and N. Of course, in general this seems to be a difficult question.

If $f \neq 0$, then recall that by the valence formula for modular forms the orders of zeros of f on the compactified Riemann surface $X_0(N) = \Gamma_0(N)\backslash\mathcal{H} \cup \mathbb{P}^1(\mathbb{Q})$ (where $\mathcal{H}$ is the complex upper half-plane) sum up to $\frac{k}{12}[\Gamma_1 : \Gamma_0(N)]$. Hence there must exist a number n in the range

$$1 \leq n \leq \frac{k}{12}[\Gamma_1 : \Gamma_0(N)]$$

such that $a(n) \neq 0$. Therefore being optimistic, one might hope for a sign change in the range

$$1 \leq n \leq \frac{k}{12}[\Gamma_1 : \Gamma_0(N)] + 1.$$

In a very special case, this indeed follows from work of Siegel [18]. To formulate the result, suppose that $k \geq 4$ is even and denote by d_k the dimension of the space $M_k(\Gamma_1)$ of modular forms of weight k on Γ_1. Recall that d_k is given by the formula

$$d_k = \begin{cases} [\frac{k}{12}] & \text{if } k \equiv 2 \pmod{12}, \\ [\frac{k}{12}] + 1 & \text{otherwise.} \end{cases}$$

Then Siegel showed that there are explicitly computable rational numbers c_n $(n = 0, 1, \ldots, d_k)$, depending on k, such that

$$\sum_{n=0}^{d_k} c_n a_f(n) = 0, \qquad \forall f \in M_k(\Gamma_1).$$

Of course, this more or less seems to follow from elementary linear algebra. However, Siegel's explicit formulas imply that if $k \equiv 2 \pmod 4$, then all the c_n are strictly positive.

Since a cusp form of weight k on Γ_1 is determined by its n-th Fourier coefficients where n runs from 1 to $d_k - 1$, we conclude immediately that if $k \equiv 2 \pmod 4$, then there must be a sign change of the $a(n)$ in the range $1 \leq n \leq d_k$. Thus using the formula for d_k given above, we see that the above optimistic expectation in this special case was justified.

Unfortunately, if $k \equiv 0 \pmod 4$ or if $N > 1$, then Siegel's arguments do not work any longer, and so one has to look for other devices.

3. Elliptic modular forms

In the following, we first look at a normalized Hecke eigenform f that is a newform of level N. Recall that "normalized" means that $a(1) = 1$ and "newform" essentially means that the exact level of f is N. In this case, the Fourier coefficients are equal to the Hecke eigenvalues.

It seems that sign changes of the $a(p)$ (where p runs through primes only), in case f is of level 1 have been studied first by Ram Murty [17].

The following result was obtained in joint work with J. Sengupta.

Theorem 3.1 ([12]). *Suppose that f is a normalized Hecke eigenform of even integral weight k and squarefree level N that is a newform. Then one has $a(n) < 0$ for some n with*

$$n \ll kN \exp\Big(c\sqrt{\log N/\log\log 3N}\Big)(\log k)^{27}, \quad (n, N) = 1.$$

Here $c > 2$ and the constant implied in $\ll$ is absolute.

Note that it is reasonable to assume that $(n, N) = 1$, since the eigenvalues $a(p)$ with $p|N$ are explicitly known by Atkin-Lehner theory.

The proof of the above result uses techniques from analytic number theory (e.g. Perron's formula and a strong convexity principle) and properties of the symmetric square L-function of f, notably the fact that the value of the latter at $s = 1$ is universally bounded from below by $\gg \frac{1}{\log(kN)}$, an important result by D. Goldfeld, J. Hoffstein and D. Lieman [8].

Recently, the above result was improved in a joint paper with H. Iwaniec and J. Sengupta, as follows.

Theorem 3.2 ([9]). *Suppose that f is a normalized Hecke eigenform of even integral weight k and level N (not necessarily squarefree) that is a newform. Then one has $a(n) < 0$ for some n with*

$$n \ll k\sqrt{N} \cdot \log^{8+\epsilon}(kN), \quad (n, N) = 1, \qquad \epsilon > 0.$$

Indeed, this immediately follows from Theorem 1 in [9]. The proof is "elementary" in the sense that it completely avoids the use of the symmetric square L-function. Instead, the Hecke relations for the eigenvalues are exploited. Let us be a bit more precise. One proves the following two Propositions.

Proposition 3.1. *One has*

$$\sum_{n \le x, (n,N)=1} \lambda(n) \log^2\left(\frac{x}{n}\right) \ll_\epsilon (k^2N)^{1/4} \log^{2+\epsilon}(kN)\sqrt{x}, \qquad x \ge 1; \epsilon > 0.$$

The proof follows in a standard way from the convexity principle in combination with Perron's formula.

Proposition 3.2. *Suppose that $\lambda(n) \ge 0$ for $1 \le n \le x$, $(n, N) = 1$. Then*

$$\sum_{n \le x, (n,N)=1} \lambda(n) \log^2\left(\frac{x}{n}\right) \gg \frac{x}{\log^2 x}, \qquad x \gg \sqrt{N}.$$

We indicate the proof in the case $N = 1$ (the general case, of course, is similar). We clearly have

$$\sum_{n \leq x} \lambda(n) \log^2 \left(\frac{x}{n}\right) \gg \sum_{n \leq x/2} \lambda(n).$$

We now restrict the summation to $n = p\ell$, where p and ℓ are primes $\leq \sqrt{x/2}$. We then find

$$\sum_{n \leq x} \lambda(n) \log^2 \left(\frac{x}{n}\right) \gg \Big(\sum_{p \leq \sqrt{x/2}} \lambda(p) \Big)^2 - \Big(\sum_{p \leq \sqrt{x/2}} 1 \Big)$$

(since $\lambda(p\ell) = \lambda(p)\lambda(\ell)$ if p and ℓ are different and $\lambda(p)^2 = \lambda(p^2) + 1$))

$$\gg \Big(\sum_{p \leq \sqrt{x/2}} 1 \Big)^2 - \Big(\sum_{p \leq \sqrt{x/2}} 1 \Big)$$

(since $\lambda(p) \geq 0$ in the given range)

$$\gg \frac{x}{\log^2 x}$$

(by the Prime Number Theorem).

It is easy to see that Propositions 3.1 and 3.2 imply Theorem 3.2.

Using the same ideas, but working a bit harder one can obtain in a similar way

Theorem 3.3 ([9]). *Suppose that f is a normalized Hecke eigenform of level N (not necessarily squarefree) and even integral weight k. Then $a(n) < 0$ for some n with*

$$n \ll (k^2 N)^{\frac{29}{60}}, \quad (n, N) = 1.$$

Note that the bound in Theorem 3.3 in weight aspect is better than the one obtained by convexity, although no sub-convexity bounds for L-functions have been used. We remark that using the recent sub-convexity bound in the case of the full modular group

$$\left| L_f \left(\frac{1}{2} + it \right) \right| \ll_\epsilon (|t| + k)^{1/3+\epsilon}, \qquad \epsilon > 0,$$

due to Jutila and Motohashi (2006, to appear) —the proof is much more difficult and involved—, one can improve the bound in Theorem 3.3 in weight aspect to $k^{2/3+\epsilon}$ if $N = 1$.

The method used in [12] can be extended in various directions. First, one can study sign changes in short intervals.

More precisely, denote by $S_f^+(x)$ and $S_f^-(x)$ the number of positive integers $n \leq x$ with $(n, N) = 1$ for which $a(n) > 0$ and $a(n) < 0$, respectively. The following result was proved in joint work with I. Shparlinski.

Theorem 3.4 ([13]). *Suppose that f is a normalized Hecke eigenform of even integral weight k and squarefree level N that is a newform. Then there are absolute constants $\eta < 1$ and $A > 0$ such that for $y = x^\eta$ one has*

$$S_f^{\pm 1}(x+y) - S_f^{\pm 1}(x) > 0$$

whenever $x \geq (kN)^A$.

In another way, one can generalize Theorem 3.1 and its method of proof to arbitrary non-zero cusp forms with real Fourier coefficients. The following result was obtained in joint work with Y.J. Choie.

Theorem 3.5 ([4]). *Let f be a non-zero cusp form of even integral weight k and squarefree level N with real Fourier coefficients $a(n)$. Then there exist $n_1, n_2 \in \mathbb{N}$ with*

$$\begin{aligned} n_1, n_2 \ll k^3 N^4 \log^{10}(kN) \cdot \exp\left(c\frac{\log(N+1)}{\log\log(N+2)}\right) \\ \cdot \max\{\psi_k(N), k^2 N^{1/2} \log^{16}(kN)\} \end{aligned}$$

such that $a(n_1) > 0$, $a(n_2) < 0$. Here $c > 0$ is an absolute constant and

$$\psi_k(N) := \prod_{p \mid N} \frac{\log(kN)}{\log p}.$$

The proof of Theorem 3.5, being a bit more technically involved, proceeds as follows. One writes f as a linear combination of a special *orthogonal* basis $\{F_\nu\}$ of Hecke eigenforms of weight k and level N and carries over to the Rankin-Selberg zeta functions $R_{F_\nu, F_\mu}(s)$ estimates partially already proved in [12] in the context of $L(\mathrm{sym}^2 F_\nu, s)$. To obtain final corresponding statements for f itself one applies Chebyshev's inequality in conjunction with uniform lower bounds for the Petersson scalar products $\langle F_\nu, F_\nu \rangle$.

The bounds obtained in this way are somewhat weaker than those in [12], being partially due to the fact that one averages over a basis of Hecke eigenforms and in this way some extra factors depending on k and N are introduced. Somewhat better bounds (using similar methods) can be obtained if one restricts to forms f, e.g. in the subspace of newforms.

We also note that in Theorem 3.1 the additional assumption $(n, N) = 1$ was made, since for Hecke eigenforms the eigenvalues $a(p)$ (p a prime, $p|N$) are explicitly known as already stated above. For arbitrary cusp forms, however, it seems unnatural to enforce this condition.

4. Siegel modular forms of genus two

Let

$$\mathcal{H}_g := \{Z \in \mathbb{C}^{g,g} \mid Z = Z',\, \Im(Z) > 0\}$$

be the Siegel upper half-space of genus g and recall that the real symplectic group $Sp_g(\mathbb{R}) \subset GL_{2g}(\mathbb{R})$ operates on $\mathcal{H}_g$ by

$$\begin{pmatrix} A & B \\ C & D \end{pmatrix} \circ Z = (AZ + B)(CZ + D)^{-1}.$$

Let $\Gamma_g := Sp_g(\mathbb{Z})$ be the group of integral symplectic matrices of size $2g$, also called the Siegel modular group of genus g.

Let F be a Siegel cusp form of integral weight k and genus g, i.e. F is a complex-valued holomorphic function on $\mathcal{H}_g$ satisfying the transformation law

$$F(M \circ Z) = \det(CZ + D)^k F(Z), \qquad \forall\, M = \begin{pmatrix} A & B \\ C & D \end{pmatrix} \in \Gamma_g$$

and having a Fourier expansion of the form

$$F(Z) = \sum_{T>0} a(T) e^{2\pi i tr(TZ)}, \qquad Z \in \mathcal{H}_g,$$

where T runs over all positive definite, symmetric half-integral matrices of size g. For basic facts on Siegel modular forms we refer e.g. to [7].

Note that

$$a(T[U]) = (-1)^k a(T), \qquad \forall\, U \in GL_g(\mathbb{Z})$$

(where $GL_g(\mathbb{Z})$ operates on $T > 0$ as above by $T[U] = U'TU$). This easily follows from the transformation formula for F applied with

$$M = \begin{pmatrix} U' & 0 \\ 0 & U^{-1} \end{pmatrix}.$$

Here as usual, for a matrix U we denote by U' its transpose.

Using the analytic properties of the Koecher-Maass Dirichlet series attached to F (cf. e.g. [11] and the literature given there) and of the Rankin-Selberg Dirichlet zeta function attached to F (for $g = 1$ cf. p. 2, l. 6; cf.

e.g. [2] in the general case), it should not be difficult to generalize the Theorem 2.1 to the situation here, i.e. if F has real Fourier coefficients and is not identically zero, then there should exist infinitely many $T > 0$ (modulo $GL_g(\mathbb{Z})$) such that $a(T) > 0$ and there should be infinitely many $T > 0$ (modulo $GL_g(\mathbb{Z})$) such that $a(T) < 0$. However, we have not checked this in detail.

Now suppose that F is an eigenfunction of all Hecke operators. Note that eigenvalues and Fourier coefficients for $g > 1$ are no longer "proportional", in any reasonable sense, and properties of the former ones in general cannot be deduced from the other ones and conversely, in an easy way. Although the Fourier coefficients remain rather mysterious and not much is known about them, the situation is a bit better for the eigenvalues, since the latter can be studied with the help of representation theory and algebraic geometry.

The situation is particularly good if $g = 2$, the easiest case after the elliptic case, and for the rest of this section we will stick to this case.

Thus in the following F will denote a cuspidal Hecke eigenform of weight k and genus 2. We will denote the linear space of all cusp forms of weight k on Γ_2 by $S_k(\Gamma_2)$.

Recall that the spinor zeta function attached to F is given by

$$Z_F(s) = \prod_p Z_{F,p}(p^{-s})^{-1}, \qquad \Re(s) \gg 1,$$

where

$$Z_{F,p}(X) := (1-\alpha_{0,p}X)(1-\alpha_{0,p}\alpha_{1,p}X)(1-\alpha_{0,p}\alpha_{2,p}X)(1-\alpha_{0,p}\alpha_{1,p}\alpha_{2,p}X)$$

and where $\alpha_{0,p}, \alpha_{1,p}, \alpha_{2,p}$ are "the" Satake p-parameters attached to F.

One has

$$\sum_{n\geq 1} \lambda_n n^{-s} = \zeta(2s-2k+4)^{-1} Z_F(s), \qquad \Re(s) \gg 1.$$

Here $\lambda(n)$ denotes the eigenvalue of F under the usual Hecke operator $T(n)$ in genus 2. The numbers $\lambda(n)$ are always real.

The completed function

$$Z_F^*(s) = (2\pi)^{-2s}\Gamma(s)\Gamma(s-k+2)Z_F(s)$$

has meromorphic continuation to $\mathbb{C}$ and is $(-1)^k$-invariant under $s \mapsto 2k-2-s$ [1]. Moreover, $Z_F^*(s)$ is entire if either k is odd or if k is even and F is contained in the orthogonal complement of the Maass subspace $S_k^M(\Gamma_2)$ [6,16].

The Maass subspace is invariant under all Hecke operators and is Hecke isomorphic to the space of elliptic cusp forms of weight $2k-2$ on Γ_1. More precisely, for a Hecke eigenform F in $S_k^M(\Gamma_2)$ there is a uniquely determined normalized cuspidal Hecke eigenform of weight $2k-2$ on Γ_1 such that the relation

$$Z_F(s) = \zeta(s-k+1)\zeta(s-k+2)L_f(s)$$

holds [5].

As a first surprise, it is not generally true that the eigenvalues of a Hecke eigenform in $S_k(\Gamma_2)$ change signs infinitely often. More precisely, S. Breulmann proved the following

Theorem 4.1 ([3]). *Suppose that k is even and let F be a Hecke eigenform in $S_k^M(\Gamma_2)$, with eigenvalues $\lambda(n)$ $(n \in \mathbb{N})$. Then $\lambda(n) > 0$ for all n.*

The proof exploits the relation between the $\lambda(n)$ and the eigenvalues of the form f corresponding to F, as stated above in terms of a relation of zeta functions, together with the fact that the latter satisfy Deligne's theorem, previously known as the Ramanujan-Petersson conjecture (actually, a simpler estimate is sufficient).

On the contrary to the above, one has the following

Theorem 4.2 ([15]). *Let F be a Hecke eigenform in $S_k(\Gamma_2)$ with Hecke eigenvalues λ_n $(n \in \mathbb{N})$. Suppose that F lies in the orthogonal complement of $S_k^M(\Gamma_2)$ if k is even. Then the sequence $(\lambda_n)_{n\in\mathbb{N}}$ has infinitely many sign changes.*

The proof is based on Landau's theorem coupled with the analytic properties of the spinor zeta function of F and a theorem of Weissauer [19] according to which the generalized Ramanujan-Petersson conjecture (saying that $|\alpha_{1,p}| = |\alpha_{2,p}| = 1$ for all p) is true for forms as in Theorem 4.2.

Of course, after Theorem 4.2 the question arises when the first sign change occurs. In this respect, very recent joint work with J. Sengupta says the following.

Theorem 4.3 ([14]). *Let F be a Siegel-Hecke eigenform in $S_k(\Gamma_2)$ and suppose either that k is odd or that k is even and F is in the orthogonal complement of $S_k^M(\Gamma_2)$. Denote by $\lambda(n)$ $(n \in \mathbb{N})$ the eigenvalues of F. Then there exists $n \in \mathbb{N}$ with*

$$n \ll k^2 \log^{20} k$$

such that $\lambda(n) < 0$. Here the constant implied in $\ll$ is absolute.

The proof follows a similar pattern as that of Theorem 3.2, with the Hecke L-function $L_f(s)$ replaced by the spinor zeta function. However, since the Hecke relations for $\lambda(n)$ are more involved in genus 2 than in the elliptic case, exploiting them naturally turns out to be more difficult. One also makes use of the result of [19].

References

1. A.N. Andrianov, Euler products corresponding to Siegel modular forms of genus 2, *Russ. Math. Surv.*, **29** (1974), 45–116.
2. S. Böcherer and S. Raghavan, On Fourier coefficients of Siegel modular forms, *J. Reine Angew. Math.*, **384** (1988), 80–101.
3. S. Breulmann, On Hecke eigenforms in the Maass space, *Math. Z.*, **232**(3) (1999), 527-530.
4. Y.J. Choie and W. Kohnen, The first sign change of Fourier coefficients of cusp forms. Preprint 2006.
5. M. Eichler and D. Zagier, The theory of Jacobi forms, *Progress in Math.*, **55** (1985), Birhäuser: Boston.
6. S.A. Evdokimov, A characterization of the Maass space of Siegel cusp forms of genus 2 (in Russian), *Mat. Sbornik*, (154)**112** (1980), 133–142.
7. E. Freitag, Siegelsche Modulformen, *Grundl. d. Math. Wiss.*, **254** (1983). Springer: Berlin Heidelberg New York.
8. J. Hoffstein and P. Lockart, Coefficients of Maass forms and the Siegel zero (with an appendix by D. Goldfeld, J. Hoffstein and D. Lieman), *Ann. of Math.*, **140** (1994), 161–180.
9. H. Iwaniec, W. Kohnen and J. Sengupta, The first negative Hecke eigenvalue. To appear in *Intern. J. Number Theory.*
10. M. Knopp, W. Kohnen and W. Pribitkin, On the signs of Fourier coefficients of cusp forms, *The Ramanujan Journal.*, **7** (2003), 269–277.
11. W. Kohnen and J. Sengupta, On Koecher-Maass series of Siegel modular forms, *Math. Z.*, **242** (2002), 149–157.
12. W. Kohnen and J. Sengupta, On the first sign change of Hecke eigenvalues of newforms, *Math. Z.*, **254** (2006), 173–184.
13. W. Kohnen and I. Shparlinski, On the number of sign changes of Hecke eigenvalues of newforms. To appear in *J. Austral. Math. Soc.*.
14. W. Kohnen and J. Sengupta, The first negative Hecke eigenvalue of a Siegel cusp form of genus two. To appear in *Acta Arithm.*.
15. W. Kohnen, Sign changes of Hecke eigenvalues of Siegel cusp forms of genus two, To appear in *Proc. AMS.*
16. T. Oda, On the poles of Andrianov L-functions, *Math. Ann.*, **256** (1981), 323–340.
17. Ram Murty, Oscillations of Fourier coefficients of modular forms, *Math. Ann.*, (4)**262** (1983), 431–446.
18. C.L. Siegel, Berechnung von Zetafunktionen an ganzzahligen Stellen, *Nachr. Akad. Wiss. Göttingen Math.-Phys. Kl. II*, (1969), 87–102.

19. R. Weissauer, The Ramanujan conjecture for genus 2 Siegel modular forms (an application of the trace formula). Preprint, Mannheim 1993.

SHIFTED CONVOLUTION SUMS OF FOURIER COEFFICIENTS OF CUSP FORMS

YUK-KAM LAU

Department of Mathematics, The University of Hong Kong,
Pokfulam Road, Hong Kong
E-mail: yklau@maths.hku.hk

JIANYA LIU*

School of Mathematics and System Sciences, Shandong University,
Jinan, Shandong 250100, China
E-mail: jyliu@sdu.edu.cn

YANGBO YE†

Department of Mathematics, The University of Iowa,
Iowa City, IA 52242-1419, U.S.A.
E-mail: yey@math.uiowa.edu

Let g be a holomorphic Hecke eigenform for $\Gamma_0(N)$ of weight l, or a Maass eigenform for $\Gamma_0(N)$ with Laplace eigenvalue $1/4+l^2$. Let $\lambda_g(n)$ be the nth Fourier coefficient of g. A shifted convolution sum of $\lambda_g(n)$ is a sum of the form $\sum_n \lambda_g(n)\overline{\lambda_g(n+h)}w(n)$, where h is a nonzero integer, and w a nice weight function. These shifted convolution sums play a crucial role in analytic number theory, and in particular, in subconvexity bound problems of automorphic L-functions. This article will survey historical developments and recent progress on estimation and analytic continuation of a type of shifted convolution sums. The techniques to be used include spectral decomposition using Poincaré series, a special choice of an orthonormal basis of Hecke eigenforms, a classical result of Good and its generalization by Krötz and Stanton, and a spectral large sieve. The shifted convolution sum will be meromorphically continued to $\Re s > -1/2$, passing through all poles from Laplace eigenvalues.

*Supported in part by the 973 Program, by NSFC Grant # 10531060, and by a Ministry of Education Major Grant Program in Sciences and Technology # 305009.

†Supported in part by the USA National Security Agency under Grant Number H98230-06-1-0075. The United States Government is authorized to reproduce and distribute reprints notwithstanding any copyright notation herein.

1. Automorphic L-functions and subconvexity problems

1.1. *The classical case: the Riemann zeta-function and Dirichlet L-functions*

In his 1859 memoir, Riemann introduced the approach of using an analytic object — Riemann zeta-function — to study the arithmetic problem of distribution of primes. Nowadays this approach has been exploited in various scopes with fruitful results. The associated artificial analytic objects are known as L-functions. They are functions defined on the complex plane under analytic/meromorphic continuation, sharing common features and conjectures with the Riemann zeta-function.

There are two important open conjectures for L-functions: Generalized Riemann Hypothesis (GRH) and Generalized Lindelöf Hypothesis (GLH). The former concerns the location of nontrivial zeros and the latter is about the size of an L-function on the critical line $\Re s = 1/2$. By standard complex analysis, it is seen that GLH follows from GRH. Though being weaker, progress towards GLH is rather slow, even for the Riemann zeta-function $\zeta(s)$. In the case of $\zeta(s)$, GLH is the assertion $\alpha = 0$ in the order estimate

$$\zeta(1/2+it) \ll_\varepsilon |t|^{\alpha+\varepsilon} \quad \text{for } |t| \geq 1. \tag{1.1}$$

The upper estimate (1.1) holds true for $\alpha = 1/4$ by the Phragmén–Lindelöf convexity principle, a robust method in complex analysis. The record to date is $\alpha = 32/205$ due to Huxley [14], but it is still far from the anticipation in GLH. Amazingly, Weyl was able to show $\alpha = 1/6$ about eighty years ago. Note that $32/205 = 1/6 - 13/1230$. The progress meanwhile is small, and it seems that a kind of obstruction at Weyl's bound is present. Below, we shall find such an obstruction occurred in other cases.

The method of convexity principle applies well to other L-functions. Naturally, the bound resulted from this principle is called a convexity bound, and we refer any improvement (usually on the exponent of the convexity bound) as a *subconvexity bound.* Besides, we call a bound *Weyl-like* if it is the 2/3-th power of the convexity bound up to an arbitrarily small $\varepsilon > 0$. For instance, the convexity bound of $\zeta(s)$ is $|t|^{1/4+\varepsilon}$ and its Weyl-like bound is $|t|^{1/6+\varepsilon}$.

The Dirichlet L-function $L(s,\chi)$ was introduced to study the primes in an arithmetic progression. In addition to the t-aspect on the critical line $s = 1/2+it$, we are interested in the aspect of conductor, that is the modulus of the character χ. For either aspect, the exponent of the convexity bound is 1/4. The best known exponent for t-aspect is 1/6 which is Weyl-like. Unlike the Riemann zeta-function, nobody can break the Weyl bound so far, though it was proven quite long time ago. On the conductor aspect, the Weyl-like bound is just achieved recently by Conrey and Iwaniec [6] for real characters. Before this, we only have 3/16 due to Burgess [4], which also remains the best for all characters. Both $\zeta(s)$ and $L(s,\chi)$ are L-functions of degree one, referring to the degree of a generic polynomial factor in p^{-s} of their Euler product factorizations. Next, we turn to higher degree examples.

1.2. *L-functions of degree two*

The L-function associated to a holomorphic Hecke eigenform or Hecke Maass eigenform is a typical example of degree two. Let $\Gamma_0(N)$ be the congruence subgroup that contains matrices in $SL_2(\mathbb{Z})$ whose lower left entry is a multiple of N. The upper half plane $\mathbb{H}$ is identified with G/K where $G = GL(2,\mathbb{R})$ and $K = O(2,\mathbb{R})$ and hence the quotient space $\Gamma_0(N)\setminus\mathbb{H} \cong \Gamma_0(N)\setminus G/K$. But instead, we consider the space $\Gamma_0(N)\setminus G$ which is regarded as the unit tangent bundle of $\Gamma_0(N)\setminus\mathbb{H}$. The Haar measure on $\Gamma_0(N)\setminus G$ is descended from

$$dg = \frac{dxdy}{y^2}\frac{d\varphi}{2\pi},$$

where an element $g \in G$ is expressed via Iwasawa decomposition as

$$g = \begin{pmatrix} 1 & x \\ & 1 \end{pmatrix} \begin{pmatrix} y^{1/2} & \\ & y^{-1/2} \end{pmatrix} \begin{pmatrix} \cos\varphi & \sin\varphi \\ -\sin\varphi & \cos\varphi \end{pmatrix},$$

mapped under the natural projection to $z = x + iy \in \mathbb{H}$. Consider the Laplace operator

$$\widetilde{\Delta} = -y^2\left(\frac{\partial^2}{\partial x^2} + \frac{\partial^2}{\partial y^2}\right) + y\frac{\partial^2}{\partial x\partial\varphi}$$

on the Hilbert space $L^2(\Gamma_0(N)\setminus G)$. A Maass eigenform with eigenvalue $1/4 + k^2$ is a square-integrable K-invariant eigenfunction of $\widetilde{\Delta}$ with eigenvalue $1/4 + k^2$. A holomorphic cusp form of weight k is a holomorphic function f on $\mathbb{H}$ such that the function in $L^2(\Gamma_0(N)\setminus G)$ corresponding to $y^{k/2}f(z)e^{ik\varphi}$ is an eigenfunction of $\widetilde{\Delta}$ with eigenvalue $(k/2)(k/2-1)$. When these forms are invariant under the Hecke operators T_n with $n \geq 1$, we call them *Hecke Maass eigenforms* and *holomorphic Hecke eigenforms*, respectively. Their associated L-functions will involve three parameters t, k and N, and the convexity bound is

$$(|t| + k)^{1/2}N^{1/4}(|t|kN)^{\varepsilon}.$$

The current best subconvexity bounds are accordingly $|t|^{1/3+\varepsilon}$ by Good [11] and Meurman [28], $k^{1/3+\varepsilon}$ by Peng [31] and Ivić [15], and $N^{1/6+\varepsilon}$ for certain forms by Conrey and Iwaniec [6]. Amazingly, all of them are only Weyl-like, that is, no further advance is achieved for L-functions of degree two. Perhaps there is a barrier behind which we can break through merely in the very special case of $\zeta(s)$.

1.3. *Rankin-Selberg L-functions*

The subconvexity problem of L-functions of degree ≥ 3 is mostly unsolved. One accessible case is the Rankin-Selberg L-function which is of degree 4. Let f be a holomorphic Hecke eigenform for $\Gamma_0(N)$ of weight k or Hecke Maass eigenform with eigenvalue $1/4+k^2$. Suppose g a fixed holomorphic or Maass cusp form of weight l or eigenvalue $1/4+l^2$ and level D. The Rankin-Selberg L-function $L(s, f\times g)$ satisfies the convexity bound

$$L(1/2+it, f\times g) \ll \big(ND(|t|+k+l)(|t|+|k-l|)\big)^{1/2+\varepsilon}. \tag{1.2}$$

We fix the level N, the form g, and t, and study the subconvexity estimate of $L(1/2+it, f\times g)$ in the k-aspect. Thus, we are seeking bounds like

$$L(1/2+it, f\times g) \ll_{N,g,t,\varepsilon} k^{\beta+\varepsilon} \tag{1.3}$$

for some $0\leq\beta<1$.

This was firstly achieved by Sarnak [34] for holomorphic f, and by Liu and Ye [25,26] for f being Maass. Progress in this direction is summarized in the following table. Throughout the paper, θ denotes a bound towards the Generalized Ramanujan Conjecture (GRC) for GL_2, for which $\theta=1/2$ is trivial, and the best bound known to date is $\theta=7/64$ due to Kim and Sarnak [20]. GRC actually predicts that $\theta=0$.

β	author(s)	the shifted convolution sum is treated by
$\frac{18}{19-2\theta}$	Sarnak [34]	spectral method
$\frac{15+2\theta}{16}$	Liu and Ye [25,26]	spectral method
$\frac{6-2\theta}{7-4\theta}$	Blomer [2]	circle method
$1-\frac{1}{8+4\theta}$	Lau, Liu, and Ye [23]	spectral method
$\frac{2}{3}$	Lau, Liu, and Ye [24]	spectral method
$\frac{2}{3}$	Jutila and Motohashi [19]	spectral method

The Weyl-like bound achieved in [24] is as follows.

Theorem 1.1. *Let f be a holomorphic Hecke eigenform for $\Gamma_0(N)$ of weight k, or a Maass Hecke eigenform for $\Gamma_0(N)$ with Laplace eigenvalue $1/4+k^2$, and correspondingly let g be a fixed holomorphic or Maass cusp form for $\Gamma_0(N)$, or for $\Gamma_0(N')$ with $(N,N')=1$. Then, for any small $\varepsilon>0$,*

$$L(1/2+it, f\times g) \ll_{N,t,g,\varepsilon} k^{2/3+\varepsilon}, \tag{1.4}$$

where the implied constant grows at most polynomially in t and N, with the degree of the polynomial growth depending on ε.

The same bound is obtained by Jutila and Motohashi [19] but for the full modular group $SL_2(\mathbb{Z})$. The result in [19] also provides a subconvexity bound in the t-aspect, when $|t|$ is suitably smaller than k. For the Rankin-Selberg L-function in question, the Weyl-like bound is only attained in the weight/spectral amongst the various aspects. The subconvexity bound is available on the level aspect N, see [29] and [13], but remains unsettled on t-aspect.

1.4. *Plan of the article*

In this article, we try to give some historical developments and recent progress on estimation and analytic continuation of a type of shifted convolution sums, and indicate some of their applications to subconvexity bounds for automorphic L-functions. In view of the huge amount of materials in these areas at hand, we will mainly mention applications to the Rankin-Selberg L-function $L(s,f\times g)$ in the weight/spectral aspect.

§2 presents some fundamentals of the shifted convolution sums. Basically, there are two methods to treat the shifted convolution sums, the circle method, and the spectral method. In §3, we describe two variants of the circle method, and their consequences in subconvexity bounds $L(s,f\times g)$. §§4–6 are devoted to the spectral method and its recent developments.

The materials are organized not always in historical order, but in logical order. For example, some results in §§4–6 are actually obtained earlier then those in §3.

1.5. *Notations*

As usual, $\tau(n),\varphi(n)$ denote, respectively, the divisor function and the Euler quotient function. For $z\in\mathbb{C}$, $\Re z$ and $\Im z$ denote, respectively, the real and imaginary part of z. Following Riemann, for the specific complex variable s, we write $\Re s = \sigma$ and $\Im s = t$; thus $s = \sigma + it$. The symbol $A \asymp B$ represents both $A \ll B$ and $B \ll A$, and $e(x) = e^{2\pi i x}$.

2. Shifted convolution sums

2.1. *Spectral theory of automorphic forms*

In 1949, Maass [27] introduced the Γ-invariant eigenfunctions of the non-Euclidean Laplacian on $\mathbb{H}$, which are now understood to be the basic elements in the harmonic analysis of $\Gamma \setminus \mathbb{H}$ with Γ being a discrete subgroup of $SL_2(\mathbb{R})$. To see the importance, one should recall that the Fourier series expansion is a kind of spectral decomposition in harmonic analysis of the Euclidean plane $\mathbb{R}^2$. More specifically, consider the group $G = \mathbb{R}^2$ acting on the plane $\mathbb{R}^2$ as translations, then the G-invariant eigenfunctions of the Euclidean Laplacian are exponential functions $e(mx + ny)$ with eigenvalues $4\pi^2(m^2 + n^2)$, where $m, n \in \mathbb{Z}$. In 1950s, Selberg wrote a couple of influential papers on harmonic analysis and discontinuous groups. In particular, he founded harmonic analysis for weakly symmetric Riemannian spaces and the Selberg trace formula generalizing the Poisson summation formula. Around the same time, there appeared important intimate works devoted by other writers.

The upper half plane $\mathbb{H}$ equipped with the hyperbolic metric is a Riemannian manifold. As the metric is invariant under $SL_2(\mathbb{Z})$ and hence under Γ, one may look for a spectral decomposition with Γ-invariant eigenfunctions of the Laplacian associated to the hyperbolic manifold. Such decomposition is possible by virtue of the theory of unbounded self-adjoint operators in Hilbert spaces. However, as Γ may not be cocompact (for example, the congruence subgroups), the Γ-invariant eigenfunctions for the discrete spectrum/eigenvalues, called Maass cusp forms, are not sufficient for complete spectral decomposition. One needs to include Eisenstein series, for the continuous spectrum, in the spectral decomposition. The spectral theory is interesting on its own. For instance, the trace formula developed by Selberg yields the Weyl's law on the discrete spectrum for congruence subgroups. Consequently it guarantees the existence of infinitely many Maass cusp forms. It is worthy of remarking that the existence problem of Maass cusp forms for general discrete subgroups remains open; the answer is likely to be negative due to the work of Phillips and Sarnak [32].

2.2. *The Rankin-Selberg method and shifted convolution sums*

One major topic in modular forms is the study of their Fourier coefficients. Rankin and Selberg independently established a method, nowadays called the Rankin-Selberg method, based on the integral

$$\int_{\Gamma\backslash\mathbb{H}} y^l g(z)\overline{g(z)}E(z,s)\,d\mu(z),$$

where $\Gamma \setminus \mathbb{H}$ is the fundamental domain of Γ, g is a modular form of even integral weight l for Γ and

$$E(z,s) = \sum_{\gamma\in\Gamma_\infty\backslash\Gamma} \Im(\gamma z)^s$$

is an Eisenstein series, and $\sigma > 1$. It is well known that $g(z)$ admits a Fourier expansion,

$$g(z) = \sum_{n\geq 1} a_g(n)e(nz). \tag{2.1}$$

By the invariance under Γ, the integral is unfolded to give

$$\begin{aligned}\int_{\Gamma\backslash\mathbb{H}} y^l g(z)\overline{g(z)}E(z,s)\,d\mu(z) &= \int_{\Gamma_\infty\backslash\mathbb{H}} y^{l+s}|g(z)|^2\,d\mu(z) \\ &= G(s)D_g(s),\end{aligned} \tag{2.2}$$

where $G(s)$ is a product of some Gamma factors, and

$$D_g(s) = \sum_{n\geq 1} \frac{|a_g(n)|^2}{n^s}.$$

A classical method with Perron's formula will yield an asymptotic formula for the summatory function

$$\sum_{n\leq x} |a_g(n)|^2$$

from $D_g(s)$, provided that $D_g(s)$ can be analytically continued to the left beyond $\sigma = 1$. With the available information on $E(z,s)$, $D_g(s)$ is meromorphically continued to the whole complex plane, and is regular on $\sigma \geq 1/2$ except for a finite number of poles lying on the segment $1/2 < s \leq 1$. This is the basic principle of the Rankin-Selberg method.

Prior to the works of Rankin and Selberg, Petersson developed an explicit formula, namely the Petersson trace formula, for the Fourier coefficients of a modular form. This formula involves the Kloosterman sum

$S(m, n, c)$ and Bessel functions, and is derived from the Poincaré series

$$P_m(z, s) = \sum_{\gamma \in \Gamma_\infty \backslash \Gamma} \Im(\gamma z)^s e(m\gamma z).$$

Apparently a good understanding of the Kloosterman sum will result in better knowledge on the Fourier coefficients. To this end, Selberg [35] investigated the series

$$Z(s, m, n) = \sum_{c \geq 1} \frac{S(m, n, c)}{c^{2s}},$$

which is a crucial component in the Fourier coefficient of $P_m(z, s)$. His method is to give $P_m(z, s)$ a spectral decomposition, regarding $P_m(\cdot, s)$ as a function in $L^2(\Gamma \setminus \mathbb{H})$. By the aforementioned spectral theory, $P_m(\cdot, s)$ is a linear combination a series of Maass cusp forms and spectral integrals of Eisenstein series. The coefficients of the discrete part, i.e. the series of Maass cusp forms, are products of gamma functions, which amount to the analytic properties of the function $P_m(z, s)$ in s. The continuous part is similar but, for simplicity, will not be further discussed here. As a result, $P_m(z, s)$ is regular for $\sigma > 1/2$ except possibly for a finite number of simple poles on $(1/2, 1]$. Replacing the Eisenstein series $E(z, s)$ in (2.2) by a Poincare series $P_m(z, s)$, the two methods can be combined to study the shifted convolution sum

$$\sum_{n \geq 1} \frac{a_g(n) a_g(n+h)}{(n + h/2)^s}.$$

This idea was pointed out by Selberg in the last section of [35], but at that time, he did not find an application for this shifted convolution sum.

During the past decades, the uses of the shifted convolution sum came up in the study of L-functions. Indeed for the classical example of degree one - the Riemann zeta-function $\zeta(s)$, the investigation of its fourth moment already leads naturally to the shifted convolution sum for the divisor function $d(n)$, which was considered by Heath-Brown [7]. One needs to handle this or similar type of sums for higher degree automorphic L-functions.

3. Variants of the circle method

Let g be a holomorphic Hecke eigenform for $\Gamma_0(N)$ of even weight l or Hecke Maass eigenform with eigenvalue $1/4 + l^2$. Then g admits the following Fourier expansions:

$$g(z) = \sum_{n \geq 1} \lambda_g(n) n^{(l-1)/2} e(nz) \tag{3.1}$$

when g is holomorphic, and

$$g(z) = y^{1/2} \sum_{n \neq 0} \lambda_g(n) K_{il}(2\pi |n| y) e(nx) \tag{3.2}$$

when g is Maass, where K_{il} is the modified Bessel function of the third kind, and $z = x + iy$. We normalize $\lambda_g(1) = 1$ in (3.1) and (3.2). In this section, we describe variants of the circle method to treat the shifted convolution sums like

$$D_g(\nu_1, \nu_2, h) = \sum_{\nu_1 m - \nu_2 n = h} \overline{\lambda_g(m)} \lambda_g(n) W(m, n) \tag{3.3}$$

uniformly in positive integers ν_1, ν_2, h, where $W : \mathbb{R} \times \mathbb{R} \to \mathbb{R}$ is a nice test function. For example, one may suppose that W is smooth, supported on $[M_1, 2M_1] \times [M_2, 2M_2]$, and satisfies

$$\|W^{(ij)}\|_\infty \ll_{i,j} M_1^{-i} M_2^{-j} \quad \text{for all } i \geq 0, j \geq 0, \tag{3.4}$$

where M_1, M_2 are real numbers greater than 1.

3.1. *The δ-symbol method*

To attack $D_g(\nu_1, \nu_2, h)$ in (3.3), Duke, Friedlander, and Iwaniec [9,10] developed the δ-symbol method, which can be viewed as a variant of the circle method. This δ-symbol method has also been used in many occasions; see for example the DFI paper series, Kowalski, Michel, and Vanderkam [21], and Michel [29]. The following description is based on [9] and Michel [30].

Let

$$\delta(n) = \begin{cases} 1 & \text{if } n = 0, \\ 0 & \text{if } n \neq 0, \end{cases} \tag{3.5}$$

be the Dirac symbol at 0 restricted to integers n; the basic idea of the δ-symbol method is to express $\delta(n)$ in terms of additive characters. One starts with a smooth, compactly supported, even function $\omega(x)$ with

$$\omega(0) = 0, \quad \sum_{r \geq 1} \omega(r) = 1.$$

Put

$$\delta_d(n) = \omega(d) - \omega\left(\frac{n}{d}\right);$$

then we have

$$\delta(n) = \sum_{d | n} \delta_d(n).$$

Now the condition $d|n$ can be detected by additive characters. Thus,

$$\begin{aligned}\delta(n) &= \sum_{d\geq 1}\frac{1}{d}\sum_{h \bmod d} e\left(\frac{hn}{d}\right)\delta_d(n)\\ &= \sum_{c\geq 1}\frac{1}{c}\sideset{}{^*}\sum_{a \bmod c} e\left(\frac{an}{c}\right)\Delta_c(n),\end{aligned}\tag{3.6}$$

where $r = (h,d), a = h/r, c = d/r$, and

$$\Delta_c(n) = \sum_{r\geq 1}\frac{1}{r}\delta_{cr}(n).$$

In practice, one applies the above identity to integers $|n| < U/2$, say, with the text function $\omega(x)$ supported on $[K/2, K]$ and whose derivative satisfy

$$\|\omega^{(j)}\|_\infty \ll K^{-j-1} \quad \text{for all } j \geq 0.$$

Then $\delta_d(n)$ vanishes save for $1 \leq d < \max(K, U/K) = K$ by choosing $K = U^{1/2}$. Hence $\Delta_c(n)$ vanishes save for $1 \leq c < K$ and $\Delta_c(n) \ll K^{-1}$.

Now applying (3.6) to the Dirac symbol $\delta(\nu_1 m - \nu_2 n - h)$ in (3.3), one therefore gets rid of the condition $\nu_1 m - \nu_2 n - h = 0$. For technical reasons, one introduces a localization factor $\phi(\nu_1 x - \nu_2 y - h)$ in $D_g(\nu_1, \nu_2, h)$, where ϕ is a smooth function compactly supported on $[-U/2, U/2]$, satisfying $\phi(0) = 1$ and

$$\|\phi^{(j)}\|_\infty \ll U^{-j} \quad \text{for all } j \geq 0.$$

Hence

$$\begin{aligned}D_g(\nu_1, \nu_2, h) = \sum_{1\leq c\leq K}\ \sideset{}{^*}\sum_{a \bmod c} e\left(-\frac{ah}{c}\right)\sum_{m,n}\overline{\lambda_g(m)}\lambda_g(n)\\ \times e\left(\frac{\nu_1 ma - \nu_2 na}{c}\right)E_c(m,n,h),\end{aligned}\tag{3.7}$$

where

$$E_c(x,y,h) = W(x,y)\phi(\nu_1 x - \nu_2 y - h)\frac{1}{c}\Delta_c(\nu_1 x - \nu_2 y - h).$$

It turns out that the derivatives of $E_c(x,y,h)$ are well controlled; in fact

$$\|E_c^{(ij)}\|_\infty \ll_{i,j} \frac{1}{(cK + |\nu_1 x - \nu_2 y - h|)}\frac{\nu_1^i \nu_2^j}{\min(M_1, M_2)^{i+j}}\left(\frac{K}{c}\right)^{i+j}. \tag{3.8}$$

Next one applies the Voronoi summation formula to both variables m and n, and the shifted convolution sum in question is transformed to

$$\begin{aligned} D_g(\nu_1, \nu_2, h) = \sum_{c \le K} \frac{(\nu_1\nu_2, c)}{c^2} \sum_{m,n} \overline{\lambda_g(m)} \lambda_g(n) S(-\nu_1' m + \nu_2' n, -h, c) \\ \times I_c(m, n, h), \end{aligned} \tag{3.9}$$

where $S(a, b, c)$ is the classical Kloosterman sum, and

$$\begin{aligned} I_c(m, n, h) =& (2\pi i^k)^2 \int_0^\infty \int_0^\infty E_c(x, y, h) J_{k-1}\left(\frac{4\pi\sqrt{mx}}{c/(\nu_1, c)}\right) \\ & \times J_{k-1}\left(\frac{4\pi\sqrt{ny}}{c/(\nu_2, c)}\right) dx dy \end{aligned} \tag{3.10}$$

with $\nu_j' = \nu_j/(\nu_j, c)$. Integrating the Bessel functions in (3.10) by parts many times, one shows that $I_c(m, n, h)$ is very small unless m and n lie in certain short ranges, and one can therefore restrict the summations of m and n in (3.9) to these short ranges. Applying Weil's bound for Kloosterman sums

$$|S(a, b, r)| \le \tau(r)\sqrt{(a, b, r)r},$$

one gets the following result.

Theorem 3.1. *Let g be a holomorphic cusp form for $\Gamma_0(N)$ of even weight l, or a Maass cusp form for $\Gamma_0(N)$ with eigenvalue $1/4 + l^2$. Then*

$$D_g(\nu_1, \nu_2, h) \ll_{N,l,\varepsilon} (\nu_1 M_1 + \nu_2 M_2)^{3/4+\varepsilon}. \tag{3.11}$$

This is proved by Duke, Friedlander, and Iwaniec [9] for the full modular group, and by Kowalski, Michel, and Vanderkam [21] for $\Gamma_0(N)$. From Theorem 3.1, one can get a subconvexity bound for $L(1/2 + it, f \times g)$.

Theorem 3.2. *Let f be a holomorphic Hecke eigenform for $\Gamma_0(N)$ of weight k, or a Maass Hecke eigenform for $\Gamma_0(N)$ with Laplace eigenvalue $1/4 + k^2$, and let g be a fixed holomorphic or Maass cusp form for $\Gamma_0(N)$, or for $\Gamma_0(N')$ with $(N, N') = 1$. Then*

$$L(1/2 + it, f \times g) \ll_{N,t,g,\varepsilon} k^{11/12+\varepsilon}. \tag{3.12}$$

Theorem 3.2 does not appear in literatures. However it can be compared with Iwaniec's bound $k^{5/12+\varepsilon}$ for single automorphic L-functions [16].

3.2. *Jutila's variant*

Let $a'/q' < a/q < a''/q''$ be three consecutive Farey fractions with denominators $\leq Q$, and

$$\mathfrak{M}\left(\frac{a}{q}\right) = \left(\frac{a+a'}{q+q'}, \frac{a+a''}{q+q''}\right].$$

Then $(0,1]$ is a disjoint union of these $\mathfrak{M}(a/q)$,

$$(0,1] = \bigsqcup_{q\leq Q} \ {\bigsqcup_{a \bmod q}}^{*} \mathfrak{M}\left(\frac{a}{q}\right).$$

Let $\delta(n)$ function defined as in (3.5). Then the circle method of Hardy and Littlewood actually starts with the following decomposition of the $\delta(n)$:

$$\delta(n) = \sum_{q\leq Q} \ {\sum_{a \bmod q}}^{*} \int_{\mathfrak{M}(a/q)} e(n\alpha)d\alpha. \tag{3.13}$$

Note that the length of the $\mathfrak{M}(a/q)$ depends on a and q, and therefore in general one cannot invert the order of the summation and the integration above. This is known as the leveling problem.

Jutila introduced another variant of the circle method in [17] and [18] to attack this leveling problem. This variant has also been used in many occasions like Harcos [12], Harcos and Michel [13], Blomer [1,2], and Blomer, Harcos, and Michel [3].

Theorem 3.3. *Let $Q \geq 1$ and $Q^{-2} \leq \delta \leq Q^{-1}$ be two parameters. Let ω be a non-negative function supported in $[Q, 2Q]$ satisfying*

$$\|\omega\|_\infty \leq 1, \qquad \sum \omega(q) > 0.$$

For $r \in \mathbb{Q}$, let $I_r(\alpha)$ be the characteristic function of the interval $[r-\delta, r+\delta]$, and define

$$\Lambda = \sum_q \omega(q)\varphi(q), \qquad \tilde{I}(\alpha) = \frac{1}{2\delta\Lambda}\sum_q \omega(q) \ {\sum_{d \bmod q}}^{*} I_{d/q}(\alpha).$$

Then $\tilde{I}(\alpha)$ is a good approximation to the characteristic function of $[0,1]$ in the sense that

$$\int_0^1 |1-\tilde{I}(\alpha)|^2 d\alpha \ll_\varepsilon \frac{Q^{2+\varepsilon}}{\delta\Lambda^2}. \tag{3.14}$$

To transform the sum in question by Jutila's variant of the circle method, we let $Q > N\nu_1\nu_2$, and $\delta = Q^{-1}$. Let $\tilde{\omega}$ be a function supported in $[Q, 2Q]$ satisfying

$$\|\tilde{\omega}\|_\infty \asymp Q^{-j} \quad \text{for all } j \geq 0,$$

and let

$$\omega(q) = \tilde{\omega}(q)\chi_{[N\nu_1\nu_2|q]}(q),$$

where $\chi_{[N\nu_1\nu_2|q]}(q)$ is the characteristic function of $N\nu_1\nu_2|q$. Then

$$\Lambda \asymp \frac{Q^2}{N\nu_1\nu_2}.$$

For simplicity, put

$$T(\alpha) = \sum_{m,n} \overline{\lambda_g(m)}\lambda_g(n)e(\nu_1 m\alpha)e(-\nu_2 n\alpha)W(m,n).$$

Then the shifted convolution sum in (3.3) can be written as

$$\begin{aligned} D_g(\nu_1,\nu_2,h) &= \int_0^1 T(\alpha)e(-\alpha h)d\alpha \\ &= \int_0^1 \tilde{I}(\alpha)T(\alpha)e(-\alpha h)d\alpha + \int_0^1 (1-\tilde{I}(\alpha))T(\alpha)e(-\alpha h)d\alpha \\ &=: D^{\mathrm{mt}} + D^{\mathrm{et}}, \end{aligned} \tag{3.15}$$

say. The error term D^{et} is, by Cauchy's inequality and Theorem 3.3,

$$\begin{aligned} D^{\mathrm{et}} &\ll \|1-\tilde{I}\|_2 \cdot \|T\|_2 \\ &\ll \frac{Q^{1+\varepsilon}}{\delta^{1/2}\Lambda} \max_{\alpha_1,\alpha_2} \left| \sum_{m,n} \overline{\lambda_g(m)}\lambda_g(n)e(\alpha_1 m)e(-\alpha_2 n)W(m,n) \right|. \end{aligned} \tag{3.16}$$

And this last quantity is acceptable by applying the estimate

$$\sum_{n\leq x} \lambda_g(n)e(m\alpha) \ll_\varepsilon Nk^{5/4}x^{1/2}(Nkx)^\varepsilon$$

uniformly in α. This is due to Wilton, but the explicit dependence on N and k is shown in [13].

The main term D^{mt} in (3.15) can be computed as

$$\begin{aligned} D^{\mathrm{mt}} &= \frac{1}{2\delta\Lambda} \sum_{N\nu_1\nu_2|q} \tilde{\omega}(q) \\ &\quad \times \sum_{d \bmod q}^{*} \int_{-\delta}^{\delta} T\left(\frac{d}{q}+\eta\right) e\left(-h\left(\frac{d}{q}+\eta\right)\right) d\eta. \end{aligned} \tag{3.17}$$

Applying Voronoi's summation formula, we find that the double sums over m, n in (3.17) is for $N\nu_1\nu_2|q$,

$$T\left(\frac{d}{q}+\eta\right) = \sum_{m,n} \overline{\lambda_g(m)}\lambda_g(n) e\left(\frac{\bar{d}(\nu_2 n - \nu_1 m)}{q}\right) \omega^*_{q,\nu_1,\nu_2,\eta}(m,n), \quad (3.18)$$

where

$$\omega^*_{q,\nu_1,\nu_2,\eta}(x_1,x_2) = \frac{4\pi^2\nu_1\nu_2}{q^2}\int_0^\infty\int_0^\infty W(t_1,t_2)e(\nu_1 t_1\eta - \nu_2 t_2\eta) \times J_{k-1}\left(\frac{4\pi\nu_1\sqrt{x_1 t_1}}{q}\right) J_{k-1}\left(\frac{4\pi\nu_2\sqrt{x_2 t_2}}{q}\right) dt_1 dt_2.$$

Inserting (3.18) into (3.17), we get

$$\sum_{N\nu_1\nu_2|q} \sum_{d \bmod q}^{*} \int_{-\delta}^{\delta} T\left(\frac{d}{q}+\eta\right) e\left(-h\left(\frac{d}{q}+\eta\right)\right) d\eta$$

$$= \int_{-\delta}^{\delta} e(-\eta h) \sum_{N\nu_1\nu_2|q} \tilde{\omega}(q) \times \sum_{m,n} S(-h, \nu_2 n - \nu_1 m, q)\overline{\lambda_g(m)}\lambda_g(n)\omega^*_{q,\nu_1,\nu_2,\eta}(m,n) d\eta$$

$$=: \int_{-\delta}^{\delta} e(-\eta h) \sum_{N\nu_1\nu_2|q} \tilde{\omega}(q) Y(m,n) d\eta, \quad (3.19)$$

say. This is where interchange of orders is needed, and this is guaranteed by the fact that the length of the intervals $[-\delta, \delta]$ is independent of the variables. The quantity $Y(m,n)$ above can be transformed as

$$Y(m,n) = \sum_{r\in\mathbb{Z}} S(-h,r,q) \sum_{\nu_2 n - \nu_1 m = r} \overline{\lambda_g(m)}\lambda_g(n)\omega^*_{q,\nu_1,\nu_2,\eta}(m,n),$$

which is similar to the inner sums in (3.9). Arguing similarly, and invoking the spectral large sieve inequality, one gets

Theorem 3.4. *Let g be a holomorphic cusp form for $\Gamma_0(N)$ of even weight l, or a Maass cusp form for $\Gamma_0(N)$ with eigenvalue $1/4 + l^2$. Then*

$$D_g(\nu_1, \nu_2, h) \ll_{N,l,\varepsilon} (\nu_1 M_1 + \nu_2 M_2)^{1/2+\theta+\varepsilon}. \quad (3.20)$$

The estimate above leads to a subconvexity bound for Rankin-Selberg L-functions.

Theorem 3.5. *Let f be a holomorphic Hecke eigenform for $\Gamma_0(N)$ of weight k, or a Maass Hecke eigenform for $\Gamma_0(N)$ with Laplace eigenvalue*

$1/4+k^2$, and correspondingly let g be a fixed holomorphic or Maass cusp form for $\Gamma_0(N)$, or for $\Gamma_0(N')$ with $(N,N')=1$. Then for any small $\varepsilon>0$, we have

$$L(1/2+it, f\times g) \ll_{N,t,g,\varepsilon} k^{(6-2\theta)/(7-4\theta)+\varepsilon}. \tag{3.21}$$

Theorems 3.4 and 3.5 are proved by Blomer [1,2].

4. The spectral method

In 2001, Sarnak [34] considered the subconvexity problem for Rankin-Selberg L-functions associated to two cusp forms with one varying weight and one fixed weight, in which the shifted convolution sum for the cusp form of fixed weight came into play. Sarnak applied Selberg's approach but made a modification of replacing $P_m(z,s)$ by

$$U_h(z,s) = \sum_{\gamma\in\Gamma_\infty\backslash\Gamma} \Im(\gamma z)^s e(-h\Re(\gamma z)),$$

where h is a positive integer. This helps simplify calculations. Together with his ingenious estimates on inner products of eigenfunctions [34], he obtained a subconvexity estimate for the Rankin-Selberg with an application to quantum unique ergodicity.

To explain the ideas of Sarnak [34], let ν_1, ν_2 be positive integers, and h an integer. For $\sigma>1$, we define

$$D_g(s,\nu_1,\nu_2,h) = \sum_{\substack{m,n>0\\ \nu_1 m-\nu_2 n=h}} \lambda_g(n)\overline{\lambda_g(m)} \left(\frac{\sqrt{\nu_1\nu_2 mn}}{\nu_1 m+\nu_2 n}\right)^{l-1} \times (\nu_1 m+\nu_2 n)^{-s} \tag{4.1}$$

when g is a holomorphic cusp form, and

$$D_g(s,\nu_1,\nu_2,h) = \sum_{\substack{m,n\neq 0\\ \nu_1 m-\nu_2 n=h}} \lambda_g(n)\overline{\lambda_g(m)} \left(\frac{\sqrt{\nu_1\nu_2 |mn|}}{\nu_1 |m|+\nu_2 |n|}\right)^{2il} \times (\nu_1 |m|+\nu_2 |n|)^{-s} \tag{4.2}$$

when g is a Maass form. To illustrate the ideas, let us consider the case of g being a holomorphic cusp form on $\Gamma_0(N)$ of weight l. Write $\Gamma=\Gamma_0(N\nu_1\nu_2)$ and

$$V(z) = y^l g(\nu_1 z)\overline{g(\nu_2 z)}.$$

Then V is a Γ-invariant function rapidly decreasing at the cusps of Γ, and $V \in L^2(\Gamma \setminus \mathbb{H})$. By the standard unfolding method, $D_g(s, \nu_1, \nu_2, h)$ can be expressed in terms of the inner product (see [34, p.444], (A7)-(A9))

$$D_g(s, \nu_1, \nu_2, h) = (2\pi)^{s+l-1}(\nu_1\nu_2)^{(l-1)/2}\frac{\langle U_h(\cdot, s), V\rangle}{\Gamma(s+l-1)}. \tag{4.3}$$

Note that V is square-integrable on $\Gamma\backslash\mathbb{H}$, because it is built from a cusp form. On the other hand, as a Poincaré series, U_h is not square integrable on $\Gamma\backslash\mathbb{H}$. However, since $\Gamma\backslash\mathbb{H}$ is of finite volume, Parseval's identity applies. Therefore

$$\begin{aligned}\langle U_h(\cdot, s), V\rangle = &\sum_{j\geq 1}\langle U_h(\cdot, s), \phi_j\rangle\overline{\langle V, \phi_j\rangle} \\ &+ \frac{1}{4\pi}\sum_{\mathfrak{a}}\int_{-\infty}^{\infty}\langle U_h(\cdot, s), E_{\mathfrak{a}}(\cdot, 1/2 + i\tau)\rangle \\ &\times\overline{\langle V, E_{\mathfrak{a}}(\cdot, 1/2 + i\tau)\rangle}\, d\tau,\end{aligned} \tag{4.4}$$

where $\mathfrak{a}$ runs over all the cusps. Note that $\langle U_h, \phi_0\rangle = 0$. In view of (4.3), one may investigate the right-side of (4.4) for the properties of $D_g(s, \nu_1, \nu_2, h)$. These inner products can be computed as follows.

Theorem 4.1. *We have*

$$\langle U_h(\cdot, s), \phi_j\rangle = \frac{\pi^{1/2-s}\overline{\rho_j(-h)}}{4|h|^{s-1/2}}\Gamma\left(\frac{s - 1/2 + it_j}{2}\right)\Gamma\left(\frac{s - 1/2 - it_j}{2}\right),$$

and

$$\begin{aligned}\langle U_h(\cdot, s), E_{\mathfrak{a}}(\cdot, 1/2 + i\tau)\rangle = &\frac{\pi^{1-s-i\tau}}{\Gamma(1/2 - i\tau)}\frac{\overline{\rho_{\mathfrak{a}}(1/2 + i\tau, -h)}}{2|h|^{s-1/2+i\tau}} \\ &\times\Gamma\left(\frac{s - 1/2 + i\tau}{2}\right)\Gamma\left(\frac{s - 1/2 - i\tau}{2}\right).\end{aligned}$$

Recall that by the Maass-Selberg theory (see Deshouillers and Iwaniec [8, p.227]), $L^2(\Gamma\backslash\mathbb{H})$ admits a spectral decomposition with respect to Δ. The spectrum of Δ consists of two components: the discrete spectrum $0 = \lambda_0 < \lambda_1 \leq \lambda_2 \leq \cdots$, and the continuous spectrum covering the segment $[1/4, \infty)$. Each eigenvalue in the discrete spectrum has finite order, and $\lambda_j \to \infty$ as $j \to \infty$. Moreover, there are two types of eigenvalues: $0 < \lambda_j < 1/4$ which are called exceptional, and $\lambda_j \geq 1/4$. The famous Selberg conjecture asserts that there is no exceptional eigenvalue for congruence groups, but the currently best known result is $\lambda_1 \geq 1/4 - \theta^2$, where $\theta = 7/64$ is the exponent of the best known bound toward the Generalized

Ramanujan Conjecture for Maass forms, due to Kim and Sarnak [20]. Write $\lambda_j = s_j(1-s_j)$ and $s_j = 1/2 + it_j$ where

$$0 < it_j \le \theta \text{ if } \lambda_j \text{ is exceptional, and } t_j \in [0,\infty) \text{ otherwise.} \tag{4.5}$$

Theorem 4.1 with (4.5) implies immediately that each summand on the right-side of (4.4) is holomorphic in $\sigma > 1/2 + \theta$. Using the estimate of individual $\langle V, \phi_j \rangle$ developed in [33], Sarnak [34, Theorem A.1] concluded that $D_g(s, \nu_1, \nu_2, h)$ extends to a holomorphic function on $\sigma > 1/2 + \theta$ and has the following upper bound estimate.

Theorem 4.2. *Let g be a holomorphic cusp form for $\Gamma_0(N)$ of even weight l, or a Maass cusp form for $\Gamma_0(N)$ of Laplace eigenvalue $1/4 + l^2$. Then $D_g(s, \nu_1, \nu_2, h)$ extends to a holomorphic function for*

$$\sigma \ge 1/2 + \theta + \varepsilon,$$

for any $\varepsilon > 0$. Moreover, in this region it satisfies

$$D_g(s, \nu_1, \nu_2, h) \ll_{N,g,\varepsilon} (\nu_1\nu_2)^{1/2+\varepsilon}|h|^{1/2+\theta+\varepsilon-\sigma}(1+|t|)^3 + \chi(g)|h|^{1-\sigma},$$

where $\chi(g) = 0$ or 1 according as g is holomorphic or Maass form.

From this, Sarnak [33] deduced the following subconvexity bound for holomorphic Hecke eigenform f. Note that it is the first subconvexity bound in the k-aspect for $L(1/2 + it, f \times g)$.

Theorem 4.3. *Let f be a holomorphic Hecke eigenform for $\Gamma_0(N)$ of weight k, and let g be a fixed holomorphic or Maass cusp form for $\Gamma_0(N)$, or for $\Gamma_0(N')$ with $(N, N') = 1$. Then for any small $\varepsilon > 0$,*

$$L(1/2 + it, f \times g) \ll_{N,t,g,\varepsilon} k^{18/(19-2\theta)+\varepsilon}. \tag{4.6}$$

Theorem 4.2 also enables Liu and Ye [25,26] to derive a subconvexity bound for $L(s, f \times g)$ with Maass eigenforms f.

Theorem 4.4. *Let f be a Maass Hecke eigenform for $\Gamma_0(N)$ with Laplace eigenvalue $1/4 + k^2$, and let g be a fixed holomorphic or Maass cusp form for $\Gamma_0(N)$, or for $\Gamma_0(N')$ with $(N, N') = 1$. Then for any small $\varepsilon > 0$,*

$$L(1/2 + it, f \times g) \ll_{N,t,g,\varepsilon} k^{(15+2\theta)/16+\varepsilon}. \tag{4.7}$$

5. The spectral method: meromorphic continuation to $\sigma > 1/2$

If we allow the occurrence of poles, $D_g(s, \nu_1, \nu_2, h)$ can indeed be meromorphically continued to a wider region. According to (4.5), if we continue $D_g(s, \nu_1, \nu_2, h)$ to $\sigma > 1/2$, the possible poles are those at $s_j = 1/2 + it_j$, where $0 < it_j \leq \theta$ with $\lambda_j = s_j(1 - s_j)$ being exceptional Laplace eigenvalues. As predicted by the GRC, these poles should not exist. Since we do not assume GRC, we will have to control the residues of these possible poles.

Furthermore, we may refine Sarnak's Theorem 4.2 in the t-aspect via the mean square estimate in Good [11] rather than the term-wise bound. Good's result was proved originally for holomorphic cusp forms of weight $l \geq 4$. In other cases, it was generalized recently by Krötz and Stanton [22]. As we will see, switching from individual bound for $\langle V, \phi_j \rangle$ to a mean square estimate provides a significant saving.

In view of (4.3), (4.4), and Theorem 4.1, we introduce the following functions

$$\begin{aligned} B_j(h, s) &= (2\pi)^{s+l-1}(\nu_1\nu_2)^{(l-1)/2}\frac{\langle U_h(\cdot, s), \phi_j\rangle}{\Gamma(s+l-1)} \\ &= (\nu_1\nu_2)^{(l-1)/2}\frac{\overline{\rho_j(-h)}}{|h|^{s-1/2}}\frac{2^{s+l-3}\pi^{l-1/2}}{\Gamma(s+l-1)} \\ &\quad \times\Gamma\left(\frac{s-1/2+it_j}{2}\right)\Gamma\left(\frac{s-1/2-it_j}{2}\right), \end{aligned} \tag{5.1}$$

$$\begin{aligned} C_{\mathfrak{a}}(h, s, \tau) &= (2\pi)^{s+l-1}(\nu_1\nu_2)^{(l-1)/2}\frac{\langle U_h(\cdot, s), E_{\mathfrak{a}}(\cdot, 1/2+i\tau)\rangle}{\Gamma(s+l-1)} \\ &= (\nu_1\nu_2)^{(l-1)/2}\frac{\overline{\rho_{\mathfrak{a}}(1/2+i\tau, -h)}}{\Gamma(1/2-i\tau)|h|^{s-1/2+i\tau}}\frac{2^{s+l-2}\pi^{l-i\tau}}{\Gamma(s+l-1)} \\ &\quad \times\Gamma\left(\frac{s-1/2+i\tau}{2}\right)\Gamma\left(\frac{s-1/2-i\tau}{2}\right), \end{aligned} \tag{5.2}$$

and denote by $R_h(s)$ the following sum over the exceptional eigenvalues,

$$\begin{aligned} R_h(s) = \frac{(\nu_1\nu_2)^{(l-1)/2}2^{s+l-3}\pi^{l-1/2}}{\Gamma(s+l-1)} &\sum_{1/2\leq s_j\leq 1/2+\theta}\frac{\overline{\rho_j(-h)}}{|h|^{s-1/2}} \\ \times\Gamma\left(\frac{s-s_j}{2}\right)\Gamma\left(\frac{s-(1-s_j)}{2}\right)&\overline{\langle V, \phi_j\rangle}. \end{aligned} \tag{5.3}$$

Note here that we include the possible nonexceptional eigenvalue $\lambda_j = 1/4$ with $s_j = 1/2$ and $t_i = 0$ in $R_h(s)$ just for technical simplicity. Then, for

$\sigma > 1$,

$$\begin{aligned} D_g(s, \nu_1, \nu_2, h) - R_h(s) = &\sum_{j:\, t_j > 0} B_j(h, s)\overline{\langle V, \phi_j\rangle} \\ &+ \frac{1}{4\pi}\sum_{\mathfrak{a}}\int_{-\infty}^{\infty} C_{\mathfrak{a}}(h, s, \tau)\overline{\langle V, E_{\mathfrak{a}}(\cdot, 1/2 + i\tau)\rangle} d\tau. \end{aligned} \tag{5.4}$$

Since $R_h(s)$ is a finite sum and

$$\langle V, \phi_j\rangle \ll \|V\|\|\phi_j\| \ll_{\nu_1,\nu_2,g} 1,$$

it follows that $R_h(s)$ is analytic in the half-plane $\sigma > 0$ except for poles at s_j and $1 - s_j$. By Sarnak [34, (A.16)], we can choose $\{\phi_j\}$ to be Hecke eigenforms such that

$$\rho_j(m) \ll_\varepsilon \frac{(mNt_j)^\varepsilon}{\sqrt{N}} \cosh\left(\frac{\pi t_j}{2}\right) m^\theta. \tag{5.5}$$

Inserting (5.5) into (5.3), and then applying Stirling's formula, we deduce, for $1/2 \le \sigma \le 2$ and $|t| \ge 1$,

$$R_h(s) \ll_{\nu_1,\nu_2,g} |h|^{1/2+\theta-\sigma+\varepsilon}. \tag{5.6}$$

However, the above estimate is not true in the region $1/2 \le \sigma \le 2$ and $|t| \le 1$, since the factor $\Gamma((s - s_j)/2)$ in (5.3) has a pole at $s = s_j = 1/2 + it_j$ with $0 < it_j \le \theta$ as in (4.5). Obviously, these poles lie in the interval $[1/2, 1/2 + \theta] \subset [1/2, 1]$. This is why we require $|t| \ge 1$ in Theorem 5.1 below.

By Theorem 4.1, $B_j(h, s)$ (when $t_j \ge 0$) and $C_{\mathfrak{a}}(h, s, \tau)$ are holomorphic in $\sigma > 1/2$. The right-side of (5.4) is analytically continued to a holomorphic function on $\sigma > 1/2$, provided that uniform convergence on compact sets is justified.

From (5.5), we infer that for $1/2 + \varepsilon \le \sigma \le 3/2$,

$$\begin{aligned} e^{-\pi t_j/2} B_j(h, s) \ll_{l,N} |h|^{1/2-\sigma+\theta+\varepsilon} t_j^\varepsilon &\frac{(1 + ||t| - t_j|)^{\sigma/2-3/4}}{(1 + |t|)^{\sigma/2+l-3/4}} \\ &\times e^{-\pi(|t-t_j|+|t+t_j|-2|t|)/4}, \end{aligned} \tag{5.7}$$

and, with the spectral large sieve in place of (5.5),

$$\begin{aligned} e^{-\pi|\tau|/2} C_{\mathfrak{a}}(h, s, \tau) \ll_{l,N} |h|^{1/2-\sigma+\varepsilon}(1 + |\tau|)^\varepsilon &\frac{(1 + ||t| - \tau|)^{\sigma/2-3/4}}{(1 + |t|)^{\sigma/2+l-3/4}} \\ &\times e^{-\pi(|t-\tau|+|t+\tau|-2|t|)/4}. \end{aligned} \tag{5.8}$$

To verify the uniform convergence of (5.4) on compact sets, we assume for instance $l \ge 4$ and invoke Good [11, Theorem 1]. The function V is of different form from that of f there; nonetheless, Good's result still covers our case. This is because his proof applies to

$$f_l(z) = y^k F(z) P_l(z)$$

where F and P_l are a cusp form and a Poincaré series for Γ, respectively; see [11, (3.2)] and [11, §4]. Note that $g(\nu_1 z)$ and $g(\nu_2 z)$ are cusp forms for Γ, and therefore $g(\nu_2 z)$ can be written as a linear combination of the Poincaré series. Hence,

$$\sum_{t_j \le T} |\langle V, \phi_j \rangle|^2 e^{\pi t_j} + \frac{1}{4\pi} \sum_{\mathfrak{a}} \int_{-T}^{T} |\langle V, E_{\mathfrak{a}}(\cdot, 1/2 + i\tau) \rangle|^2 e^{\pi|\tau|}\, d\tau \ll T^{2l}. \tag{5.9}$$

The estimate (5.9) is also valid for other cases, by Krötz and Stanton [22]. Plainly $|t-\tau| + |t+\tau| - 2|t| \ge |\tau|$ if $|\tau| \ge 2|t|$.

Thus, by (5.7) and Weyl's law

$$\#\{j : t_j \le T\} = cT^2 + O(T \log T) \tag{5.10}$$

for $T \ge 2|t|$, we have

$$\begin{aligned} \sum_{t_j \ge T} e^{-\pi t_j} |B_j(h,s)|^2 &\ll |h|^{1+2\theta-2\sigma+\varepsilon} (1+|t|)^{3/2-\sigma-2l} \sum_{t_j \ge T} e^{-\pi t_j/4} \\ &\ll |h|^{1+2\theta-2\sigma+\varepsilon} e^{-3T/4}. \end{aligned} \tag{5.11}$$

Also, by (5.8), we have, for $T \ge 2|t|$,

$$\begin{aligned} \int_{|\tau| \ge T} e^{-\pi|\tau|} |C_{\mathfrak{a}}(h,s,\tau)|^2\, d\tau &\ll |h|^{1-2\sigma+\varepsilon} e^{-3T/4} \\ &\ll |h|^{1+2\theta-2\sigma+\varepsilon} e^{-3T/4}. \end{aligned} \tag{5.12}$$

Now assume $T_0 \ge 2|t|$. Dividing dyadically and applying the Cauchy-Schwarz inequality, we obtain

$$\begin{aligned} &\sum_{j:\, T_0 < t_j \le T} |B_j(h,s) \overline{\langle V, \phi_j \rangle}| \\ &\le \sum_{0 \le r \le \log T} \sum_{j:\, Y_r < t_j \le 2Y_r} |B_j(h,s) \overline{\langle V, \phi_j \rangle}| \\ &\le \sum_{0 \le r \le \log T} \Bigg(\sum_{j:\, Y_r < t_j \le 2Y_r} e^{-\pi t_j} |B_j(h,s)|^2 \Bigg)^{1/2} \\ &\qquad \times \Bigg(\sum_{j:\, Y_r < t_j \le 2Y_r} e^{\pi t_j} |\langle V, \phi_j \rangle|^2 \Bigg)^{1/2}. \end{aligned} \tag{5.13}$$

Here Y_r denotes $2^r T_0$. The sums in the brackets of the last line of (5.13) are $\ll |h|^{1-2\sigma+2\theta+\varepsilon} e^{-3Y_r/4}$ and $\ll Y_r^{2l}$, respectively, by (5.11) and (5.9). Therefore (5.13) becomes

$$\begin{aligned}\sum_{j:\, T_0<t_j\le T} |B_j(h,s)\overline{\langle V,\phi_j\rangle}| &\ll |h|^{1/2+\theta-\sigma+\varepsilon} \sum_{0\le r\le \log T} e^{-3Y_r/8} Y_r^l \\ &\ll |h|^{1/2+\theta-\sigma+\varepsilon} e^{-T_0/4} \end{aligned} \tag{5.14}$$

for any $T \ge T_0$.

Similarly it follows from Cauchy-Schwarz's inequality that

$$\begin{aligned} &\int_{T_0\le|\tau|\le T} \left|C_{\mathfrak{a}}(h,s,\tau)\overline{\langle V, E_{\mathfrak{a}}(\cdot,1/2+i\tau)\rangle}\right| d\tau \\ &\ll \sum_{0\le r\le \log T} \left(\int_{Y_r\le|\tau|\le 2Y_r} e^{-\pi|\tau|} |C_{\mathfrak{a}}(h,s,\tau)|^2\, d\tau \right)^{1/2} \\ &\qquad \times \left(\int_{Y_r\le|\tau|\le 2Y_r} e^{\pi|\tau|} |\langle V, E_{\mathfrak{a}}(\cdot,1/2+i\tau)\rangle|^2\, d\tau \right)^{1/2} \\ &\ll |h|^{1/2+\theta-\sigma+\varepsilon} e^{-T_0/4}. \end{aligned} \tag{5.15}$$

Consequently, for any fixed compact subset K in the half plane $\sigma > 1/2$, (5.14) and (5.15) hold uniformly for $s \in K$ when $T_0 \ge 2\max_{s\in K}|t|$. By the Cauchy criterion, uniform convergence of (4.4) on compact sets is verified.

It remains to bound $D_g(s,\nu_1,\nu_2,h)$ in (5.4). Let us consider $\sigma \ge 1/2+\varepsilon$ and $|t| \ge 1$. By (5.14) and (5.15) with $T_0 = 2|t|$, we see that the tail part in (5.4) yields

$$\begin{aligned} &\sum_{j:\, t_j\ge 2|t|} B_j(h,s)\overline{\langle V,\phi_j\rangle} + \frac{1}{4\pi}\sum_{\mathfrak{a}} \int_{|\tau|\ge 2|t|} C_{\mathfrak{a}}(h,s,\tau)\overline{\langle V, E_{\mathfrak{a}}(\cdot,1/2+i\tau)\rangle}\, d\tau \\ &\quad \ll |h|^{1/2+\theta-\sigma+\varepsilon} e^{-|t|/2}. \end{aligned} \tag{5.16}$$

Since $|t-t_j| + |t+t_j| - 2|t|$ is always nonnegative, we may apply (5.7) to

$B_j(h,s)$, to get

$$\begin{aligned}
&\sum_{j:\, t_j\le 2|t|} \left|B_j(h,s)\overline{\langle V,\phi_j\rangle}\right| \\
&\ll |h|^{1/2+\theta-\sigma+\varepsilon}|t|^{3/4-\sigma/2-l+\varepsilon} \sum_{t_j\le 2|t|} |\langle V,\phi_j\rangle| e^{\pi t_j/2}(1+|t-t_j|)^{\sigma/2-3/4} \\
&\ll |h|^{1/2+\theta-\sigma+\varepsilon}|t|^{3/4-\sigma/2-l+\varepsilon}\left(\sum_{t_j\le 2|t|} |\langle V,\phi_j\rangle|^2 e^{\pi t_j}\right)^{1/2} \\
&\qquad \times\left(\sum_{t_j\le 2|t|}(1+|t-t_j|)^{\sigma-3/2}\right)^{1/2} \\
&\ll |h|^{1/2+\theta-\sigma+\varepsilon}|t|^{1+\varepsilon}
\end{aligned}$$

by (5.9) and Weyl's law (5.10). We remark that the precise form of (5.10) is needed to evaluate the last bracket.

Similarly, by (5.9),

$$\begin{aligned}
&\sum_{\mathfrak{a}}\int_{-2|t|}^{2|t|} |C_{\mathfrak{a}}(h,s,\tau)\overline{\langle V,E_{\mathfrak{a}}(1/2+i\tau)\rangle}|\,d\tau \\
&\ll |h|^{1/2-\sigma+\varepsilon}|t|^{3/4-\sigma/2-l+\varepsilon}\sum_{\mathfrak{a}}\int_{-2|t|}^{2|t|} |\langle V,E_{\mathfrak{a}}(1/2+i\tau)\rangle| e^{\pi|\tau|/2} \\
&\qquad\qquad \times (1+|t-|\tau||)^{\sigma/2-3/4}\,d\tau \\
&\ll |h|^{1/2-\sigma+\varepsilon}|t|^{3/4-\sigma/2-l+\varepsilon}\sum_{\mathfrak{a}}\left(\int_{-2|t|}^{2|t|} |\langle V,E_{\mathfrak{a}}(1/2+i\tau)\rangle|^2 e^{\pi|\tau|}\,d\tau\right)^{1/2} \\
&\qquad\qquad \times\left(\int_{-2|t|}^{2|t|}(1+|t-|\tau||)^{\sigma-3/2}\,d\tau\right)^{1/2} \\
&\ll |h|^{1/2-\sigma+\varepsilon}|t|^{1/2+\varepsilon}.
\end{aligned}$$

Hence,

$$\begin{aligned}
&\sum_{j:\, t_j\le 2|t|} B_j(h,s)\overline{\langle V,\phi_j\rangle} + \frac{1}{4\pi}\sum_{\mathfrak{a}}\int_{|\tau|\le 2|t|} C_{\mathfrak{a}}(h,s,\tau)\overline{\langle V,E_{\mathfrak{a}}(\cdot,1/2+i\tau)\rangle}\,d\tau \\
&\quad \ll |h|^{1/2+\theta-\sigma+\varepsilon}|t|^{1+\varepsilon}.
\end{aligned}$$

Inserting this, (5.16), and (5.6) into (5.4), we [23] prove the following result.

Theorem 5.1. *With notation as in Theorem 4.2, the function $D_g(s,\nu_1,\nu_2,h)$ admits an analytic continuation to a meromorphic function on $\sigma>1/2$, with at most a finite number of poles $s_j\in(1/2,1/2+\theta]$*

due to possible exceptional eigenvalues $\lambda_j = s_j(1-s_j)$ *of the Laplacian* Δ. *Moreover, for* $\sigma \geq 1/2+\varepsilon$ *and* $|t| \geq 1$, *we have*

$$D_g(s,\nu_1,\nu_2,h) \ll_{N,\nu_1,\nu_2,g,\varepsilon} |h|^{1/2+\theta-\sigma+\varepsilon}|t|^{1+\varepsilon} + \chi(g)|h|^{1-\sigma}. \tag{5.17}$$

We note that in (5.17), the implied constants depend possibly on ν_1 and ν_2, while in Theorem 4.2, the dependence of the bounds for $D_g(s,\nu_1,\nu_2,h)$ in the ν_1,ν_2 aspect can be controlled explicitly. The main reason that the estimates in Theorem 5.1 lack this ν_1,ν_2 aspect is that this dependence is unavailable in (5.9) by [11] and [22]. The dependence on ν_1 and ν_2 plays a crucial role in subconvexity problems in the level aspect (cf. Cogdell [5]), and therefore it is an interesting question to estimate $\langle V,\phi_j\rangle$ not individually but on average while keeping track of its dependence on ν_1,ν_2.

The meromorphic continuation of $D_g(s,\nu_1,\nu_2,h)$ to $\sigma > 1/2$ in Theorem 5.1 enabled the authors [23] to prove

Theorem 5.2. *Let* f *be a holomorphic Hecke eigenform for* $\Gamma_0(N)$ *of weight* k, *or a Maass Hecke eigenform for* $\Gamma_0(N)$ *with Laplace eigenvalue* $1/4+k^2$, *and let* g *be a fixed holomorphic or Maass cusp form for* $\Gamma_0(N)$, *or for* $\Gamma_0(N')$ *with* $(N,N')=1$. *Then for any small* $\varepsilon > 0$,

$$L(1/2+it, f\times g) \ll_{N,t,g,\varepsilon} k^{1-1/(8+4\theta)+\varepsilon}. \tag{5.18}$$

Even if we take the trivial $\theta = 1/2$, (5.18) still yields a subconvexity bound $k^{9/10+\varepsilon}$ for $L(1/2+it, f\times g)$.

6. Spectral method: meromorphic continuation to $\sigma > -1/2$

To reach the Weyl bound as in (1.4), we need to meromorphically continue $D_g(s,\nu_1,\nu_2,h)$ further to the left.

6.1. *Further meromorphic continuation to* $\sigma > -1/2$

First, let us look at $R_h(s)$ in (5.3). As $R_h(s)$ is a finite sum and

$$\langle V,\phi_j\rangle \ll \|V\|\|\phi_j\| \ll_{\nu_1,\nu_2,g} 1,$$

$R_h(s)$ is analytic in the complex plane except for poles lying on the real axis, which arise from the two gamma functions. In particular, on the half-plane $\sigma > 0$, there are only finitely many poles at s_j and $1-s_j$ lying in the interval $[1/2-\theta, 1/2+\theta] \subset [0,1]$. Using Stirling's formula, we deduce from

(5.5) in the same way as we deduce (5.6) that, for $|\sigma| \le A_0$ and $|t| \ge 1$,

$$\begin{aligned} R_h(\sigma+it) \ll_{A_0} & \frac{|\rho_j(-h)|}{|h|^{\sigma-1/2}|\Gamma(\sigma+l-1+it)|} \\ & \times \left|\Gamma\left(\frac{s-s_j}{2}\right)\Gamma\left(\frac{s-(1-s_j)}{2}\right)\right| |\langle V,\phi_j\rangle| \\ \ll & \ |h|^{1/2-\sigma+\theta+\varepsilon}|t|^{-l} \ll |h|^{1/2-\sigma+\theta+\varepsilon}. \end{aligned} \tag{6.1}$$

Now let us turn to the first sum on the right side of (5.4). Recall (5.1) and (5.2) and write

$$\begin{aligned} B_j(s) = (\nu_1\nu_2)^{(l-1)/2}\frac{2^{s+l-3}\pi^{l-1/2}}{\Gamma(s+l-1)}\Gamma\left(\frac{s-1/2+it_j}{2}\right) \\ \times\Gamma\left(\frac{s-1/2-it_j}{2}\right), \end{aligned} \tag{6.2}$$

$$\begin{aligned} C_{\mathfrak{a}}(s,\tau) = (\nu_1\nu_2)^{(l-1)/2}\frac{2^{s+l-2}\pi^{l-i\tau}}{\Gamma(s+l-1)}\Gamma\left(\frac{s-1/2+i\tau}{2}\right) \\ \times\Gamma\left(\frac{s-1/2-i\tau}{2}\right). \end{aligned} \tag{6.3}$$

Then from (5.4), (5.1), (5.2), (6.2), and (6.3) we have

$$\begin{aligned} & D_g(s,\nu_1,\nu_2,h) - R_h(s) \\ & = \sum_{j:t_j>0} \frac{\overline{\rho_j(-h)}}{|h|^{s-1/2}} B_j(s)\overline{\langle V,\phi_j\rangle} \\ & \quad + \frac{1}{4\pi}\sum_{\mathfrak{a}}\int_{-\infty}^{\infty}\frac{\overline{\rho_{\mathfrak{a}}(1/2+i\tau,-h)}}{|h|^{s-1/2+i\tau}}\frac{C_{\mathfrak{a}}(s,\tau)}{\Gamma(1/2-i\tau)}\overline{\langle V,E_{\mathfrak{a}}(\cdot,1/2+i\tau)\rangle}\,d\tau. \end{aligned} \tag{6.4}$$

We deduce from (6.2) that

$$B_j(s) \ll_{l,\nu_1,\nu_2,\varepsilon} T \tag{6.5}$$

for $0 \le t_j \le 2T$ and $-1/2 \le \sigma \le 3/2$. Similarly, from (6.3) we derive that

$$C_{\mathfrak{a}}(s,\tau) \ll_{l,\nu_1,\nu_2,\varepsilon} T$$

for $|\tau| \le 2T$ and $|s-(1/2 \pm it_j)| \ge \varepsilon$ or $|s-(1/2\pm i\tau)| \ge \varepsilon$. Besides, we may deduce that

$$\begin{aligned} \int_{|t|\asymp T}\left|B_j\left(\frac{1}{2}+\varepsilon+it\right)\right|^2 dt \ \text{ and } \int_{|t|\asymp T}\left|C_{\mathfrak{a}}\left(\frac{1}{2}+\varepsilon+it,\tau\right)\right|^2 dt \\ \ll_{l,\nu_1,\nu_2,\varepsilon} T^{1-2l} \end{aligned} \tag{6.6}$$

and

$$\int_{|t|\asymp T}\left|B_j\Big(-\frac{1}{2}+it\Big)\right|^2 dt, \int_{|t|\asymp T}\left|C_{\mathfrak{a}}(-\frac{1}{2}+it,\tau\Big)\right|^2 dt \\ \ll_{l,\nu_1,\nu_2} T^{2-2l}. \tag{6.7}$$

6.2. *Illustration for the proof of Theorem 1.1*

The proof of Theorem 1.1 follows the line of arguments in [34], and our salient point is a delicate study on the Mellin transform of the shifted convolution sum against an oscillatory function. We need to give it a good upper estimate. To do so, we decompose spectrally the shifted convolution sum. The oscillatory function is given by an exponential integral, to which we apply the stationary phase method to extract the main part. Our desired estimate then follows from the spectral large sieve inequality and an estimate of Good on inner products of eigenfunctions. The spectral decomposition of shifted convolution sum is powerful and interesting on its own. It plays a key role in [34] as well, but there, Sarnak considered only for his purpose the analytic continuation to the plane $\sigma > 1/2+\theta$. We need a more precise form so that the meromorphic continuation is carried out to the wider region $\sigma > -1/2$.

To illustrate the crucial roles played by the meromorphic continuation to $\sigma > -1/2$ and bounds in (6.7), let us look at [24, (9.25)]:

$$\Sigma''_d(C,T)_0^\ell = \frac{1}{2\pi i}\int_{\ell''}\sum_{j:\,0<t_j\le 2T}\overline{\langle V,\phi_j\rangle} B_j(s)\sum_{|h|\sim CT}\frac{\overline{\rho_j(-h)}}{|h|^{1/2}}S_{h,0}(s)\,ds, \tag{6.8}$$

where $\ell''=\ell''_-\cup\ell''_+$ is the union of the two vertical line segments joining $-1/2-ic_0'T$ to $-1/2-ic_0T$, and $-1/2+ic_0T$ to $-1/2+ic_0'T$, and $S_{h,0}(s)$ is a nice function defined in [24, (9.7)]. We apply the Cauchy-Schwarz inequality to the integral in (6.8) to get rid of $B_j(s)$. In this case $\sigma=-1/2$, and we have by (6.7),

$$\int_{\ell''}|B_j(s)|^2\,|ds| \ll T^{2-2l}. \tag{6.9}$$

Define

$$\mathcal{I}_0(\sigma,\rho)=\int_{\ell''}\left|\sum_{|h|\sim CT}\frac{\overline{\rho(-h)}}{|h|^{1/2}}S_{h,0}(s)\right|^2|ds|. \tag{6.10}$$

Then

$$\begin{aligned}
&\Sigma''_d(C,T)_0^\ell \\
&\ll T^{1-l} \sum_{j:\,0<t_j\le 2T} |\langle V,\phi_j\rangle|\,\mathcal{I}_0\Big(-\frac{1}{2},\rho_j\Big)^{1/2} \\
&\ll T^{1-l}\Big(\sum_{0\le t_j\le 2T} |\langle V,\phi_j\rangle|^2 e^{\pi t_j}\Big)^{1/2}\Big(\sum_{0\le t_j\le 2T} e^{-\pi t_j}\mathcal{I}_0\Big(-\frac{1}{2},\rho_j\Big)\Big)^{1/2}
\end{aligned}$$

by the Cauchy-Schwarz inequality for the sum. Then, we apply Good's estimate (5.9) to the first factor, and apply the spectral large sieve estimate to the second factor (see [24, pp.51-54] for details). Finally, we get a bound for $\Sigma''_d(C,T)_0^\ell$, which eventually leads to our Weyl-like bound (1.4) in Theorem 1.1.

References

1. V. Blomer, Shifted convolution sums and subconvex bounds for automorphic L-functions, *Internat. Math. Res. Notices*, **73** (2004), 3905–3926.
2. V. Blomer, Rankin-Selberg L-functions on the critical line, *Manuscripta Math.*, **117** (2005), 111–133.
3. V. Blomer, G. Harcos, and P. Michel, A Burgess-like subconvex bound for twisted L-functions, *Forum Math.*, **19** (2007), 61–105.
4. D.A. Burgess, On character sums and L-series, I, *Proc. London Math. Soc.*, **12** (1962), 193–206.
5. J.W. Cogdell, On sums of three squares, *J. Théorie Nombres Bordeaux*, **15** (2003), 33–44.
6. J.B. Conrey and H. Iwaniec, The cubic moment of central values of automorphic L-functions, *Ann. Math.*, **151** (2000), 1175–1216.
7. D.R. Heath-Brown, The fourth power moment of the Riemann zeta function, *Proc. London Math. Soc.*, **38** (1979), 385–422.
8. J.-M. Deshouillers and H. Iwaniec, Kloosterman Sums and Fourier Coefficients of Cusp Forms, *Invent. Math.*, **70** (1982), 219–288.
9. W. Duke, J. Friedlander, and H. Iwaniec, Bounds for automorphic L-functions, *Invent. Math.*, **112** (1993), 1–8.
10. W. Duke, J. Friedlander, and H. Iwaniec, Bounds for automorphic L-functions, II, *Invent. Math.*, **115** (1994), 219–239.
11. A. Good, The square mean of Dirichlet series associated with cusp forms, *Mathematika*, **29** (1982), 278–295.
12. H. Harcos, An additive problem in the Fourier coefficients of Maass forms, *Math. Ann.*, **326** (2003), 347–365.
13. G. Harcos and P. Michel, The subconvexity problem for Rankin-Selberg L-functions and equidistribution of Heegner points. II, *Invent. Math.*, **163** (2006), 581–655.

14. M. N. Huxley, Exponential sums and the Riemann zeta function. V. *Proc. London Math. Soc.*, **90** (2005), 1–41.
15. A. Ivić, On sums of Hecke series in short intervals, J. Théor. *Nombres Bordeaux,* **13** (2001), 453–468.
16. H. Iwaniec, The spectral growth of automorphic L-functions, *J. Reine Angew. Math.*, **428** (1992), 139–159.
17. M. Jutila, Transformation of exponential sums, *Proceedings of the Amalfi Conference on Analytic Number Theory, Univ. Salerno, Salerno*, 1992, 263–270.
18. M. Jutila, *A variant of the circle method, in: Sieve methods, exponential sums and their applications in number theory*, Cambridge Press, 1996, 245–254.
19. M. Jutila and Y. Motohashi, Uniform bounds for Rankin-Selberg L-Functions, in: Multiple Dirichlet Series, Automorphic Forms, and Analytic Number Theory, edited by Friedberg, Bump, Goldfeld, and Hoffstein, *Proc. Symp. Pure Math.*, **75** (2006), 243–256.
20. H. Kim, P. Sarnak, Appendix 2: Refined estimates towards the Ramanujan and Selberg conjectures, *J. Amer. Math. Soc.*, **16** (2003), 175–181.
21. E. Kowalski, P. Michel, and J. VanderKam, Rankin-Selberg L-functions in the level aspect, *Duke Math. J.*, **114** (2002), 123–191.
22. B. Krötz and R. J. Stanton, Holomorphic extensions of representations (I): automorphic functions, *Ann. Math.*, **159** (2004), 641–724.
23. Y.-K. Lau, J. Liu, and Y. Ye, Subconvexity bounds for Rankin-Selberg L-functions for congruence subgroups, *J. Number Theory,* **121** (2006), 204–223.
24. Y.-K. Lau, J. Liu, and Y. Ye, A new bound $k^{2/3+\varepsilon}$ for Rankin-Selberg L-functions for Hecke congruence subgroups, *Intern. Math. Res. Papers,* **7**(2006), Article ID 35090, 1–78.
25. J. Liu and Y. Ye, Subconvexity for Rankin-Selberg L-functions of Maass forms, *Geom. Funct. Anal.*, **12** (2002), 1296–1323.
26. J. Liu and Y. Ye, Petersson and Kuznetsov trace formulas, in: *Lie Groups and Automorphic Forms*, Amer. Math. Soc. and International Press, 2005, 147–168.
27. H. Maass, Über eine neue Art von nichtanalytischen automorphen Funktionen und die Bestimmung Dirichletscher Reihen durch Funktionalgleichungen, *Math. Ann.*, **121** (1949), 141–183.
28. T. Meurman, On the order of the Maass L-function on the critical line, in: K. Györy, G. Halász (Eds.), *Number Theory, Vol. I (Budapest, 1987)*, Colloquium of Mathematical Society Janos Bolyai, Vol. 51, North-Holland, Amsterdam, 1990, 325–354.
29. P. Michel, The subconvexity problem for Rankin-Selberg L-functions and equidistribution of Heegner points, *Ann. Math.*, **160** (2004), 185–236.
30. P. Michel, *Analytic number theory and families of automorphic L-functions*, IAS/Park City Mathematics Series, manuscript.
31. Z. Peng, *Zeros and central values of automorphic L-functions*, Ph.D thesis, Princeton University, Princeton, 2001.
32. R.S. Phillips, P. Sarnak, On cusp forms for co-finite subgroups of PSL$(2, R)$, *Invent. Math.*, **80** (1985), 339–364.

33. P. Sarnak, Integrals of products of eigenfunctions, *Internat. Math. Res. Notices*, **6** (1994), 251–260.
34. P. Sarnak, Estimates for Rankin-Selberg L-functions and Quantum Unique Ergodicity, *J. Funct. Anal.*, **184** (2001), 419–453.
35. A. Selberg, On the estimation of Fourier coefficients of modular forms, in: *Proc. Sympos. Pure Math.*, Vol. VIII, Amer. Math. Soc., Providence, R.I, 1965, 1–15.

TWO EXPOSITIONS ON ARITHMETIC OF CUBICS

KATSUYA MIYAKE

Department of Mathematical Science,
School of Science and Engineering,
Waseda University,
3-4-1 Okubo, Shinjuku-ku, Tokyo 169-8555, Japan
E-mail: miyakek@aoni.waseda.jp

In the former half of the article we introduce cubic *generic polynomials* $R(t;X)$ and $Q(s;X)$ for the symmetric group of degree 3 and for the cyclic group of order 3, respectively, in a down-to-earth way. As an application we review the main result of Kishi and the author [8] on the divisibility of the class numbers of quadratic fields by three. In the latter half we introduce two kinds of families of those elliptic curves over the rational number field $\mathbb{Q}$ whose sets of rational points over $\mathbb{Q}$ are described by certain subsets of the related cubic fields. The analysis of the cubic polynomials $R(t;X)$ and $Q(s;X)$ in the former half is fully utilized.

1. Introduction

This article consists of two parts. Part I is devoted to the study of cubic generic polynomials

$$R(t;X) = X^3 + tX + t \tag{1.1}$$

and

$$Q(s;X) = X^3 - (s-3)X^2 - sX - 1 \tag{1.2}$$

for the symmetric group S_3 of degree 3 and the cyclic group C_3 of order 3, respectively, over a field of characteristic other than 3 and 2. The contents are restricted to the cubic cases and appear rather elementary. They do not seem, however, to be found in usual text books in spite of the fact that the cubic case is certainly the simplest gateway to step out into non-Abelian lands. It may be better to note that some results were obtained or checked by using computers. As an application of the 'genericness' of the polynomial $R(t;X)$, we review an elegant result on the divisibility of class numbers of quadratic fields by 3 obtained by Yasuhiro Kishi and the author [8] in §2.5; indeed, all unramified cyclic cubic extensions of quadratic fields are covered by a cubic polynomial with two integral parameters satisfying simple conditions as the splitting fields over $\mathbb{Q}$.

In Part II, we study two types of families of elliptic curves whose sets of rational points over $\mathbb{Q}$ are described by certain subsets of suitable cubic fields. One of the two families of elliptic curves is given by

$$w^3 = u^3 + au^2 + bu + c$$

with $a, b, c \in \mathbb{Q}$ and may look rather special because they admit complex multiplication by the third roots of unity. Their short forms are Mordell curves of the form

$$y^2 = x^3 + k.$$

The other family $\tilde{H}(\mu, t)$ of our concern consists of twists of the Hessian family of elliptic curves

$$H_\mu : U^3 + V^3 + W^3 = 3\mu UVW, \qquad \mu \in \mathbb{Q} - \{1\}$$

over the splitting fields $\tilde{K}_t$ of those cubic polynomials $R(t; X)$ with $t \in \mathbb{Q}$ which are irreducible over $\mathbb{Q}$. We also treat twists of the arising curves over the field K_t generated by a root of $R(t; X)$, and find that they are the quadratic twists of the original Hessian elliptic curves. It should be noted that the twists $\tilde{H}(\mu, t)$ of H_μ over $\tilde{K}_t$ are not ones as elliptic curves but as algebraic curves. This is because they may not have any rational points over $\mathbb{Q}$ even though the curves themselves are defined over $\mathbb{Q}$. The two-parameter family, however, contains infinitely many types of one-parameter subfamilies each of which has 'parametrized' rational points over $\mathbb{Q}$ and hence consists of elliptic curves defined over $\mathbb{Q}$. In the special case of $\mu = 0$, we will show a necessary and sufficient condition on t for $\tilde{H}(0, t)$ to have a rational point over $\mathbb{Q}$; (Proposition 3.12 in the last Subsection 3.5).

We will not give any detailed proofs to our results in this article. Interested readers may find them in the author's papers (partly with coauthors) [4,8,11–14].

2. Part I: Generic polynomials of degree 3

First of all, the author recommends the readers the basic textbook [6] by C. U. Jensen, A. Ledet and N. Yui on generic polynomials. Here in this article we do not show any general results on them other than the simplest cubic cases.

2.1. *Generic polynomials of degree* 3

Let us start with an irreducible cubic polynomial in X

$$P(X) = X^3 + aX^2 + bX + c$$

over a field k. We suppose that the base field is of characteristic other than 3 and contains infinitely many elements. Our basic interest is the case of $k = \mathbb{Q}$. For the moment, however, we handle a general field k with the afore-mentioned conditions. Let ξ be a root of $P(X)$ in the fixed algebraic closure of k, and $K := k(\xi)$ be the cubic extension of k generated by ξ.

Proposition 2.1. *By an affine transformation of X over k, the polynomial $P(X)$ is reduced to either $X^3+s, s \in k$, or $X^3+tX+t, t \in k$ up to a multiple by a cube of an element of k.*

Indeed, by putting $Y = X + a/3$, we have $P(X) = Y^3 + b'Y + c'$ with

$$\begin{aligned} b' &= -a^2/3 + b, \\ c' &= 2a^3/27 - ab/3 + c. \end{aligned}$$

Since $P(X)$ is irreducible over k, c' is not equal to 0. If $b' = 0$, then we have $P(X) = Y^3 + c'$, while if $b' \neq 0$, putting $Y = \frac{b'}{c'} \cdot X$, we see

$$\left(\frac{b'}{c'}\right)^3 \cdot P(X) = Y^3 + \frac{b'^3}{c'^2} \cdot Y + \frac{b'^3}{c'^2}.$$

Proposition 2.2. *Every cubic extension K of k is obtained as $K = k(\xi)$ by a root ξ of the cubic polynomial $R(t; X)$ of* (1.1) *with a suitable $t \in k$.*

Indeed, if $K = k(\sqrt[3]{s})$, we take $\eta := \sqrt[3]{s} + \sqrt[3]{s}^{-1} \in K$. Then it is a root of $X^3 - 3X - (s + s^{-1}) = 0$; and we also have $K = k(\eta)$.

Now let t be an independent variable over k. Then $R(t; X)$ is a polynomial over the rational function field $k(t)$. As we see below, the Galois group of the splitting field of $R(t; X)$ over $k(t)$ is isomorphic to the symmetric group S_3 of degree 3. Then it follows from Proposition 2.2 that each of those Galois extensions $\tilde{K}$ over k whose group $\mathrm{Gal}(\tilde{K}/k)$ are isomorphic to S_3 is obtained by some specialization of $t := a, a \in k$ as the splitting field of $R(a; X)$ over k. Hence we say that

the cubic polynomial $R(t; X)$ is *parametric* for S_3 over k.

Furthermore, $R(t; X)$ is parametric for S_3 not only over k but also over each of those fields F which contain k; namely, $R(t; X)$ belongs to the polynomial ring $F[t]$ and is parametric over F as long as t is an independent variable over F. This property is known as

the cubic polynomial $R(t; X)$ is *generic* for S_3 over k.

Actually, it is generic for S_3 over the prime field k_0 contained in k. It should be noted, however, that, when a Galois extension L/k has a non-Abelian Galois group like S_3, the base field k cannot be a finite field.

Next let C_3 be the cyclic group of order 3. Then the following proposition is also well known:

Proposition 2.3. *The cubic polynomial $Q(s; X)$ of (1.2) over the rational function field $k(s)$ of one variable s is generic for C_3 over k.*

The polynomial $Q(s; X)$ is characterized by the following lemma:

Lemma 2.1. *Under the linear fractional transformation $-X/(X+1)$, we have*

$$Q\left(s; \frac{-X}{X+1}\right) = \frac{-1}{(X+1)^3} Q(s; X).$$

Since the linear fractional transformation $-X/(X+1)$ is of order 3, the Galois group of the splitting field of $Q(s; X)$ over $k(s)$ is isomorphic to C_3. The genericness of the cyclic polynomial is assured by

Proposition 2.4. *Suppose that K/k is a cyclic extension of degree 3, and let σ be a generator of the Galois group $\mathrm{Gal}(K/k)$. Then there exists an element $\xi \in K$ which satisfies the condition $\xi^\sigma = -\xi/(\xi+1)$. For such an element $\xi \in K$, we have $K = k(\xi)$ and $Q(s; \xi) = 0$ for some $s \in k$.*

This result for the case of $k = \mathbb{Q}$ is generalized for a cyclic group of odd order n over the maximal real subfield of the n-th cyclotomic field in the author's paper [10]. D. Shanks [19] may be the first who published on the polynomial $Q(s; X)$. He investigated arithmetic of the cyclic cubic fields given by the polynomial with integral coefficients.

Recently Yuuichi Rikuna found a simple way to prove Proposition 2.3 in the case where the characteristic of the base field k is other than 2 and 3, and -3 is not a square in k.

Let $P(X) := X^3 + aX^2 + bX + c$ be an irreducible cyclic polynomial in $k[X]$. Then the discriminant $D := a^2b^2 - 4b^3 - 4a^3c + 18abc - 27c^2$ of $P(X)$ is a square in k (cf. Proposition 2.6 below): $D = d^2, d \in k$. Note that $a^2 - 3b \neq 0$ because $a^2 - 3b = 0$ implies, by assumption, that $D = -(a^3 - 27c)^2/27$ is not a square in k.

Proposition 2.5 (Y. Rikuna). *The notation and the assumptions being as above, let ξ be a root of $P(X) = 0$, and put*

$$\eta := \frac{a^2 - 3b}{d}\xi + \frac{ab - 9c - d}{2d}.$$

Then the minimal polynomial of η over k is $Q(s; X) = X^3 - (s-3)X^2 - sX - 1$ with

$$s = -\frac{2a^3 - 9ab + 27c - 3d}{2d}.$$

Moreover, if we take $-d$ in place of d, then we have a root γ of $Q(u; X) = 0$ with $u = 3 - s$.

Remark 2.1. In case of $a^2 - 3b = 0$, we have

$$27P(X) = (3X + a)^3 - (a^3 - 27c).$$

Hence if we assume further that D is a square in k, then k contains $\sqrt{-3}$ and the third root of unity. Even in this case we may easily find by the method given just after Proposition 2.2 an element $\eta \in k(\sqrt[3]{a^3 - 27c})$ which is a root of $Q(s; X) = 0$ with a suitable $s \in k$.

2.2. *The Splitting Field of $R(t; X)$*

In this section, we suppose that the characteristic of the base field k is not equal to 3 nor to 2. We consider a quantity t which generates a field $k(t)$ with infinitely many elements, and fix an algebraic closure $\overline{k(t)}$. We also suppose that the cubic polynomial $R(t; X)$ is irreducible over $k(t)$, and denote the splitting field of $R(t; X)$ over $k(t)$ by $\tilde{K}_t$.

Let $\xi \in \overline{k(t)}$ be a root of $R(t; X)$, and

$$K_t := k(t, \xi) \tag{2.1}$$

be the field generated by ξ over $k(t)$.

Lemma 2.2. *The discriminant of $R(t;X)$ is $-t^2(4t+27)$.*

We put

$$d_t := -(4t+27). \tag{2.2}$$

Proposition 2.6. *The splitting field $\tilde{K}_t$ of the cubic polynomial $R(t;X)$ over $k(t)$ contains $\sqrt{d_t}$, $d_t = -(4t+27) \in k(t)$, and $\tilde{K}_t = K_t(\sqrt{d_t})$. Furthermore, the following two statements hold:*
(1) $[\tilde{K}_t : k(t)] = 3$ if and only if d_t is a square in $k(t)$; if this is the case, we have $\tilde{K}_t = K_t$; the Galois group $\mathrm{Gal}(\tilde{K}_t/k(t))$ is isomorphic to C_3.
(2) $[\tilde{K}_t : k(t)] = 6$ if and only if d_t is not a square in $k(t)$; if this is the case, $\tilde{K}_t$ contains the quadratic extension $k(t, \sqrt{d_t})$ of $k(t)$, and $\tilde{K}_t = K_t(\sqrt{d_t})$; the Galois group $\mathrm{Gal}(\tilde{K}_t/k(t))$ is isomorphic to S_3.

The proposition immediately follows from the following lemma. Note, hereby, that the Galois group is not abelian in the case (2) of the proposition. Indeed, if it were abelian, then the subfield K_t of $\tilde{K}_t$ should be normal over $k(t)$ and hence coincide with $\tilde{K}_t$.

Lemma 2.3. *The notation and the assumptions being as above, we have $\tilde{K}_t = k(t, \xi, \alpha, \beta)$ where α and β are the roots of the quadratic polynomial $X^2 + \xi X + \xi^2 + t$. We also have*

$$R(t;X) = (X-\xi)\,(X^2 + \xi X + \xi^2 + t), \qquad t = \frac{-\xi^3}{\xi+1}.$$

The discriminant d of the quadratic factor is given with d_t of (2.4) by

$$d = -(3\xi^2 + 4t) = d_t \left(\frac{\xi}{2\xi+3}\right)^2.$$

2.3. *A Criterion for $K_t \simeq K_{t'}$*

In this section whose details can be found in the paper [4] by Akinari Hoshi and the author, we still suppose that the characteristic of the base field k is not equal to 3 nor to 2. We suppose, furthermore, that the quantities t and t' generate the same field $k(t) = k(t')$, and that the cubic polynomials $R(t;X)$ and $R(t';X)$ are irreducible over $k(t)$.

Let ξ and ξ' be roots of $R(t;X)$ and $R(t';X)$, respectively, in the algebraically close field $\overline{k(t)}$. If $K_t = k(t,\xi)$ is isomorphic to $K_{t'} = k(t',\xi')$ over $k(t) = k(t')$, then one of the conjugates of ξ' over $k(t)$ has to belong to K_t. Hence we replace ξ' with the conjugate of it in K_t and may assume $\xi' \in K_t$. Since $[K_t : k(t)] = 3$, K_t has no non-trivial subfields. Therefore we have $K_t = K_{t'}$ in this case.

Proposition 2.7. *The notation and the assumptions being as above, $K_t = K_{t'}$ if and only if there exists an element $u \in k(t)$ such that*

$$t' = t(u^2 + 9u - 3t)^3/(u^3 - 2tu^2 - 9tu - 2t^2 - 27t)^2.$$

Since the condition $K_t = K_{t'}$ is reciprocal in t and t', so should the latter condition of the proposition be. Indeed, let T and U be two independent variables, and put

$$T' := T(U^2 + 9U - 3T)^3/(U^3 - 2TU^2 - 9TU - 2T^2 - 27T)^2,$$
$$U' := -(U^2 + 3T)(U^2 + 9U - 3T)/(U^3 - 2TU^2 - 9TU - 2T^2 - 27T).$$

Then we have a rational endomorphism of the rational function field $k(T,U)$ of two variables, $\Phi : k(T,U) \to k(T,U)$, by assigning (T',U') to (T,U).

Proposition 2.8. *The endomorphism Φ is involutive; that is, $\Phi \circ \Phi$ is equal to the identity map. Hence Φ is an automorphism of $k(T,U)$ and an involutive Cremona transformation of dimension 2.*

2.4. *Notes on the Reducible Cases*

It is tautologically true that $R(t;X)$ is reducible over $k(t)$ if and only if t is of the form $t = -r^3/(r+1)$ with some $r \in k(t) - \{-1\}$. In this case it is convenient to put $\xi := r$ and $K_t := k(t)$ for later use in Part II. The splitting field $\tilde{K}_t$ is either a quadratic extension of $k(t)$ or $k(t)$ itself. We now have

$$R(t;X) = (X - r)\left(X^2 + rX + \frac{r^2}{r+1}\right).$$

The discriminant D of the quadratic factor is given as

$$D = \frac{r^2(r-3)}{r+1}.$$

Proposition 2.9. *The cubic polynomial $R(t;X)$ is reducible over $k(t)$ if and only if t is of the form $t = -r^3/(r+1)$ with some $r \in k(t) - \{-1\}$. Furthermore in this case, $\tilde{K}_t = k(t, \sqrt{(r-3)/(r+1)})$, so that, $\tilde{K}_t = k(t)$ if and only if $(r-3)/(r+1)$ is a square in $k(t)$.*

Remark 2.2. It is clear that $d = (r-3)/(r+1)$ if and only if $r = -(d+3)/(d-1)$. For a fixed field $k_0(= k(t))$ and its quadratic extension $k_0(\sqrt{d})$, therefore, there always exist r and $t = -r^3/(r+1)$ in k_0 which give rise to $\tilde{K}_t = k_0(\sqrt{d})$. If k_0 is an infinite field, then we have infinitely many such pairs of r and t because m^2d with $m \in k_0$ gives the same quadratic extension $k_0(\sqrt{d})$.

Remark 2.3. Non-trivial subgroups of the symmetric group S_3 are the cyclic group C_3 of order 3 and the cyclic group C_2 of order 2. Hence Proposition 2.6, (2), and Proposition 2.9 with Remark 2.2 confirm that the cubic polynomial $R(t;X)$ is *descent-generic* over k; cf. G. Kemper [7].

2.5. *An Application: Parametrization of Unramified Cyclic Cubic Extensions of Quadratic Fields*

In this subsection, we review the main theorem of Yasuhiro Kishi and the author [8]. The base field is the rational number field: $k = \mathbb{Q}$. We start with a basic fact on an unramified Abelian extension of a quadratic field.

Proposition 2.10. *An unramified Abelian extension of a quadratic field is a normal extension over $\mathbb{Q}$.*

This fact is an outcome of the two special conditions; one is that the base field is a quadratic field, and the other is that the ideal class group of $\mathbb{Q}$ is trivial. Let d be an integer which is not a square in $\mathbb{Q}$, and put $K := \mathbb{Q}(\sqrt{d})$. Let L/K be an unramified Abelian extension of K. We denote the Hilbert class field of K by $\tilde{L}$; this is the maximal unramified Abelian extension of K, and is normal over $\mathbb{Q}$. Put $G := \mathrm{Gal}(\tilde{L}/\mathbb{Q})$, $H := \mathrm{Gal}(\tilde{L}/K)$ and $N := \mathrm{Gal}(\tilde{L}/L)$. It is clear that H is a normal subgroup of G. We denote the absolute class group of K by Cl_K. By class field theory, there is a canonical isomorphism $\varphi : H \to \mathrm{Cl}_K$ which is the inverse isomorphism of the Artin map. Take an element $\sigma \in G$ which is not trivial on K. Then the inner automorphism of G defined by σ induces an automorphism of H. It is independent of the choice of such σ because the subgroup H of G is Abelian. Furthermore, we have $\varphi(\sigma^{-1}\tau\sigma) = \varphi(\tau)^\sigma$ for $\tau \in H$ by class field theory; here the right-hand side comes from the natural action of σ on ideals of K. Take a prime ideal $\mathfrak{p}$ of K from the class $\varphi(\tau)$ which is unramified in $K/\mathbb{Q}$ and of degree 1. Since the class group of $\mathbb{Q}$ is trivial, we see that $\mathfrak{p}^\sigma \cdot \mathfrak{p}$ is a principal ideal. Hence we have $\varphi(\tau)^\sigma = \varphi(\tau)^{-1}$ and $\sigma^{-1}\tau\sigma = \tau^{-1}$. Therefore, every subgroup of H is normal in G. Hence in particular, so is the group $N = \mathrm{Gal}(\tilde{L}/L)$.

As for class field theory and related fundamental elements of algebraic number theory, the readers are recommended to consult recent books J. Neukirch [17] and G. Gras [3].

Corollary 2.1. *An unramified cyclic cubic extension of a quadratic field is normal over $\mathbb{Q}$. Its Galois group over $\mathbb{Q}$ is isomorphic to the symmetric group S_3.*

The first part of the corollary is assured by Proposition 2.10. Regarding the second part, we note that the Galois group is of order 6. Hence if it were isomorphic to C_6, it should be Abelian, and the cyclic cubic extension of the quadratic field contains a cyclic cubic extension of $\mathbb{Q}$. Then we see by Minkowski's theorem and the Hilbert theory that a prime number must fully ramify here. Since the ramification index should be 3 in this case, a prime ideal of the quadratic field over the prime number must also ramify in the original cyclic extension of the quadratic field; this contradicts to the presupposition.

It follows from the corollary that every unramified cyclic cubic extension of a quadratic field is the Galois closure of a non-Galoisian cubic extension of $\mathbb{Q}$. That is to say, it is the splitting field of an irreducible cubic polynomial $R(t;X)$ over $\mathbb{Q}$ with a suitable $t \in \mathbb{Q}$. We can now use the results of the paper [9] of P. Llorente and E. Nart. To utilize them, we have to form a cubic polynomial with integral coefficients. First put $t = -v/u, u, v \in \mathbb{Z}, (u,v) = 1$ and $Y := uX$. Then we have

$$f(Y) = Y^3 - uvY - u^2v = 0$$

from the equation $R(t;X) = 0$. Then we have $v = w^3, w \in \mathbb{Z}$ from [9] if we assume that the splitting field $\tilde{K}_t$ of $R(t;X)$ is unramified over the quadratic subfield. Hence, putting $Y = wZ$, we have a cubic equation

$$g(Z) := Z^3 - uwZ - u^2 = 0, \qquad u, w \in \mathbb{Z}, (u, w) = 1 \tag{2.3}$$

which is equivalent to the original $R(t; X) = 0$. The discriminant of $g(Z)$ is given by

$$D := u^2 d, \qquad d := 4uw^3 - 27u^2.$$

Hence the quadratic field in $\tilde{K}_t$ is $\mathbb{Q}(\sqrt{d})$.

Theorem 2.1 (Kishi and Miyake [8]). *The notation being as above, suppose that $g(Z)$ is irreducible over $\mathbb{Q}$ and that one of the following three conditions is satisfied:*

$$\begin{aligned} &\text{(i)} && 3 \nmid w; \\ &\text{(ii)} && 3 \mid w, \quad uw \not\equiv 3 \pmod 9, \quad u \equiv w \pm 1 \pmod 9; \\ &\text{(iii)} && 3 \mid w, \quad uw \equiv 3 \pmod 9, \quad u \equiv w \pm 1 \pmod{27}. \end{aligned} \tag{2.4}$$

Then the splitting field of $g(Z)$ over $\mathbb{Q}$ is an unramified cyclic cubic extension of the quadratic field $\mathbb{Q}(\sqrt{d})$. Conversely, every unramified cyclic cubic extension of a quadratic field $\mathbb{Q}(\sqrt{d})$ (if exists) is given in this way as the splitting field of $g(Z)$ over $\mathbb{Q}$ by a suitable pair of integers u and w which satisfies $d = 4uw^3 - 27u^2$ and one of the three conditions (2.4).

3. Part II: Some Families of Elliptic Curves related with Cubic Fields

In Part II, the base field is fixed to be the rational number field: $k = \mathbb{Q}$. The contents of this part mainly come from the papers [13,14] of the author, the first half from the former and the second half from the latter. The two other papers [11,12] also deal with related topics. Basic references on arithmetic aspects of elliptic curves are J. Silverman [20] and [21], and J. Silverman and J. Tate [22].

3.1. *Elliptic Curves of the Form $w^3 = u^3 + au^2 + bu + c$*

For an irreducible cubic polynomial over $\mathbb{Q}$

$$P(u) = u^3 + au^2 + bu + c, \qquad a, b, c \in \mathbb{Q}, \tag{3.1}$$

define an elliptic curve by

$$E = E(P) : w^3 = P(u) = u^3 + au^2 + bu + c. \tag{3.2}$$

Indeed, $P(u)$ being irreducible over $\mathbb{Q}$, E is of genus 1; since the leading coefficient of $P(u)$ is 1, just one of its three points at infinity is rational over $\mathbb{Q}$. Therefore E is an elliptic curve defined over $\mathbb{Q}$. We denote the set of rational points of E over $\mathbb{Q}$ including the point at infinity by $E[\mathbb{Q}]$.

Take a root ξ of $P(u)$ in the complex number field $\mathbb{C}$, and let $K = \mathbb{Q}(\xi)$ be the cubic field generated by ξ. Put

$$\mathcal{W}(\xi) = \{\alpha = q\xi + r \mid \mathrm{N}_{K/\mathbb{Q}}(\alpha) = 1, q, r \in \mathbb{Q}\}. \tag{3.3}$$

Then there is a bijective map from $\mathcal{W}(\xi)$ onto $E[\mathbb{Q}]$ which maps $1 \in \mathcal{W}(\xi)$ to the point at infinity in $E[\mathbb{Q}]$. Note that the subset of $K^\times$ depends on ξ or, we may say, on the polynomial $P(u)$; indeed, the isomorphism between the cubic fields K and $\mathbb{Q}(\eta)$ bijectively maps $\mathcal{W}(\xi)$ to $\mathcal{W}(\eta)$ if we take another root η of $P(u)$. In this way we can completely describe the set of all rational points of E over $\mathbb{Q}$ by the subset $\mathcal{W}(\xi)$ of the cubic field K. We state the fact as

Proposition 3.1 ([13, Proposition 2]). *Let the notation be as above. Then there exists a bijective map from $\mathcal{W}(\xi)$ onto $E[\mathbb{Q}]$ which maps $1 \in \mathcal{W}(\xi)$ to the point at infinity in $E[\mathbb{Q}]$.*

It is clear that $\mathcal{W}(\xi) = \mathcal{W}(q\xi + r)$ for $q, r \in \mathbb{Q}, q \neq 0$. Then by Proposition 2.1 in §2.1, we reduce the curve $E(P)$ to one of the two typical elliptic curves $F_0(s)$ and $E_0(t)$ which are defined by

$$F_0(s) : w^3 = u^3 + s,$$
$$E_0(t) : w^3 = R(t; u) := u^3 + tu + t,$$

and parametrized by s and t in $\mathbb{Q}^\times$, respectively. The first $F_0(s)$ is isomorphic to the well known pure cubic twist $X^3 + Y^3 + sZ^3 = 0$ of the cubic Fermat curve. It was extensively studied, for examples, by J. W. S. Cassels [1] and E. Selmer [18]. Its short form is a Mordell curve $M_5(s)$ defined by

$$M_5(s) : y^2 = x^3 - 2^4 3^3 s^2$$

(cf. e.g. Cassel [1]). L. J. Mordell investigated this type of curves in [15]. Because of this, J. Gebel, A. Pethő and H. G. Zimmer used the term 'Mordell's equations' in their title of [2].

We can show by a standard manner that a short form of $E_0(t)$ is given by

$$M_7(T) : y^2 = x^3 - T^2(T + 3), \quad T = \frac{4t}{9}.$$

(Cf. [13, §6]. Note that we use the indices of curves here in such a way as they are to be consistent with the ones in the papers [11] and [12].)

Next let us consider the cubic polynomial $P(c;u) := u^3 + u^2 + c$ with $c \in \mathbb{Q}$. A short form of the elliptic curve

$$E_1(c) := E(P(c;u)) : w^3 = u^3 + u^2 + c$$

is obtained in [11] as

$$M_1(c) : y^2 = x^3 - 2^4 c(27c + 4).$$

The cubic field generated by a root of $P(c;u)$ is isomorphic to the one obtained by $R(t;u)$ with $t = c^{-1}$. Indeed, if ξ is a root of $P(c;u) = 0$, then ξ^{-1} is a root of $R(c^{-1};u) = 0$. This relation between $P(c;u)$ and $R(c^{-1};u)$ may give us a morphism between the two algebraic curves $E_1(c)$ and $E_0(t)$ for $t = c^{-1}$. If we try to pick up such a morphism, however, it may not be defined over $\mathbb{Q}$ as we don't see a natural correspondence between $\mathcal{W}(\xi)$ and $\mathcal{W}(\xi^{-1})$. Hence to see a relation defined over $\mathbb{Q}$ between these two families of elliptic curves, we should carry out the reduction process we did in §2.1 and reduce a general cubic polynomial to our $R(t;X)$. Then we obtain the following proposition.

Proposition 3.2. *Suppose $27c+2 \neq 0$. Then $E_1(c)$ is isomorphic to $E_0(t)$ for $t = -27/(27c+2)^2$ over $\mathbb{Q}$. In this case, indeed, the defining equation*

$$w^3 = u^3 + u^2 + c$$

of $E_1(c)$ is equivalent to

$$\left(\frac{-9w}{27c+2}\right)^3 = \left(\frac{-3(3u+1)}{27c+2}\right)^3 + t\left(\frac{-3(3u+1)}{27c+2}\right) + t.$$

It should be noted that $c \in \mathbb{Q}$ may not always be obtained from t in $\mathbb{Q}$ through the relation $t = -27/(27c+2)^2$. Hence the family of elliptic curves $E_1(c), c \in \mathbb{Q}$ covers a strict subfamily of $E_0(t), t \in \mathbb{Q}$.

If we pick up the generic polynomial $Q(s;u) = u^3 - (s-3)u^2 - su - 1$ for C_3 of Proposition 2.3 in §2.1, then it corresponds to the cyclic cubic field defined by the roots of $Q(s;u)$. Let us denote the elliptic curve $E(Q(s;u))$ as $E_4(s)$:

$$E_4(s) : w^3 = u^3 - (s-3)u^2 - su - 1.$$

Then we see

Proposition 3.3. *The short form of* $E_4(s)$ *is*

$$M_8(s): y^2 = x^3 + 2^4(s^2 - 3s + 9)^2.$$

Let θ *be a root of* $Q(s;u)$ *and* $K = \mathbb{Q}(\theta)$. *Then* $-\theta - 1$ *belongs to*

$$\mathcal{W}(\theta) = \{\alpha = q\theta + r \mid \mathrm{N}_{K/\mathbb{Q}}(\alpha) = 1, q, r \in \mathbb{Q}\}.$$

The rational point $(w, u) = (1, -1)$ *of* $E_4(s)$ *corresponding to* $-\theta - 1$ *is mapped to the rational point*

$$(x, y) = (0,\ 4(s^2 - 3s + 9))$$

of $M_8(s)$. *This is a torsion point of order* 3.

3.2. *Some facts on the Hessian curves*

In this latter half of Part II, we explain some results in [14].

In 1840's, L. O. Hesse (1811–74) investigated plane curves in a series of papers, and found an interesting family of elliptic curves of the form

$$H_\mu : U^3 + V^3 + W^3 = 3\mu UVW \tag{3.4}$$

on the projective plane $\mathbb{P}^2(U : V : W)$. If $\mu \neq 1$, H_μ is non-singular, and is an elliptic curve with points of order 3 defined over $\mathbb{Q}(\mu)$.

In this section we quote some facts on the Hessian curves from the text-books of D. Husemöller [5, Ch.4], and L. J. Mordell [16, Ch.3].

Proposition 3.4.
(1) *If* $\mu \neq 1$, H_μ *is an elliptic curve defined over* $\mathbb{Q}(\mu)$. *If we take, in this case, the point* $P_\infty = (1 : -1 : 0)$ *as the origin of the addition on the elliptic curve, then all of its points of order* 3 *are given by the list (excluding* P_∞*),*

$$\begin{array}{lll} (0:-1:1), & (1:0:-1), & (1:-1:0) = P_\infty, \\ (0:-\omega:1), & (\omega:0:-\omega^2), & (-1:\omega^2:0), \\ (0:-\omega^2:1), & (\omega^2:0:-\omega), & (-1:\omega:0), \end{array}$$

where ω *is the third root of unity.*
(2) *If* $\mu = 1$, *then* H_μ *is singular and consists of a line and a quadratic curve:*

$$U^3+V^3+W^3-3UVW = \frac{-1}{2}(U+V+W)\{(U+V+W)^2-3(U^2+V^2+W^2)\}.$$

Proposition 3.5. *The affine elliptic curve*

$$A_\mu : y^2 + 3\mu xy + y = x^3$$

is a 3-isogeny of an affine model

$$u^3 + v^3 = 1 - 3\mu uv$$

of $H_\mu, (u = -U/W, v = -V/W)$. *The isogeny is given by* $x = -uv$, $y = -v^3$.

Theorem 3.1. *Suppose that* 3μ *is a rational integer.*

(1) *Case of* $3\mu \neq -1, 3, 5$*:*
 (1-1) H_μ *has only three* $\mathbb{Q}$*-rational points, or*
 (1-2) H_μ *has infinitely many* $\mathbb{Q}$*-rational points.*

(2) *Case of* $3\mu = -1$*:* H_μ *has six* $\mathbb{Q}$*-rational points*

$$(0 : -1 : 1), (1 : 0 : -1), (1 : -1 : 0), (1 : 1 : -1), (1 : -1 : 1), (-1 : 1 : 1).$$

(3) *Case of* $3\mu = 5$*:* H_μ *has six* $\mathbb{Q}$*-rational points*

$$(0 : -1 : 1), (1 : 0 : -1), (1 : -1 : 0), (1 : 1 : 2), (1 : 2 : 1), (2 : 1 : 1).$$

(4) *Case of* $3\mu = 3$*:* H_μ *is singular, and has infinitely many* $\mathbb{Q}$*-rational points.*

Remark 3.1. Mordell uses the tangential method to show the contents of this theorem. Here it is essential to assume that 3μ is an integer.

3.3. *Twists of Hessian Elliptic Curves (1)*

Our main concern in this subsection is to review some results in [14] on twists of the curves H_μ with μ in $\mathbb{Q} - \{1\}$ over quadratic fields and the Galois closures of cubic fields, and, in the latter case, to describe the set of the rational points of the curve over $\mathbb{Q}$ by a certain subset of the related cubic field. Here we mean by twists of the elliptic curves twists as algebraic curves because the twists we construct may not have any rational points over $\mathbb{Q}$ even though the curves themselves are defined over the field.

We utilize the cubic polynomial $R(t; X) = X^3 + tX + t$ with $t \in \mathbb{Q}$ to construct the twists. Note that it parametrizes not only all cubic fields but also all quadratic fields as we stated in §2.4; we exclude the values 0 and $-27/4$ of the parameter t to save $R(t; X)$ from having multiple roots;

indeed, the discriminant of the polynomial in X is $-t^2(4t+27)$. As before let ξ be a root of $R(t;X)=0$ in the complex number field $\mathbb{C}$ and put $K_t := \mathbb{Q}(\xi)$; if $R(t;X)=0$ has a root r in $\mathbb{Q}$, then we pick it up as $\xi = r$ and have $K_t = \mathbb{Q}$; in this case, obviously, we have $t = -r^3/(r+1)$. Let $\tilde{K}_t$ be the splitting field of the cubic polynomial $R(t;X)$ over $\mathbb{Q}$. If t runs over all those rational values for which we have $K_t = \mathbb{Q}$, then $\tilde{K}_t$ covers all quadratic fields as $\mathbb{Q}(\sqrt{-(4t+27)})$ beside $\mathbb{Q}$.

We define the twists $\tilde{H}(\mu,t)$ of H_μ by making use of 3×3 matrices. Put

$$\Xi = \begin{pmatrix} 0 & 1 & 0 \\ 0 & 0 & 1 \\ -t & -t & 0 \end{pmatrix}.$$

The characteristic polynomial of the matrix Ξ is equal to the cubic polynomial $R(t;X)$.

To define our twist of the Hessian curve H_μ, take independent variables x,y,z and put

$$M(x,y,z) := x1_3 + y\Xi + z\Xi^2$$

where 1_3 is the unit matrix of size 3. The curve $\tilde{H}(\mu,t)$ is defined on the projective plane $\mathbb{P}^2(x:y:z)$ by

$$\tilde{H}(\mu,t) : \operatorname{Tr}\left(M(x,y,z)^3\right) = 3\mu \operatorname{Det}(M(x,y,z))$$

where Tr and Det denote the trace and the determinant of matrices, respectively.

It should be noted again that, in general, the curve $\tilde{H}(\mu,t)$ may not have any rational points over $\mathbb{Q}$ even though it is defined over $\mathbb{Q}$.

Proposition 3.6. *For $\mu,t\in\mathbb{Q}, \mu\neq 1, t\neq 0,-27/4$, the curve $\tilde{H}(\mu,t)$ is isomorphic to H_μ over $\tilde{K}_t$.*

In the case where $R(t;X)$ is irreducible over $\mathbb{Q}$, the set of rational points of $\tilde{H}(\mu,t)$ over $\mathbb{Q}$ is described by a certain subset of the cubic field K_t.

Let $\tilde{H}(\mu,t)[\mathbb{Q}]$ denote the set of all $\mathbb{Q}$-rational points on the curve $\tilde{H}(\mu,t)$.

Define a subset $S_{\mu,t}$ of the cubic field K_t by

$$S_{\mu,t} = \{\eta\in K_t^\times \mid \operatorname{Tr}_{K_t/\mathbb{Q}}\left(\eta^3\right) = 3\mu\,\mathrm{N}_{K_t/\mathbb{Q}}(\eta)\}.$$

Here $\operatorname{Tr}_{K_t/\mathbb{Q}}$ and $\mathrm{N}_{K_t/\mathbb{Q}}$ are the trace map and the norm map of K_t over $\mathbb{Q}$, respectively. It is clear that the multiplicative group $\mathbb{Q}^\times$ naturally acts on the set $S_{\mu,t}$. We denote the set of orbits by $S_{\mu,t}/\sim_{\mathbb{Q}^\times}$.

Proposition 3.7. *Let the notation and the assumptions be as above. Then there exists a canonical one-to-one and onto correspondence between* $S_{\mu,t}/\sim_{\mathbb{Q}^\times}$ *and* $\tilde{H}(\mu,t)[\mathbb{Q}]$.

We may now transfer the well-known tangential method on elliptic curves into our $S_{\mu,t}$. For an element η in the cubic field K_t, we denote its conjugates over $\mathbb{Q}$ by η, η', η''. Note that η' and η'' may not belong to K_t in general.

Proposition 3.8. *If* $\eta \in K_t$ *belongs to* $S_{\mu,t}$, *then*

$$\sqrt{-(4t+27)}\,\eta\left(\eta'^3 - \eta''^3\right)$$

also belongs to $S_{\mu,t}$.

3.4. *Twists of Hessian Elliptic Curves (2)*

We now give a twist $H_{\mu,t}$ of $\tilde{H}(\mu,t)$ over K_t which is a quadratic twist of our original Hessian curve H_μ.

Let $H_{\mu,t}$ be the curve on the projective plane $\mathbb{P}^2(u:v:w)$ defined by

$$H_{\mu,t} : 2u^3 + 6d_0uv^2 + w^3 = 3\mu\,(u^2 - d_0v^2\,)w, \quad d_0 = -(4t+27).$$

Proposition 3.9. *Suppose that* $\mu, t \in \mathbb{Q}, \mu \neq 1, t \neq 0, -27/4$.
(1) *The curve* $\tilde{H}(\mu,t)$ *is isomorphic to* $H_{\mu,t}$ *over* K_t.
(2) *The curve* $H_{\mu,t}$ *is a twist of the Hessian curve* H_μ *over* $\mathbb{Q}(\sqrt{d_0}) = \mathbb{Q}(\sqrt{-(4t+27)})$, *and is an elliptic curve defined over* $\mathbb{Q}$ *with a* $\mathbb{Q}$*-rational point* $(u:v:w) = (0:1:0)$. *The isomorphism is given by*

$$\begin{cases} U = u + \sqrt{d_0}\,v, \\ V = u - \sqrt{d_0}\,v, \\ W = w, \end{cases} \qquad \begin{cases} u = (U+V)/2, \\ v = (U-V)/(2\sqrt{d_0}), \\ w = W. \end{cases}$$

In Proposition 3.5 in §3.2, we saw a simple affine model A_μ of the 3-isogeny of the Hessian elliptic curve H_μ for $\mu \neq 1$; namely,

$$A_\mu : y^2 + 3\mu xy + y = x^3.$$

We now give an affine model $A_{\mu,t}$ of a 3-isogeny of the elliptic curve $H_{\mu,t}$ for $\mu \neq 0, 1$. This affine curve is a quadratic twist of A_μ over $\mathbb{Q}(\sqrt{d_0})$. Define $A_{\mu,t}$ by

$$A_{\mu,t} : d_0 Y^2 = \left(\frac{2X-1}{3\mu}\right)^3 + X^2, \qquad d_0 = -(4t+27).$$

It has a $\mathbb{Q}$-rational point $(X : Y : Z) = (0 : 1 : 0)$ at infinity.

Proposition 3.10. *Suppose that* $\mu, t \in \mathbb{Q}, \mu \neq 0, 1, t \neq 0, -27/4$.
(1) *The elliptic curve* $A_{\mu,t}$ *is an affine model of a 3-isogeny of the elliptic curve* $H_{\mu,t}$*; the isogeny is defined by*

$$\begin{cases} X = \left(\dfrac{u}{-w}\right)^3 + 3d_0\left(\dfrac{u}{-w}\right)\left(\dfrac{v}{-w}\right)^2, \\[2ex] Y = 3\left(\dfrac{u}{-w}\right)^2\left(\dfrac{v}{-w}\right) + d_0\left(\dfrac{v}{-w}\right)^3. \end{cases}$$

(2) *If* d_0 *is not a square in* $\mathbb{Q}$, $A_{\mu,t}$ *is the quadratic twist of* A_μ *over the quadratic field* $\mathbb{Q}(\sqrt{d_0})$.

3.5. *Some cases of non-empty* $\tilde{H}(\mu, t)[\mathbb{Q}]$

We now present two propositions to demonstrate some cases of non-empty $\tilde{H}(\mu, t)[\mathbb{Q}]$ before we close the article.

Proposition 3.11. *Suppose that the cubic polynomial* $R(t; X)$ *is irreducible over* $\mathbb{Q}$ *for* $t \in \mathbb{Q} - \{0, -27/4\}$. *Then* $\tilde{H}(\mu, t)$ *has a rational point over* $\mathbb{Q}$ *in each of the following three cases:* (1) *in case of* $\mu = t/3 + 1 \neq 1$, *the point* $(t : 0 : 1)$ *belongs to* $\tilde{H}(\mu, t)[\mathbb{Q}]$, (2) *in case of* $\mu = 1 - 3t \neq 1$, *the point* $(1 : 1 : 0)$ *belongs to* $\tilde{H}(\mu, t)[\mathbb{Q}]$, *and* (3) *in case of* $\mu = 1 - 2t/3 \neq 1$, *the point* $(0 : 0 : 1)$ *belongs to* $\tilde{H}(\mu, t)[\mathbb{Q}]$.

In the special case of $\mu = 0$ we give a necessary and sufficient condition on t for $\tilde{H}(0, t)$ to have a rational point over $\mathbb{Q}$.

Proposition 3.12. *In case of* $\mu = 0$, *the curve* $\tilde{H}(0, t)$ *has a rational point over* $\mathbb{Q}$ *for* $t \in \mathbb{Q} - \{0, -27/4\}$ *if and only if either* $t = -r^3/(r+1)$

with $r \in \mathbb{Q} - \{0, -1\}$, *or* $\tilde{H}(0,t)$ *is isomorphic over* $\mathbb{Q}$ *to* $\tilde{H}(0,t')$ *with* $t' = h^3/(2h-1)^2, h \in \mathbb{Q} - \{0, 1/2\}$.

References

1. J. W. S. Cassels, The Rational Solutions of the Diophantine Equation, $y^2 = x^3 - D$. *Acta Math.*, **82** (1950), 243–273.
2. J. Gebel, A. Pethő and H. G. Zimmer, On Mordell's Equation. *Comp. Math.*, **110** (1998), 335–367.
3. G. Gras, *Class Field Theory, from Theory to Practice.* Springer Monographs in Mathematics, Springer-Verlag, New York/Berlin/Heidelberg/London/Paris/Tokyo, 1st ed., 2003. Corr. 2nd printing, 2005.
4. A. Hoshi and K. Miyake, Tschirnhausen Transformation of a Cubic Generic Polynomial and a 2-dimensional Involutive Cremona Transformation. *Proc. Japan Acad.*, **83** Ser. A, (2007), 21–26.
5. D. Husemöller, *Elliptic Curves* with an Appendix by Ruth Lawrence, Springer-Verlag, New York/Berlin/Heidelberg/London/Paris/Tokyo, 1987.
6. C. U. Jensen, A. Ledet and N. Yui, *Generic Polynomials, Constructive Aspects of the Inverse Galois Problem*, Cambridge U.P., Cambridge, 2002.
7. G. Kemper, Generic polynomials are descent-generic. *Manuscripta math.*, **105** (2001), 139–141.
8. Y. Kishi and K. Miyake, Parametrization of the Quadratic Fields whose Class Numbers are divisible by Three. *Jour. Number Theory*, **80** (2000), 209–217.
9. P. Llorente and E. Nart, Effective determination of the decomposition of the rational prime in a cubic field. *Proc. American Math. Soc.*, **87** (1983), 579–585.
10. K. Miyake, Linear Fractional Transformations and Cyclic Polynomials. *Advanced Studies in Contemporary Mathematics* **1**, ed. by Han-Soo Kim and Toru Nakahara, *Memoirs of the Jangjun Math. Soc.*, 1999, Gu-Duk Publisher, Pusan, pp.137–142.
11. K. Miyake, Some Families of Mordell Curves associated to Cubic Fields. *Jour. Comp. and Applied Math. Sciences*, **160** (2003), 217–231.
12. K. Miyake, An Introduction to Elliptic Curves and their Diophantine Geometry — Mordell Curves —. *Compte Rendus de la conférence internationale Maroc-Québec* (Mai 2003) "Théorie des nombres et applications" 2004, pp. 173–185, *Annales des Sciences Mathématiques du Québec*, 2004.
13. K. Miyake, Cubic Fields and Mordell Curves. *Number Theory: Tradition and Modernization,* ed. by W. Zhang and Y. Tanigawa, pp. 175–183, Springer Science + Buisiness Media, Inc., 2006.
14. K. Miyake, Twists of Hessian Elliptic Curves and Cubic Fields, submitted to *The Proceedings of Congrés International, Algèbre, Théorie des nombres et leurs Applications, Université Mohammed I, Oujda-Saidia, Maroc, du 11 au 14 mai, 2006.*
15. L. J. Mordell, The Diophantine Equation $y^2 - k = x^3$. *Proc. London Math. Soc.*, **13** (1914), 60–80.

16. L. J. Mordell, *Diophantine Equations.* Acad. Press, London and New York, 1969.
17. J. Neukirch, *Algebraic Number Theory.* Grundlehren der mathematischen Wissenschaften, Vol.322, Springer-Verlag, New York/Berlin/Heidelberg/London/Paris/Tokyo, 1999.
18. E. S. Selmer, The Diophantine Equation $ax^3 + by^3 + cz^3 = 0$. *Acta Math.*, **85** (1951), 203–362; Completion of the Table. *Ibid.*, **92** (1954), 191–197.
19. D. Shanks, The Simplest Cubic Fields, *Math. Comp.*, **28** (1974), 1137–1152.
20. J. H. Silverman, *The Arithmetic of Elliptic Curves.* **GTM 106**, Springer-Verlag, New York, 1986.
21. J. H. Silverman, *Advanced Topics in the Arithmetic of Elliptic Curves.* **GTM 151**, Springer-Verlag, New York, 1994.
22. J. H. Silverman and J. Tate, *Rational Points on Elliptic Curves.* **UTM**, Springer-Verlag, New York, 1992.

DISTRIBUTION OF POINTS ON MODULAR HYPERBOLAS

IGOR E. SHPARLINSKI

Department of Computing, Macquarie University
Sydney, NSW 2109, Australia
E-mail: igor@ics.mq.edu.au

We give a survey of several recent results and pose several open problems about the distribution and some geometric properties of points (x, y) on modular hyperbolas $xy \equiv a \pmod m$. We also outline a very diverse range of applications of such results and discuss multivariate generalisations.

1. Introduction

1.1. *Modular Hyperbolas and Kloosterman Sums*

For a positive integer m and an arbitrary integer a with $\gcd(a, m) = 1$, we consider the set of points (x, y) on the modular hyperbola

$$\mathcal{H}_{a,m} = \{(x, y) \ : \ xy \equiv a \pmod m\}.$$

We give a survey of various results about the distribution and some geometric properties of points on $\mathcal{H}_{a,m}$. We also point out several very surprising links of these results with a wide variety of other number theoretic questions. Finally, we briefly consider some multidimensional generalisations, which often require very different techniques. Several open problems are formulated as well.

Given two sets of integers $\mathcal{X}$ and $\mathcal{Y}$, we write

$$\mathcal{H}_{a,m}(\mathcal{X}, \mathcal{Y}) = \{(x, y) \in \mathcal{H}_{a,m} \ : \ x \in \mathcal{X}, \ y \in \mathcal{Y}\}.$$

Since the case of $a = 1$ is of special interest we also write

$$\mathcal{H}_m = \mathcal{H}_{1,m}, \qquad \text{and} \qquad \mathcal{H}_m(\mathcal{X}, \mathcal{Y}) = \mathcal{H}_{1,m}(\mathcal{X}, \mathcal{Y}).$$

Obtaining precise asymptotic formulas for and establishing the positivity of $\#\mathcal{H}_{a,m}(\mathcal{X}, \mathcal{Y})$ for various "interesting" sets $\mathcal{X}$ and $\mathcal{Y}$ have been the central themes of many works in this direction. Certainly the problem becomes harder and far more interesting when the sets $\mathcal{X}$ and $\mathcal{Y}$ become "thinner".

Let

$$\mathbf{e}_m(z) = \exp(2\pi i z/m).$$

One immediately observes that the well known bound of *Kloosterman sums*

$$K_m(r,s) = \sum_{\substack{(x,y)\in\mathcal{H}_m\\ 1\leqslant x,y\leqslant m}} \mathbf{e}_m\left(rx+sy\right) \leqslant \left(m\gcd(r,s,m)\right)^{1/2+o(1)}, \qquad (1.1)$$

see [39, Corollary 11.12], can be used to study the points on $\mathcal{H}_{a,m}$. One only needs to recall some standard tools which link exponential sums with uniformity of distribution which we present in §2.2.

In this survey, we do not present complete proofs but sometimes explain their underlying ideas and specific ingredients. Typically the error terms in the asymptotic formulas we give contain a factor $m^{o(1)}$. In most cases it can be replaced by a more explicit function. Furthermore, when $m = p$ it can usually (but not always) be replaced by just some low power of $\log m$.

There is a large number of papers in this area which rather routinely study seemingly distinct, but in fact closely related, problems about $\mathcal{H}_{a,m}$ on the case by case basis. Here, we explain some standard principles which can be used to derive these and many other results of similar spirit about the points on $\mathcal{H}_{a,m}$ as simple corollaries of just one general result about the uniformity of distribution of points on $\mathcal{H}_{a,m}$ in certain domains. In §3.1, such a result is presented in Theorem 3.1 and derived in a very straightforward fashion from (1.1), using some standard arguments.

We remark that most of the results which rely only on (1.1) do not appeal to anything specific about the congruence $xy \equiv a \pmod m$, and at least when $m = p$ prime they can be extended to the distribution of solutions to more general congruences $f(x,y) \equiv 0 \pmod p$, with a polynomial f with integer coefficients. In the case of prime $m = p$, the Bombieri bound [9] of exponential sums along a curve replaces the bound (1.1). Although, as we have mentioned, we present such a generic result in §3.1, our main purpose is to outline some more intricate arguments, which use special properties of the congruence $xy \equiv a \pmod m$ and cannot be generalised to other congruences. In particular, in §3.2 we discuss the behaviour of points on $\mathcal{H}_{a,m}$ on average over a. We also describe some geometric properties of the set $\mathcal{H}_{a,m}$ in §§4.1–4.3.

Before presenting results about the properties of $\mathcal{H}_{a,m}$, we give a short overview of the surprisingly diverse variety of number theoretic tools which have been used in the study of $\mathcal{H}_{a,m}$ and their multivariate generalisations, see §§2.1–2.3.

Finally, we outline various applications of the results on the distribution of points on $\mathcal{H}_{a,m}$. Some of them are quite natural with very transparent connections to $\mathcal{H}_{a,m}$, see §5.1. However, there are also several less obvious and thus less expected applications, see §§5.2–5.8. An especially striking example of such unexpected applications is given by a result of [54] on torsion of elliptic curves, see §5.5.

1.2. *Notation*

Throughout the paper, any implied constants in symbols O, $\ll$ and $\gg$ may occasionally depend, where obvious, on the real positive parameter ε and are absolute otherwise. We recall that the notations $U = O(V)$, $U \ll V$ and $V \gg U$ are all equivalent to the statement that $|U| \leqslant cV$ holds with some constant $c > 0$.

We use p, with or without a subscript, to denote a prime number and use m to denote a positive integer.

We denote by $\mathbb{Z}/m\mathbb{Z}$ the residue ring modulo m. Typically we assume that the set $\{0, \dots, m-1\}$ is used to represent the elements of $\mathbb{Z}/m\mathbb{Z}$. Accordingly, we often consider the following subset of $\mathcal{H}_{a,m}$

$$\overline{\mathcal{H}}_{a,m} = \mathcal{H}_{a,m} \cap [0, m-1]^2.$$

We also put

$$\overline{\mathcal{H}}_{m} = \mathcal{H}_{m} \cap [0, m-1]^2.$$

As usual, $\omega(k)$, $\tau(k)$ and $\varphi(k)$ denote the number of distinct prime divisors, the number of positive integer divisors and the Euler function of $k \geqslant 1$, respectively.

Finally, $\mu(k)$ denotes the Möbius function. We recall that $\mu(1) = 1$, $\mu(k) = 0$ if $k \geqslant 2$ is not squarefree and $\mu(k) = (-1)^{\omega(k)}$ otherwise.

1.3. *Acknowledgements*

The author is very grateful to the organisers of the 4th China-Japan Seminar on number theory Jianya Liu and Shigeru Kanemitsu for their kind invitation to this meeting and help with preparation of this manuscript.

The author would like to thank Arne Winterhof for careful reading of the manuscript and making many valuable suggestions; in particular, the equation (4.6) is due to him.

This work was supported in part by ARC grant DP0556431.

2. Number Theory Background

2.1. *Exponential and Character Sums*

We have already mentioned the prominent role of Kloosterman sums and the bound (1.1) in particular.

Most of the works also use the identity

$$\frac{1}{m}\sum_{r\in\mathbb{Z}/m\mathbb{Z}} \mathbf{e}_m(rv) = \begin{cases} 1 & \text{if } v \equiv 0 \pmod m, \\ 0 & \text{if } v \not\equiv 0 \pmod m \end{cases} \tag{2.1}$$

to express various characteristic functions and thus relating various counting questions to exponential sums.

It is very often complemented by the bound

$$\sum_{z=W+1}^{W+Z} \mathbf{e}_m(rz) \ll \min\{Z, m/|r|\} \tag{2.2}$$

which holds for any integers r, W and $Z \geqslant 1$ with $0 < |r| \leqslant m/2$, see [39, Bound (8.6)].

We need the estimate from [63] of exponential sums with rational functions of special type, which generalises the bound (1.1) of Kloosterman sums.

Lemma 2.1. *Let $n_1, \ldots, n_s$ be nonzero pairwise distinct integers. Then the bound*

$$\max_{\gcd(a_1,\ldots,a_s,m)=d} \sum_{\substack{z=1\\ \gcd(z,m)=1}}^{m} \mathbf{e}_m(a_1 z^{n_1} + \ldots + a_s z^{n_s}) \ll d^{1/s} m^{1-1/s+o(1)}$$

holds, where the implied constant depends only on $n_1, \ldots, n_s$.

However, in many cases using bounds of multiplicative character sums yields stronger results.

Let Φ_m be the set of all $\varphi(m)$ multiplicative characters modulo m. We have the following analogue of (2.1). For any integer r,

$$\frac{1}{\varphi(m)}\sum_{\chi\in\Phi_m} \chi(r) = \begin{cases} 1 & \text{if } r \equiv 1 \pmod m, \\ 0 & \text{otherwise.} \end{cases} \tag{2.3}$$

We also use χ_0 to denote the principal character.

The following result is a combination of the Polya-Vinogradov bound (for $\nu = 1$) and Burgess (for $\nu \geqslant 2$) bounds, see [39, Theorems 12.5 and 12.6].

Lemma 2.2. *For any integers W and Z with $1 \leqslant Z \leqslant m$, the bound*

$$\max_{\substack{\chi \in \Phi_m \\ \chi \neq \chi_0}} \left| \sum_{z=W+1}^{W+Z} \chi(z) \right| \leqslant Z^{1-1/\nu} m^{(\nu+1)/4\nu^2+o(1)}$$

holds with $\nu = 1, 2, 3$ for any m and with arbitrary positive integer ν if $m = p$ is a prime.

The identity (2.3) immediately implies that for $1 \leqslant Z \leqslant m$

$$\sum_{\chi \in \Phi_m} \left| \sum_{z=W+1}^{W+Z} \chi(z) \right|^2 = \varphi(m) \sum_{\substack{z=W+1 \\ \gcd(z,m)=1}}^{W+Z} 1 \leqslant \varphi(m) Z \qquad (2.4)$$

which has been used in many works on $\mathcal{H}_{a,m}$.

Furthermore, it turns out, that sometimes one gets better results using the following 4th moment estimate from [3] (for prime $m = p$) and [28] (for arbitrary m), see also [32].

Lemma 2.3. *For an arbitrary integer W, if $m = p$ is a prime, and for $W = 0$ for arbitrary m, and an arbitrary positive integer $Z \leqslant m$, the bound*

$$\sum_{\chi \in \Phi_m} \left| \sum_{z=W+1}^{W+Z} \chi(z) \right|^4 \leqslant m^{1+o(1)} Z^2$$

holds.

2.2. *Theory of Uniform Distribution*

For a finite set $\mathcal{F} \subseteq [0,1]^s$ of the s-dimensional unit cube, we define its *discrepancy with respect to a domain* $\Xi \subseteq [0,1]^s$ as

$$\Delta(\mathcal{F}, \Xi) = \left| \frac{\#\{\mathbf{f} \in \mathcal{F} : \ \mathbf{f} \in \Xi\}}{\#\mathcal{F}} - \lambda(\Xi) \right|,$$

where λ is the Lebesgue measure on $[0,1]^s$.

We now define the *discrepancy* of $\mathcal{F}$ as

$$D(\mathcal{F}) = \sup_{\Pi \subseteq [0,1]^s} \Delta(\mathcal{F}, \Pi),$$

where the supremum is taken over all boxes $\Pi = [\alpha_1, \beta_1) \times \ldots \times [\alpha_s, \beta_s) \subseteq [0, 1]^s$.

A link between the discrepancy and exponential sums is provided by the celebrated *Koksma–Szüsz inequality*, see [22, Theorem 1.21]. However, for points of $\mathcal{H}_{a,m}$, due to the discrete structure of the problem, one can immediately establish such a link directly by the identity (2.1).

For example, one can consider the points

$$\left(\frac{x}{m}, \frac{y}{m}\right) \in [0, 1]^2, \qquad (x, y) \in \overline{\mathcal{H}}_{a,m},$$

and apply the bound (1.1) to estimate their discrepancy, which in turn is equivalent to studying points of $\mathcal{H}_{a,m}(\mathcal{X}, \mathcal{Y})$ where $\mathcal{X}$ and $\mathcal{Y}$ are sets of consecutive integers.

Moreover, the *Koksma–Hlawka inequality*, see [22, Theorem 1.14], allows to estimate average values of various functions on the points $(x, y) \in \mathcal{H}_{a,m}$.

Lemma 2.4. *For any continuous function $\psi(\mathbf{z})$ on the unit cube $z \in [0, 1]^s$ and a finite set $\mathcal{F} \subseteq [0, 1]^s$ of discrepancy $D(\mathcal{F})$, the following bound holds:*

$$\frac{1}{\#\mathcal{F}} \sum_{\mathbf{f} \in \mathcal{F}} \psi(\mathbf{f}) = \int_{[0,1]^s} \psi(\mathbf{z}) d\mathbf{z} + O\left(D(\mathcal{F})\right)$$

where the implied constant depends only on s and the function ψ.

To study $\mathcal{H}_{a,m} \cap \mathcal{W}$ for more general sets $\mathcal{W}$ some additional tools are required from the theory of uniform distribution.

As usual, we define the distance between a vector $\mathbf{u} \in [0, 1]^s$ and a set $\Xi \subseteq [0, 1]^s$ by

$$\operatorname{dist}(\mathbf{u}, \Xi) = \inf_{\mathbf{w} \in \Xi} \|\mathbf{u} - \mathbf{w}\|,$$

where $\|\mathbf{v}\|$ denotes the Euclidean norm of $\mathbf{v}$. Given $\varepsilon > 0$ and a domain $\Xi \subseteq [0, 1]^s$ we define the sets

$$\Xi_\varepsilon^+ = \{\mathbf{u} \in [0, 1]^s \backslash \Xi \ : \ \operatorname{dist}(\mathbf{u}, \Xi) < \varepsilon\}$$

and

$$\Xi_\varepsilon^- = \{\mathbf{u} \in \Xi \ : \ \operatorname{dist}(\mathbf{u}, [0, 1]^s \backslash \Xi) < \varepsilon\}.$$

Let $h(\varepsilon)$ be an arbitrary increasing function defined for $\varepsilon > 0$ and such that

$$\lim_{\varepsilon \to 0} h(\varepsilon) = 0.$$

As in [47,56], we define the class $\mathcal{S}_h$ of domains $\Xi \subseteq [0,1]^s$ for which

$$\lambda\left(\Xi_\varepsilon^+\right) \leqslant h(\varepsilon) \qquad \text{and} \qquad \lambda\left(\Xi_\varepsilon^-\right) \leqslant h(\varepsilon).$$

A relation between $D(\mathcal{F})$ and $\Delta(\mathcal{F},\Xi)$ for $\Xi \in \mathcal{S}_h$ is given by the following inequality of [47] (see also [56]).

Lemma 2.5. *For any domain $\Xi \in \mathcal{S}_h$, we have*

$$\Delta(\mathcal{F},\Xi) \ll h\left(s^{1/2}D(\mathcal{F})^{1/s}\right).$$

Finally, the following bound, which is a special case of a more general result of H. Weyl [74] shows that if Ξ has a piecewise smooth boundary such that $\Xi \in \mathcal{S}_h$ for some linear function $h(\varepsilon) = C\varepsilon$.

Lemma 2.6. *For any domain $\Xi \in \mathcal{S}_h$ with piecewise smooth boundary, we have*

$$\lambda\left(\Xi_\varepsilon^\pm\right) = O(\varepsilon).$$

To use the above results for the study of points on $\mathcal{H}_{a,m}$, one usually considers points

$$\left(\frac{x}{m},\frac{y}{m}\right) \in [0,1]^2, \qquad (x,y) \in \mathcal{H}_{a,m},\ 1 \leqslant x,y \leqslant m. \tag{2.5}$$

2.3. *Arithmetic Functions, Divisors, Prime Numbers*

Certainly some elementary bounds such as

$$\varphi(k) \gg \frac{k}{\log\log(k+2)}$$

and

$$2^{\omega(k)} \leqslant \tau(k) \ll \exp\left((\log 2 + o(1))\,\frac{\log k}{\log\log k}\right), \tag{2.6}$$

see [71, Section I.5.2 and I.5.4] appear at various stages of the proofs of relevant results.

The following well known consequence of the *sieve of Eratosthenes* (essentially of the inclusion-exclusion principle expressed via the Möbius function) is very often needed to estimate the main terms of various asymptotic formulas (see, for example, [67,68]).

Lemma 2.7. *For any integers $m, Z \geqslant 1$ and $W \geqslant 0$,*

$$\sum_{\substack{z=W+1\\ \gcd(z,m)=1}}^{W+Z} 1 = \frac{\varphi(m)}{m}Z + O(2^{\omega(m)}).$$

For an infinite monotonically increasing sequence of positive integers $\mathcal{A} = (a_n)_{n=1}^{\infty}$, we define

$$H(x, y, z; \mathcal{A}) = \#\{n \leqslant x \ : \ \exists\, d | a_n \text{ with } y < d \leqslant z\}.$$

For $\mathcal{A} = \mathbb{N}$, the set of natural numbers, the order of magnitude of $H(x, y, z; \mathbb{N})$ for all x, y, z has been determined in [25], see also [38]. Also in [25], one can find upper bounds for $H(x, y, z; \mathcal{P}_b)$ of the expected order of magnitude, where $\mathcal{P}_b = \{p + b \ : \ p \text{ prime}\}$ is a set of so-called shifted primes. However, for the problem of studying $\mathcal{H}_{a,m}$, we need analogous results where n is restricted to an arithmetic progression. More precisely, let us define the sequences

$$\mathcal{T}_k = \{mk - 1 \ : \ m \in \mathbb{N}\} \qquad \text{and} \qquad \mathcal{U}_k = \{pk - 1 \ : \ p \text{ prime}\}.$$

It has been shown in [26] that the arguments of [25] imply the following estimates.

It is usual that in questions of this kind, the constant

$$\kappa = 1 - \frac{1 + \log\log 2}{\log 2} = 0.086071\ldots. \tag{2.7}$$

plays an important role, see also [38].

Lemma 2.8. *Uniformly for* $100 \leqslant y \leqslant x^{0.51}$, $1.1y \leqslant z \leqslant y^{1.1}$, $1 \leqslant k \leqslant \log x$, *we have*

$$H(x, y, z; \mathcal{T}_k) \ll x \frac{k}{\varphi(k)} u^{\kappa} (\log(1/u))^{-3/2},$$
$$H(x, y, z; \mathcal{U}_k) \ll x \frac{k}{\varphi(k)} u^{\kappa} (\log(1/u))^{-3/2},$$

where $z = y^{1+u}$.

A certain result of [26] relies on the existence of infinitely many primes p with a prescribed structure of divisors of $p-1$, which is done using a very deep results of [10] concerning the *Bombieri-Vinogradov theorem*.

For an integer $k \geqslant 1$ we write

$$T(k) = \max_{i=1,\ldots,\tau(m)-1} \frac{d_{i+1}}{d_i}$$

where $1 = d_1 < \ldots < d_{\tau(m)} = m$ are the positive divisors of m.

By [60, Theorem 1], we have:

Lemma 2.9. *Uniformly in* $z \geqslant t \geqslant 2$,

$$\frac{z \log t}{\log z} \gg \#\{k \leqslant z \ : \ T(k) \leqslant t\} \gg \frac{z \log t}{\log z}.$$

Finally, we remark, that several interesting results about the distribution of points on $\mathcal{H}_{a,m}(\mathcal{X},\mathcal{Y})$ on average over a, for some special sets $\mathcal{X}$ and $\mathcal{Y}$, such as intervals

$$\mathcal{X} = \mathcal{Y} = \{z \ : \ 1 \leqslant z \leqslant m/2\},$$

are based on various asymptotic formulas for average values of Dirichlet L-functions, see, for example, [48,75,86].

3. Distribution of Points on $\mathcal{H}_{a,m}$

3.1. *Points on $\mathcal{H}_{a,m}$ in Intervals for All a*

A classical conjecture asserts that for any fixed $\varepsilon > 0$ and a sufficiently large p, for every integer a there are integers x and y with $|x|, |y| \leqslant p^{1/2+\varepsilon}$ and such that $xy \equiv a \pmod p$, see [30,33–35] and references therein. The question has probably been motivated by the following observation. Using the Dirichlet pigeon-hole principle, one can easily show that for every integer a there are integers x and y with $|x|, |y| \leqslant 2p^{1/2}$ with $y/x \equiv a \pmod p$.

Unfortunately, this is known only with $|x|, |y| \geqslant Cp^{3/4}$ for some absolute constant $C > 0$, which is shown in [31]. Several modifications of this bound, for example for composite m, are also known, see [44]. These results are based on the bound (1.1) of Kloosterman sums (and its more precise form in the case when $m = p$ is a prime) combined with some other standard arguments. The same arguments also produce the following estimate which is a slight generalisation of several previously known results, see [5,29] and references therein.

Theorem 3.1. *Let $\mathcal{X} = \{U+1, \dots, U+X\}$, where $X \geqslant 1$ and $U \geqslant 0$ are arbitrary integers. Suppose that for every $x \in \mathcal{X}$ we are given a set $\mathcal{Y}_x = \{V_x+1, \dots, V_x+Y\}$ where $Y \geqslant 1$ and $V_x \geqslant 0$ are arbitrary integers. Then for any integer $m \geqslant 1$ and a with $\gcd(a,m) = 1$, we have*

$$\sum_{\substack{(x,y)\in\mathcal{H}_{a,m}\\ x\in\mathcal{X},\, y\in\mathcal{Y}_x}} 1 = XY\frac{\varphi(m)}{m^2} + O(m^{1/2+o(1)}).$$

Proof. Using (2.1) we write

$$\begin{aligned}\sum_{\substack{(x,y)\in\mathcal{H}_{a,m}\\ x\in\mathcal{X},\, y\in\mathcal{Y}_x}} 1 &= \frac{1}{m^2}\sum_{\substack{(x,y)\in\mathcal{H}_{a,m}\\ 1\leqslant x,y\leqslant m}}\ \sum_{w\in\mathcal{X}}\ \sum_{z\in\mathcal{Y}_w}\ \sum_{r,s\in\mathbb{Z}/m\mathbb{Z}} \mathbf{e}_m\left(r(x-w)+s(y-z)\right)\\ &= \frac{1}{m^2}\sum_{r,s\in\mathbb{Z}/m\mathbb{Z}} K_m(r,as)\sum_{w\in\mathcal{X}}\mathbf{e}_m(-rw)\sum_{z\in\mathcal{Y}_w}\mathbf{e}_m(-sz).\end{aligned}$$

We now separate the main term which corresponds to $r = s = 0$ and is equal to

$$\frac{XY}{m^2} \sum_{\substack{(x,y)\in\mathcal{H}_{a,m} \\ 1\leqslant x,y\leqslant m}} 1 = XY\frac{\varphi(m)}{m^2}.$$

For the error term $\mathbf{E}$, for each divisor $d|m$, we collect together pairs (r, s) with the same value $\gcd(r, s, m) = d$.

Applying the bounds (1.1) and (2.2) we obtain

$$\begin{aligned}\mathbf{E} &\ll m^{1/2+o(1)} \sum_{\substack{d|m \\ d<m}} d^{1/2} \sum_{\substack{-(m-1)/2\leqslant r,s\leqslant m/2 \\ \gcd(r,s,m)=d}} \frac{1}{(|r|+1)(|s|+1)} \\ &\ll m^{1/2+o(1)} \sum_{\substack{d|m \\ d<m}} d^{1/2} \left(\sum_{-(m-1)/2d\leqslant t\leqslant m/2d} \frac{1}{d|t|+1} \right)^2 \\ &\ll m^{1/2+o(1)} \sum_{\substack{d|m \\ d<m}} d^{-3/2} \ll m^{1/2+o(1)}\tau(m).\end{aligned}$$

Now the bound (2.6) leads to the desired statement. □

For example if $V_x = V$ for all $x \in \mathcal{X} = \{U+1, \dots, U+X\}$ and $\mathcal{Y} = \{V+1, \dots, V+Y\}$ then Theorem 3.1 yields

$$\#\mathcal{H}_{a,m}(\mathcal{X}, \mathcal{Y}) = XY\frac{\varphi(m)}{m^2} + O(m^{1/2+o(1)}). \tag{3.1}$$

It seems that improving Theorem 3.1 or even just the asymptotic formula (3.1) and making them nontrivial for $XY < m^\alpha$ with some fixed $\alpha < 3/2$ is out of reach at the present time. However there are some easier questions which could be more feasible to answer.

Question 3.1. Improve the asymptotic formula (3.1) for some special moduli m, such as primes or prime powers.

Question 3.2. Improve the asymptotic formula (3.1) on average over the moduli $m \leqslant M$.

Clearly, Theorem 3.1 can be viewed as the bound $O(m^{-1/2+o(1)})$ on the discrepancy of the points (2.5). Thus one can now apply Lemmas 2.5 and (2.6) to study the distribution of points on $\mathcal{H}_{a,m}$ in more complicated domains than boxes $[U+1, U+X] \times [V+1, V+Y]$ covered by Theorem 3.1. For example, we immediately deduce that

$$\#\{(x,y) \in \overline{\mathcal{H}}_{a,m} \ : \ x^2 + y^2 \leqslant r^2\} = \frac{\pi r^2 \varphi(m)}{4m^2} + O(m^{3/4+o(1)}).$$

Furthermore, the asymptotic formula (3.1), combined with Lemma 2.4, provides the most direct way to the following asymptotic formula

$$\sum_{(x,y)\in\overline{\mathcal{H}}_{a,m}} (x-y)^{2\nu} = \frac{1}{(2\nu+1)(\nu+1)} m^{2\nu}\varphi(m) + O\left(m^{2\nu+1/2+o(1)}\right) \quad (3.2)$$

which has been given in [82] (in a slightly more precise form, which can also be obtained within our elementary arguments).

Similarly Theorem 3.1 and Lemma 2.4, imply that for any real positive $\Delta < m/2$

$$\sum_{\substack{(x,y)\in\overline{\mathcal{H}}_{a,m} \\ |x-y|\leqslant\Delta}} 1 = \frac{\Delta(2m-\Delta)\varphi(m)}{m^2} + O\left(m^{1/2+o(1)}\right) \quad (3.3)$$

which is a version of a result of [83].

As we have mentioned, Theorem 3.1 uses very little specific information about the congruence $xy \equiv a \pmod m$ and can be extended to many other congruences. For prime $m = p$ one can use the Bombieri bound [9] instead of (1.1) and obtain exactly the same results in much more general settings. For example, this has been done for solutions of polynomial congruences modulo p and also for joint distribution of inverses modulo p of s linear forms $a_j x + b_j$, $j = 1, \ldots, s$, with integer coefficients, see [14,16,18–21,36, 72,77,85,88] and references therein.

3.2. *Points on $\mathcal{H}_{a,m}$ in Intervals on Average Over a*

It is natural to expect that one can get stronger results than Theorem 3.1 on average over a.

This indeed is true, and it has been shown in the series of works [30, 33–35] that the congruence $a \equiv xy \pmod p$ is solvable for all but $o(m)$ values of $a = 1, \ldots, m-1$, with x and y significantly smaller than $m^{3/4}$. In particular, in [34], this is proved for x and y in the range $1 \leqslant x, y \leqslant m^{1/2}(\log m)^{1+\varepsilon}$. Certainly this result is very sharp. Indeed, it has been noticed in [30] that well known estimates for integers with a divisor in a given interval (see [25,38]) immediately imply that for any $\varepsilon > 0$ almost all residue classes modulo m are not of the form $xy \pmod m$ with $1 \leqslant x, y \leqslant m^{1/2}(\log m)^{\kappa-\varepsilon}$ where κ is given by (2.7).

One can also derive from [26] (which in turn makes use of Lemma 2.8) that for any $\varepsilon > 0$ the inequality

$$\max\{|x|, |y| \ : \ xy \equiv 1 \pmod m\} \geqslant m^{1/2}(\log m)^{\kappa/2}(\log\log m)^{3/4-\varepsilon}$$

holds:

- for all positive integers $m \leqslant M$, except for possibly $o(M)$ of them,
- for all prime $m = p \leqslant M$ except for possibly $o(M/\log M)$ of them.

Similar questions about the ratios x/y, have also been studied, see [30, 34,61].

The result of [34] shows that almost all reduced classes modulo m can be represented as xy with $1 \leqslant x, y \leqslant m^{1/2+\varepsilon}$. However, it does not imply that these products are uniformly distributed in reduced residue classes, which is some times required in applications.

In this respect, the following bound is a minor modification of a result of [68] and gives the desired uniformity of distribution for $1 \leqslant x \leqslant X$, $V+1 \leqslant y \leqslant V+Y$ provided that $X, Y \geqslant m^{1/2+\varepsilon}$ for a fixed $\varepsilon > 0$ and sufficient large integer m. In turn, it is based on some ideas from [4].

Theorem 3.2. *Let $\mathcal{X} = \{1, \ldots, X\}$ and $\mathcal{Y} = \{V+1, \ldots, V+Y\}$ where $X, Y \geqslant 1$ and $V \geqslant 0$ are arbitrary integers. Then for any integer $m \geqslant 1$,*

$$\sum_{\substack{a=1\\ \gcd(a,m)=1}}^{m} \left| \#\mathcal{H}_{a,m}(\mathcal{X}, \mathcal{Y}) - XY\frac{\varphi(m)}{m^2} \right|^2 \ll X(X+Y)m^{o(1)}.$$

Proof. We rewrite the congruence $xy \equiv a \pmod m$ as $y \equiv ax^{-1} \pmod m$ (where the inversion is taken modulo m). Using the identity (2.1), we write

$$\begin{aligned}\#\mathcal{H}_{a,m}(\mathcal{X}, \mathcal{Y}) &= \frac{1}{m} \sum_{\substack{x=1\\ \gcd(x,m)=1}}^{X} \sum_{y=V+1}^{V+Y} \sum_{-(m-1)/2 \leqslant r \leqslant m/2} \mathbf{e}_m\left(r(ax^{-1} - y)\right)\\ &= \frac{1}{m} \sum_{-(m-1)/2 \leqslant r \leqslant m/2} \mathbf{e}_m(-rV) \sum_{\substack{x=1\\ \gcd(x,m)=1}}^{X} \mathbf{e}_m\left(arx^{-1}\right) \sum_{y=1}^{Y} \mathbf{e}_m(-ry).\end{aligned}$$

By Lemma 2.7, the main term corresponding to $r = 0$ is

$$\frac{1}{m} \sum_{\substack{x=1\\ \gcd(x,m)=1}}^{X} \sum_{y=1}^{Y} 1 = XY\frac{\varphi(m)}{m^2} + O\left(Ym^{-1+o(1)}\right).$$

Hence

$$\#\mathcal{H}_{a,m}(\mathcal{X},\mathcal{Y}) - XY\frac{\varphi(m)}{m^2} \ll \frac{1}{m}\mathbf{E}_{a,m}(X,Y) + Ym^{-1+o(1)},$$

where

$$\mathbf{E}_{a,m}(X,Y) = \sum_{1\leqslant |r|\leqslant m/2}\left|\sum_{\substack{x=1\\ \gcd(x,m)=1}}^{X} \mathbf{e}_m\left(arx^{-1}\right)\right|\left|\sum_{y=1}^{Y}\mathbf{e}_m(-ry)\right|.$$

Using the Cauchy inequality we have

$$\begin{aligned}\sum_{a=1}^{m}&\left|\#\mathcal{H}_{a,m}(\mathcal{X},\mathcal{Y}) - XY\frac{\varphi(m)}{m^2}\right|^2\\ &\qquad\qquad \leqslant \frac{1}{m^2}\sum_{a=1}^{m}\mathbf{E}_{a,m}(X,Y)^2 + Y^2m^{-1+o(1)}.\end{aligned} \tag{3.4}$$

We now put $J = \lfloor \log(Y/2)\rfloor$ and define the sets

$$\begin{aligned}\mathcal{R}_0 &= \left\{r \ :\ 1\leqslant |r|\leqslant \frac{m}{Y}\right\},\\ \mathcal{R}_j &= \left\{r \ :\ e^{j-1}\frac{m}{Y} < |r| \leqslant e^{j}\frac{m}{Y}\right\}, \qquad j=1,\ldots,J,\\ \mathcal{R}_{J+1} &= \left\{r \ :\ e^{J}\frac{m}{Y} < |r|\leqslant m/2\right\}\end{aligned}$$

(we can certainly assume that $J\geqslant 1$ since otherwise the bound is trivial).

Applying the Cauchy inequality again, we deduce

$$\mathbf{E}_{a,m}(X,Y)^2 \leqslant (J+2)\sum_{j=0}^{J+1}\mathbf{E}_{a,m,j}(X,Y)^2, \tag{3.5}$$

where

$$\mathbf{E}_{a,m,j}(X,Y) = \sum_{r\in\mathcal{R}_j}\left|\sum_{\substack{x=1\\ \gcd(x,m)=1}}^{X} \mathbf{e}_m\left(arx^{-1}\right)\right|\left|\sum_{y=1}^{Y}\mathbf{e}_m(-ry)\right|.$$

Using (2.2), we conclude that that

$$\sum_{y=1}^{Y}\mathbf{e}_m(-ry) \ll e^{-j}Y.$$

for $r\in\mathcal{R}_j$, $j=0,\ldots,J+1$. Thus

$$\mathbf{E}_{a,m,j}(X,Y) \ll e^{-j}Y\left|\sum_{r\in\mathcal{R}_j}\vartheta_r\sum_{\substack{x=1\\ \gcd(x,m)=1}}^{X}\mathbf{e}_m\left(arx^{-1}\right)\right|, \qquad j=0,\ldots,J+1,$$

for some complex numbers ϑ_r with $|\vartheta_r| \leqslant 1$ for $|r| \leqslant m/2$. Therefore,

$$\sum_{a=1}^{m} \mathbf{E}_{a,m,j}(X,Y)^2 \ll e^{-2j}Y^2 \sum_{a=1}^{m} \left| \sum_{r\in\mathcal{R}_j} \vartheta_r \sum_{\substack{x=1\\ \gcd(x,m)=1}}^{X} \mathbf{e}_m\left(arx^{-1}\right) \right|^2$$

$$= e^{-2j}Y^2 \sum_{r_1,r_2\in\mathcal{R}_j} \vartheta_{r_1}\vartheta_{r_2} \sum_{\substack{|x_1|,|x_2|\leqslant X\\ \gcd(x_1x_2,m)=1}}^{X} \sum_{a=1}^{m} \mathbf{e}_m\left(a\left(r_1x_1^{-1} - r_2x_2^{-1}\right)\right).$$

Clearly the inner sum vanishes if $r_1x_1^{-1} \not\equiv r_2x_2^{-1} \pmod m$ and is equal to m otherwise. Therefore

$$\sum_{a=1}^{m} \mathbf{E}_{a,m,j}(X,Y)^2 \ll e^{-2j}Y^2 mT_j, \tag{3.6}$$

where T_j is the number of solutions to the congruence

$$r_1x_2 \equiv r_2x_1 \pmod m, \qquad r_1,r_2 \in \mathcal{R}_j,\ |x_1|,|x_2| \leqslant X,\ \gcd(x_1x_2,m)=1.$$

We now see that if r_1 and x_2 are fixed, then r_2 and x_1 are such that their product $s = r_2x_1 \ll e^j mX/Y$ belongs to a prescribed residue class modulo m. Thus there are at most $O\left(e^jX/Y+1\right)$ possible values of s and for each fixed $s \ll e^j mX/Y$ there are $\tau(s) = m^{o(1)}$ values of r_1 and x_2 with $s = r_2x_1$, see (2.6). Therefore

$$T_j \leqslant X\#\mathcal{R}_j\left(e^jX/Y+1\right)m^{o(1)} = \frac{e^{2j}X^2m^{1+o(1)}}{Y^2} + \frac{e^jXm^{1+o(1)}}{Y}$$

and after substitution into (3.6) we get

$$\sum_{a=1}^{m} \mathbf{E}_{a,m,j}(X,Y)^2 \ll e^{-2j}Y^2mT_j = X^2m^{2+o(1)} + e^{-j}XYm^{2+o(1)}.$$

Substituting this bound in (3.5) and recalling (3.4), we conclude the proof. □

We note that the proof of Theorem 3.2 can easily be extended to arbitrary sets $\mathcal{X} \subseteq \{1,\ldots,X\}$, see [68]. However, it breaks down if x runs through a short interval away from the origin. An alternative approach has been suggested in [32] and is based on bounds of the 4th moment of multiplicative character sums, see Lemma 2.3. If $m = p$ is prime, it can handle such shifted intervals $\mathcal{X} = \{U+1,\ldots,U+X\}$ (but not arbitrary sets $\mathcal{X} \subseteq \{1,\ldots,X\}$ as that of [68]). Furthermore, the technique of [32] leads to more explicit expressions instead of $m^{o(1)}$ in the error term. Thus, although the approaches of [32] and [68] complement each other they still leave some natural open questions.

Question 3.3. Extend Theorem 3.2 to sets $\mathcal{X} = \{U+1, \ldots, U+X\}$ with arbitrary U.

As we have mentioned, if $m = p$, Question 3.3 is addressed in [32], however the some of the necessary ingredients are not known for composite m.

We also note that several results "on average" related to various modifications of the Lehmer problem are given in [48,73,75,86,87].

Finally, we remark that $\#\mathcal{H}_{a,m}(\mathcal{X}, \mathcal{Y})$ has been studied in [17] for the same sets as in Theorem 3.1, that is, for $\mathcal{X} = \{1, \ldots, X\}$ and $\mathcal{Y} = \{V + 1, \ldots, V + Y\}$, but on average over V. It is shown in [17] that in this case one can also obtain stronger bounds than that of Theorem 3.1.

3.3. *Points on $\mathcal{H}_{a,m}$ in Sets with Arithmetic Conditions*

Theorems 3.1 and 3.2 consider the case when x and y belong to sets of consecutive integers. However, studying points on $\mathcal{H}_{a,m}$ in other sets is of ultimate interest as well.

We start with a very simple observation that no general result of the type of Theorems 3.1 and 3.2 applying to arbitrary sets $\mathcal{X}$ and $\mathcal{Y}$ is possible (even for very massive sets $\mathcal{X}$ and $\mathcal{Y}$). For example, if $m = p$ is a prime and $\mathcal{X} = \mathcal{Y}$ consist of all $(p-1)/2$ quadratic residues modulo p, then $\mathcal{H}_{a,p}(\mathcal{X}, \mathcal{Y}) = \varnothing$ for every quadratic nonresidue a.

The problem of distribution of pairs of primes $(p, q) \in \overline{\mathcal{H}}_{a,m}$ has been considered in [24]. Unfortunately, it seems that even the *Extended Riemann Hypothesis* is not powerful enough to get a satisfactory answer to this question, see [24] for details. However it seems that the method of [64] can be used to study points $(x, y) \in \overline{\mathcal{H}}_{a,m}$ with squarefree x and y.

Question 3.4. Obtain an asymptotic formula for

$$\#\{(x, y) \in \mathcal{H}_{a,m}(\mathcal{X}, \mathcal{Y}) \ : \ x \text{ and } y \text{ are squarefree}\}$$

where $\mathcal{X} = \{1, \ldots, X\}$ and $\mathcal{Y} = \{1, \ldots, Y\}$ and $1 \leqslant X, Y \leqslant m$ are arbitrary integers.

One can also study the distribution of points $(x, y) \in \mathcal{H}_m(\mathcal{X}, \mathcal{Y})$ with some prescribed structure of prime factors. For example, let $P_+(k)$ and $P_-(k)$ denote the largest and the smallest prime divisors of an integer $k \geqslant 1$.

Question 3.5. Obtain an asymptotic formula for

$$\#\{(x,y)\in\mathcal{H}_{a,m}(\mathcal{X},\mathcal{Y})\ :\ P_+(xy)\leqslant R\}$$

and

$$\#\{(x,y)\in\mathcal{H}_{a,m}(\mathcal{X},\mathcal{Y})\ :\ P_-(xy)\geqslant r\}$$

where $\mathcal{X}=\{1,\ldots,X\}$ and $\mathcal{Y}=\{1,\ldots,Y\}$ and $1\leqslant X,Y\leqslant m$ are arbitrary integers with R and r in reasonably large ranges.

We remark that using elementary sieving arguments one can extend the result of Theorem 3.1 to counting $(x,y)\in\mathcal{H}_{a,m}$ such that x is of the largest possible multiplicative order modulo m (and thus so is y) which is given by the *Carmichael function* $\lambda(n)$. In particular, when $m=p$ is a prime, this addresses the problem of counting $(x,y)\in\mathcal{H}_{a,m}$ where x is a primitive root modulo p, for example, see [5,44]. Proofs of these results usually follow the same standard lines as the proof of Theorem 3.1, except that instead of (1.1) one uses the bound of the same strength on Kloosterman sums twisted with multiplicative characters

$$\sum_{\substack{(x,y)\in\mathcal{H}_m\\1\leqslant x,y\leqslant m}}\chi(x)\mathbf{e}_m\left(rx+sy\right)\ll\left(m\gcd(r,s,m)\right)^{1/2+o(1)}.$$

There are still some delicate issues of getting the $m^{o(1)}$ term as small as possible.

Finally, we note that several very interesting results have recently been obtained in [55] about points $(x,y)\in\overline{\mathcal{H}}_{a,m}$ such that x and y have restricted g-ary expansions to some fixed base $g\geqslant 2$.

4. Geometric Properties of $\mathcal{H}_{a,m}$

4.1. *Distances*

We observe that the asymptotic formulas (3.2) and (3.3) have a natural interpretation as the bounds on the power moments and the distribution function of the distances between an element $x\in\{1,\ldots,m\}$ with $\gcd(x,m)=1$ and its modular inverse. Several results about the average (over a) value of power moments can be found in [48–51], see also references therein.

We now define the *width* $w_{a,m}$ of the set $\overline{\mathcal{H}}_{a,m}$:

$$w_{a,m}=\max\left\{|x-y|\ :\ (x,y)\in\overline{\mathcal{H}}_{a,m}\right\}.$$

We also put

$$w_m=w_{1,m}$$

for the width of $\overline{\mathcal{H}}_m$, which has been the main object of study of [26,43,44].

Using Theorem 3.1 one easily derives that

$$w_{a,m} = m + O(m^{3/4+o(1)}). \tag{4.1}$$

Using the same arguments as in [44], one can obtain a more precise expression for the factor $m^{o(1)}$.

On the other hand, it has been noticed in [43] that

$$m - w_m \geq \lceil 2\sqrt{m-1} \rceil$$

with equality for all m of the form

$$m = k^2 + \ell k + 1 \tag{4.2}$$

with integers k and ℓ such that $k > 0$, $0 \leq \ell < 2\sqrt{k} + 1$ and hence

$$\liminf_{m\to\infty} \frac{m - w_m}{\sqrt{m}} = 2. \tag{4.3}$$

Question 4.1. Show that there are infinitely many primes $m = p$ of the form (4.2) with $0 \leq \ell < 2\sqrt{k} + 1$.

As a curiosity, we recall the following, it has been noted in [26], that $m - w_m \leqslant \sqrt{8m}$ for all positive integers $m = 2^s$ with $s \in \mathbb{Z}$. Indeed, if s is even, then $m = (2^{s/2} - 1)^2 + 2(2^{s/2} - 1) + 1$ is of the form (4.2), if s is odd, then this follows from

$$(2^{(s+1)/2} - 1)(2^s - 2^{(s+1)/2} - 1) \equiv 1 \pmod{2^s}.$$

In the opposite direction it is shown in [26] that

$$\limsup_{m\to\infty} \frac{m - w_m}{\sqrt{m}} = \infty. \tag{4.4}$$

Furthermore, analogues of (4.3) and (4.4) also hold for prime values $m = p$:

$$\liminf_{p\to\infty} \frac{p - w_p}{\sqrt{p}} = 2 \qquad \text{and} \qquad \limsup_{p\to\infty} \frac{p - w_p}{\sqrt{p}} = \infty,$$

which follow from the following two results given in [26].

Theorem 4.1. *For infinitely many primes p, we have*

$$p - w_p \leqslant 2\sqrt{p} + \frac{\sqrt{p}}{\log p}.$$

Proof. Let $\varepsilon = 1/(4\log Q)$. Using [10] one can show that for sufficiently large Q, there is a prime in the interval $((1-\varepsilon)Q, Q]$ such that $p-1$ has a divisor d in the interval $((1-2\varepsilon)\sqrt{Q}, (1-\varepsilon)\sqrt{Q}]$. If we write $p-1 = df$, then $w_p \geqslant p - f - d$. But, if Q is so large that $\varepsilon \leqslant 0.01$, then

$$f + d = \frac{p-1}{d} + d \leqslant \frac{x}{(1-2\varepsilon)\sqrt{x}} + (1-\varepsilon)\sqrt{x} \leqslant (2+3\varepsilon)\sqrt{p},$$

which implies the desired result. □

Theorem 4.2. *Let $f(M)$ be any positive function tending monotonically to zero as $M \to \infty$. Then the inequality*

$$m - w_m \geqslant m^{1/2}(\log m)^{\kappa/2}(\log\log m)^{3/4} f(m)$$

holds:

- *for all positive integers $m \leqslant M$, except for possibly $o(M)$ of them,*
- *for all prime $m = p \leqslant M$ except for possibly $o(M/\log M)$ of them.*

Proof. Let x be large and set

$$z = (\log M)^{\kappa/2}(\log\log M)^{3/4} f(M/2).$$

It suffices to show $m - w_m \leqslant zm^{1/2}$ for $o(M)$ of integers m between $M/2$ and M. Without loss of generality, suppose $f(M) \geqslant 1/\log\log M$ for all $M > 10$. We define $\mathcal{J}_k$ to be the set of positive integers $m \in (M/2, M]$ for which $m - w_m \leqslant ym^{1/2}$ and such that there are $(x, y) \in \overline{\mathcal{H}}_m$ with $d_m = y - x$ and $x(m-y) = km - 1$.

By the arithmetic-geometric mean inequality, for every $m \in \mathcal{J}_k$, we have

$$\frac{m - w_m}{2} = \frac{m - y + x}{2} \geq \sqrt{x(m-y)} = \sqrt{km-1}. \qquad (4.5)$$

Thus $\mathcal{J}_k = \varnothing$ for $k \geqslant z^2 + 1$. Suppose $1 \leqslant k < z^2 + 1$, $m \in \mathcal{J}_k$, $(x, y) \in \overline{\mathcal{H}}_m$ and $x(m-y) = km - 1$. Then

$$\sqrt{kM/2 - 1} \leqslant \max(x, m-y) \leqslant z\sqrt{M}.$$

By Lemma 2.8,

$$\#\mathcal{J}_k \leqslant H(M, \sqrt{kM/2-1}, z\sqrt{M}; \mathcal{T}_k) \ll \frac{kM(\log(3z^2/k))^{\kappa}}{\varphi(k)(\log M)^{\kappa}(\log\log M)^{3/2}}$$

which after simple calculations leads to the estimate

$$\sum_{1 \leqslant k < z^2+1} \#\mathcal{J}_k = o(M)$$

and proves the first part of the theorem.

The proof of the second part is completely analogous. □

The following questions have been formulated as conjectures in [26] where one can also find some heuristic arguments in their support.

Question 4.2. Let $g(M)$ be any positive function tending monotonically to ∞ as $M \to \infty$. Show that then the inequality

$$m - w_m \leqslant m^{1/2}(\log m)^{\kappa/2}(\log\log m)^{3/4}g(m)$$

holds:

- for all positive integers $m \leqslant M$, except for possibly $o(M)$ of them,
- for all prime $m = p \leqslant M$ except for possibly $o(M/\log M)$ of them.

Question 4.3. Prove that $m - w_m \leqslant m^{1/2}(\log m)^{\kappa/2+1/2+o(1)}$ for $m \to \infty$.

Even a weaker form of Question 4.3 with just $m^{1/2+o(1)}$ is already of great interest and would have some interesting applications, see [45].

Besides the extreme and average values of the distances between an element and its modular inverse it is also interesting to study the number of distinct differences between x and y for $(x, y) \in \overline{\mathcal{H}}_{a,m}$.

Question 4.4. Estimate the cardinality of the set

$$\mathcal{D}_{a,m} = \{|x - y| \ : \ (x, y) \in \overline{\mathcal{H}}_{a,m}\}.$$

Clearly Question 4.4 is equivalent to counting $u = 0, \dots, m-1$ for which the quadratic congruence

$$x(x + u) \equiv a \pmod m$$

has a solution x with $1 \leqslant x < m - u$.

For prime $m = p$ this question has been studied in [69] where an explicit formula

$$\#\mathcal{D}_{a,p} = \frac{1}{4}\left(p + 1 + \left(\frac{a}{p}\right)\left(1 + (-1)^{(p-1)/2}\right)\right).$$

is given, where (a/p) denotes the *Legendre symbol* of a modulo p. In [69] the same question is also studied when x and y ranges over intervals, however the method of that work does not immediately apply to the case of composite moduli m.

A similar question can also be asked about Euclidean distances between distinct points of $\overline{\mathcal{H}}_{a,m}$ and also between the origin and points of $\overline{\mathcal{H}}_{a,m}$.

Question 4.5. Estimate the cardinality of the sets

$$\mathcal{E}_{a,m} = \{\sqrt{(x_1 - x_2)^2 + (y_1 - y_2)^2} \ : \ (x_1, y_1), (x_2, y_2) \in \overline{\mathcal{H}}_{a,m}\}$$

and

$$\mathcal{F}_{a,m} = \{\sqrt{x^2 + y^2} \ : \ (x, y) \in \overline{\mathcal{H}}_{a,m}\}.$$

The author is grateful to Arne Winterhof for the observation that if $m = p$ is a prime then $\#\mathcal{F}_{a,p} = (p-1)/2$. Indeed if $u \equiv x^2 + y^2 \pmod p$ for $(x, y) \in \overline{\mathcal{H}}_{a,p}$ then the congruence $Z^4 - uZ^2 + a^2 \equiv 0 \pmod p$ has exactly four roots $x, p-x, y, p-y$ (in particular, we have two double roots if $x = y$). Since $(p-x)^2 + (p-y)^2 \neq x^2 + y^2$ we have

$$\#\mathcal{F}_{a,p} = \frac{1}{2}\#\overline{\mathcal{H}}_{a,p} = \frac{p-1}{2}. \tag{4.6}$$

Clearly, Theorem 3.1 can be used to obtain some lower bounds on $\#\mathcal{D}_{a,m}$ and $\#\mathcal{E}_{a,m}$. For example, an easy modification of an argument which leads to the bound (4.1) yields

$$\#\mathcal{D}_{a,m} \geqslant m^{1/4+o(1)}.$$

However we are mostly interested in more precise results which should certainly be based on some additional ideas.

Finally, we ask a question of a different flavour, which is about the number of possible directions on the Euclidean plane defined by the pairs of distinct points $(x_1, y_1), (x_2, y_2) \in \overline{\mathcal{H}}_{a,m}$.

Question 4.6. Estimate the cardinality of the set

$$\mathcal{L}_{a,m} = \left\{ \frac{x_1 - x_2}{y_1 - y_2} \ : \ (x_1, y_1), (x_2, y_2) \in \overline{\mathcal{H}}_{a,m}, \ (x_1, y_1) \neq (x_2, y_2) \right\}.$$

Obviously Questions 4.5 and 4.6 are influenced by the *Erdős* and *Kakeya* problems, respectively, see [13, Sections 5.3 and 7.1] and also surveys [12,42]. They can also be asked for points of $\mathcal{H}_{a,m}(\mathcal{X}, \mathcal{Y})$ for various sets $\mathcal{X}$ and $\mathcal{Y}$.

Finally, motivated by [6,7] and some other works one can also ask various questions about the distribution of the angles of elevation $\arctan(y/x)$ of points $(x, y) \in \mathcal{H}_{a,m}(\mathcal{X}, \mathcal{Y})$ over the horisontal line.

4.2. *Convex Hull*

We consider the convex closure $\mathcal{C}_m$ of the point set $\overline{\mathcal{H}}_m$. It is not hard to see that $\mathcal{C}_m$ is always a convex polygon with nonempty interior, except when $m = 2, 3, 4, 6, 8, 12, 24$, see [45].

Following [45], we denote by $v(m)$ the number of vertices of C_m and by $V(M)$ its average value,

$$V(M) = \frac{1}{M-1} \sum_{m=2}^{M} v(m).$$

Using Theorem 3.1, one can easily derive

$$v(m) \leqslant m^{3/4+o(1)}. \tag{4.7}$$

Naturally, exactly as Theorem 3.1, the bound (4.7) can be extended to sets of solutions of many other congruences.

More interestingly, one can obtain another bound, which is much better in some cases and relies on more specific properties of $\overline{\mathcal{H}}_m$

$$v(m) \leqslant T(m-1)m^{o(1)}, \tag{4.8}$$

see [45], where $T(k)$ is defined in §2.3.

The lower bound

$$v(m) \geqslant 2(\tau(m-1)-1) \tag{4.9}$$

is also given in [45]. Furthermore, it is shown in [45] that in fact $v(m) = 2(\tau(m-1)-1)$ whenever $T(m-1) \leqslant 5$. Thus, this and the bound (4.8) explain why Lemma 2.9 comes into play.

One can also find in [45] several efficient algorithms for computing $v(m)$, together with their complexity analysis.

Numerical calculations show that while the behaviour of $v(m)$ is not adequately described by any of the above bounds, the lower bound (4.9) seems to be more precise than (4.7) and (4.8).

It is quite natural to view the points of $\overline{\mathcal{H}}_m$ as being randomly distributed in the square $[0, m] \times [0, m]$ (which is supported by the theoretic results which we have presented in §3.1 and 3.2) and then appeal to the following result of Rényi and Sulanke [57, Satz 1]. Let $\mathcal{R}$ be a convex polygon in the plane with r vertices and let P_i, $i = 1, \ldots, n$, be n points chosen at random in $\mathcal{R}$ with uniform distribution. Let X_n be the number of vertices of the convex closure of the points P_i, and let $E(X_n)$ be the expectation of X_n. Then

$$E(X_n) = \frac{2}{3}r(\log n + \gamma) + c_{\mathcal{R}} + o(1), \tag{4.10}$$

where $\gamma = 0.577215\ldots$ is the Euler constant, and $c_{\mathcal{R}}$ depends on $\mathcal{R}$ and is maximal when $\mathcal{R}$ is a regular r-gon or is affine equivalent to a regular

r-gon. In particular, for the unit square $\mathcal{R} = [0,1]^2$ we have

$$c_{\mathcal{R}} = -\frac{8}{3}\log 2.$$

Using (4.10) with $r = 4$, it seems plausible to conjecture that for most m

$$v(m) \approx h(m),$$

where

$$h(m) = \frac{8}{3}(\log\varphi(m) + \gamma - \log 2).$$

However, surprisingly enough, the numerical results of [45] show that $V(M)$ deviates from

$$H(M) = \frac{1}{M-1}\sum_{m=2}^{M}\frac{8}{3}(\log\varphi(m)+\gamma-\log 2) = \frac{8}{3}(\log M+\gamma+\eta-1-\log 2),$$

where

$$\eta = \sum_{p \text{ prime}} \frac{\log(1-1/p)}{p} = -0.580058\ldots,$$

quite significantly, and is apparently larger than $H(M)$ by a fixed factor. Some partial explanation to this phenomenon has been given in [45] and suggests that for each m, the convex hull $\mathcal{C}_m$, besides some "random" points, also contains a "regular" component with $2(\tau(m-1)-1)$ points associated with divisors of $m-1$ whose average contribution

$$\frac{1}{M-1}\sum_{m=2}^{M} 2(\tau(m-1)-1) \sim 2\log M$$

is of the same order of magnitude as the size of $H(M)$.

It is also shown [45] that this affect is specific to the points of $\overline{\mathcal{H}}_m$ and disappears for the convex hull of points on a "generic" curve which behaves in much better agreement with (4.10) than $v(m)$.

Clearly since $(1,1),(m-1,m-1) \in \overline{\mathcal{H}}_m$ the diameter of $\overline{\mathcal{H}}_m$ takes the largest possible value $\sqrt{2}(m-2)$. However for the other values of a the question about the diameter of $\overline{\mathcal{H}}_{a,m}$ is more interesting.

Question 4.7. Estimate the diameter

$$\Delta_{a,m} = \max\{\sqrt{(x_1-x_2)^2+(y_1-y_2)^2} \ : \ (x_1,y_1),(x_2,y_2) \in \overline{\mathcal{H}}_{a,m}\}.$$

4.3. *Visible Points*

For two sets of integers $\mathcal{X}$ and $\mathcal{Y}$ we denote

$$\mathcal{G}_{a,m}(\mathcal{X},\mathcal{Y}) = \#\{(x,y) \in \mathcal{H}_{a,m}(\mathcal{X},\mathcal{Y}) \ : \ \gcd(x,y) = 1\}$$

and also, following our usual agreement, we put $\mathcal{G}_m(\mathcal{X},\mathcal{Y}) = \mathcal{G}_{1,m}(\mathcal{X},\mathcal{Y})$.

Clearly $\mathcal{G}_{a,m}(\mathcal{X},\mathcal{Y})$ is the set of points $(x,y) \in \mathcal{H}_{a,m}(\mathcal{X},\mathcal{Y})$ which are "visible" from the origin (that is, which are not "blocked" by other points with integer coordinates).

The following estimate is obtained in [64] for $\mathcal{G}_m(\mathcal{X},\mathcal{Y})$ but its extension to the general case is immediate and we present it here (in fact we also simplify the argument).

Theorem 4.3. *Let $\mathcal{X} = \{1,\ldots,X\}$ and $\mathcal{Y} = \{1,\ldots,Y\}$ where $1 \leqslant X, Y \leqslant m$ are arbitrary integers. For all integers m, we have*

$$\#\mathcal{G}_{a,m}(\mathcal{X},\mathcal{Y}) = \frac{6}{\pi^2}\cdot\frac{XY}{m}\prod_{p|m}\left(1+\frac{1}{p}\right)^{-1} + O\left(X^{1/2}Y^{1/2}m^{-1/4+o(1)}\right),$$

where the product is taken over all prime numbers $p \mid m$.

Proof. For an integer z we let

$$\mathcal{H}_{a,m}(z;\mathcal{X},\mathcal{Y}) = \{(x,y) \in \mathcal{H}_{a,m}(\mathcal{X},\mathcal{Y}) \ : \ z \mid \gcd(x,y)\}.$$

By the inclusion-exclusion principle, we write

$$\#\mathcal{G}_{a,m}(\mathcal{X},\mathcal{Y}) = \sum_{z=1}^{\infty}\mu(z)\#\mathcal{H}_{a,m}(z;\mathcal{X},\mathcal{Y}).$$

Clearly

$$\mathcal{H}_{a,m}(z;\mathcal{X},\mathcal{Y}) = \emptyset$$

if $\gcd(z,m) > 1$ or $z > m$. For $\gcd(z,m) = 1$, writing

$$x = zs \qquad \text{and} \qquad y = zt, \tag{4.11}$$

we have

$$\mathcal{H}_{a,m}(z;\mathcal{X},\mathcal{Y}) = \{(zs,zt) \ : \ st \equiv az^{-2} \pmod m, \\ 1 \leqslant s \leqslant X/z,\ 1 \leqslant t \leqslant Y/z\}.$$

Now, we define

$$R = \left\lceil X^{1/2}Y^{1/2}m^{-3/4}\right\rceil \qquad \text{and} \qquad Q = \left\lceil X^{1/2}Y^{1/2}m^{-1/2}\right\rceil,$$

and note that

$$XY/Q^2 \leqslant m.$$

A variant of Theorem 3.1 gives

$$\#\mathcal{H}_{a,m}(z;\mathcal{X},\mathcal{Y}) = \frac{XY\varphi(m)}{z^2m^2} + O\left(m^{1/2+o(1)}\right),$$

which we apply for "small" $z \leqslant R$.

We also note that for each z, the product $r = st \leqslant XY/z^2$, where s and t are given by (4.11), belongs to a fixed residue class modulo m and thus can take at most $XY/z^2m + 1$ possible values. For each fixed $r \leqslant XY/z^2 \leqslant XY \leqslant m^2$, there are $\tau(r) = r^{o(1)} = m^{o(1)}$ pairs (s,t) of integers s and t with $r = st$, see (2.6). Therefore,

$$\#\mathcal{H}_{a,m}(z;\mathcal{X},\mathcal{Y}) \leqslant \left(\frac{XY}{z^2m} + 1\right) m^{o(1)},$$

which we apply for "medium" z with $Q \geqslant z > R$.

Finally, we note that Lemma 2.1 with gives the bound

$$\sum_{\substack{z=1\\ \gcd(z,m)=1}}^{m} \mathbf{e}_m(Az^{-2} + Bz) \ll \gcd(A,B,m)^{1/2} m^{1/2+o(1)}.$$

Using the same arguments as in the proof of Theorem 3.1, we deduce that for the number of positive integers $z \leqslant Z$ with $az^{-2} \equiv w \pmod m$ with some $w \leqslant W$ is $ZW + O(m^{1/2+o(1)})$.

We now note that $az^{-2} \equiv r \pmod m$ where, as before,

$$r = st \leqslant XY/z^2 \leqslant XY/Q^2 \leqslant m.$$

Furthermore, for every z the value of r is uniquely defined and leads to at most $\tau(r) = m^{o(1)}$ possible pairs (s,t). Hence, the total contribution from "large" $z \geqslant Q$ can be estimated as

$$\begin{aligned}\sum_{m \geqslant z \geqslant Q} \#\mathcal{H}_{a,m}(z;\mathcal{X},\mathcal{Y}) &\leqslant \sum_{\nu=0}^{\lceil 2\log m\rceil} \sum_{2^{\nu+1}Q > z \geqslant 2^{\nu}Q} \#\mathcal{H}_{a,m}(z;\mathcal{X},\mathcal{Y}) \\ &\leqslant \sum_{\nu=0}^{\lceil 2\log m\rceil} \left(2^{\nu+1}Q \cdot \frac{XY}{(2^{\nu}Q)^2} + m^{1/2+o(1)}\right) m^{o(1)} \\ &\leqslant \sum_{\nu=0}^{\lceil 2\log m\rceil} \left(\frac{XY}{2^{\nu}Q} + m^{1/2+o(1)}\right) m^{o(1)} = \frac{XY}{Q} m^{o(1)}.\end{aligned}$$

Combining the above bounds and recalling our choice of R and Q (which optimises the error term), we derive the desired result. □

In particular, under the conditions of Theorem 4.3

$$\#\mathcal{G}_{a,m}(\mathcal{X},\mathcal{Y}) \sim \frac{6}{\pi^2} \cdot \frac{XY}{m} \prod_{p|m} \left(1+\frac{1}{p}\right)^{-1}$$

provided that $XY \geqslant m^{3/2+\varepsilon}$ for some fixed $\varepsilon > 0$.

There is little doubt that our approach can also be used to obtain asymptotic formulas for $(x,y) \in \mathcal{H}_{a,m}(\mathcal{X},\mathcal{Y})$ with arbitrary integer a, and also the sums

$$\sum_{(x,y)\in\mathcal{H}_m(\mathcal{X},\mathcal{Y})} |\mu(xy)| \qquad \text{and} \qquad \sum_{(x,y)\in\mathcal{H}_m(\mathcal{X},\mathcal{Y})} |\mu(x)\mu(y)| \tag{4.12}$$

under the same conditions on the sets $\mathcal{X}$ and $\mathcal{Y}$ as in Theorem 4.3, see also Question 3.4. However, we do not have any approaches to the following problem.

Question 4.8. Extend Theorem 4.3 to intervals of the form

$$\mathcal{X} = \{U+1,\ldots,U+X\} \qquad \text{and} \qquad \mathcal{Y} = \{V+1,\ldots,V+Y\}$$

with arbitrary U and V.

It is also interesting to study sums of other arithmetic functions on $(x,y) \in \mathcal{H}_m(\mathcal{X},\mathcal{Y})$.

Question 4.9. For intervals

$$\mathcal{X} = \{U+1,\ldots,U+X\} \qquad \text{and} \qquad \mathcal{Y} = \{V+1,\ldots,V+Y\}$$

of length $X, Y \leqslant m$,

- obtain nontrivial bounds for the sums

$$\sum_{(x,y)\in\mathcal{H}_m(\mathcal{X},\mathcal{Y})} \mu(xy) \qquad \text{and} \qquad \sum_{(x,y)\in\mathcal{H}_m(\mathcal{X},\mathcal{Y})} \left(\frac{x}{y}\right),$$

  where (x/y) is the *Jacobi symbol* of x modulo y, which we also extend to even values of y by simply putting $(x/y) = 0$ in this case;
- obtain asymptotic formulas for the sums

$$\sum_{(x,y)\in\mathcal{H}_m(\mathcal{X},\mathcal{Y})} \varphi(|x-y|) \qquad \text{and} \qquad \sum_{(x,y)\in\mathcal{H}_m(\mathcal{X},\mathcal{Y})} \omega(|x-y|).$$

5. Applications

5.1. *Lehmer Problem*

One of the most natural and immediate applications of the uniformity of distribution results outlined in §3.1 is a positive solution to the *Lehmer problem*, see [37, Problem F12], about the joint distribution of the parity of x and y for $(x, y) \in \mathcal{H}_{a,m}(\mathcal{X}, \mathcal{Y})$ with some intervals

$$\mathcal{X} = \{U+1, \ldots, U+X\} \qquad \text{and} \qquad \mathcal{Y} = \{V+1, \ldots, V+Y\}.$$

This distribution is naturally expected to be close to uniform (that is, each parity combination is taken about in 25% of the cases) for any odd m and sufficiently large X and Y. The Lehmer problem can easily be reformulated as a question about the cardinality of $\#\mathcal{H}_{a,m}(\widetilde{\mathcal{X}}, \widetilde{\mathcal{Y}})$ for some a and some other sets $\widetilde{\mathcal{X}}$ and $\widetilde{\mathcal{Y}}$ of about $X/2$ and $Y/2$ consecutive integers, respectively. Indeed, we are interested in solutions to the congruence

$$(2\widetilde{x} + \vartheta_1)(2\widetilde{y} + \vartheta_2) \equiv 1 \pmod m$$

with some fixed $\vartheta_1, \vartheta_2 \in \{0, 1\}$ and

$$\frac{U+1-\vartheta_1}{2} \leqslant \widetilde{x} \leqslant \frac{U+X-\vartheta_1}{2}, \qquad \frac{V+1-\vartheta_2}{2} \leqslant \widetilde{y} \leqslant \frac{V+Y-\vartheta_2}{2}.$$

It remains to notice that the above congruence is equivalent to

$$(\widetilde{x} + 2^{-1}\vartheta_1)(\widetilde{y} + 2^{-1}\vartheta_2) \equiv 4^{-1} \pmod m$$

(where the inversion is taken modulo m), which after a shift of variables takes the desired shape.

Close links between the Lehmer problem and bounds of Kloosterman sums has been first observed in [79,80]. The question and the above approach have been extended in several directions. Accordingly, instead of the bound (1.1) more general bounds of incomplete or multiple Kloosterman sums are used, see [2,19,20,48,49,51,52,75,81–84,86] and references therein. However, it has turned out that for multivariate analogues of the Lehmer problem, and a number of similar questions, bounds of character sums provide a more efficient tool than Kloosterman sums, see [67], where Lemmas 2.2 and 2.3 and the bound (2.4) have been used to improve several previous results.

5.2. *Distribution of Angles in Some Point Sets*

A version of Theorem 3.1 has been used in [7] to study the distribution of angles between visible points (viewed from the origin) in a dilation of a certain plain region $\Omega \in [0,1]^2$. More precisely, for $\Omega \in [0,1]^2$ and a sufficiently large Q, we define

$$\Omega_Q = \{(Q\alpha, Q\beta) \ : \ (\alpha, \beta) \in \Omega)\}.$$

We now consider $\#\mathcal{F}_\Omega(Q)$ angles between the horisontal line and the points of the set

$$\mathcal{F}_\Omega(Q) = \{(x,y) \in \Omega_Q \cap \mathbb{Z}^2 \ : \ \gcd(x,y) = 1\}.$$

The distribution of this angles is studied in [7] and shown to exhibit somewhat unexpected behaviour.

A slight generalisation of Theorem 3.1 has also played an important role in a study of angles defined by some other point set [6], see also [8].

5.3. *Sums of Divisor Functions over Arithmetic Progressions*

For example, the strength of estimates on the error terms in asymptotic formulas for sums of some divisor functions over an arithmetic progression is closely related to the precision of our knowledge of $\#\mathcal{H}_{a,m}(\mathcal{X}, \mathcal{Y})$, where $\mathcal{X}$ and $\mathcal{Y}$ are sets of consecutive integers, see [4,27,28]. This link becomes transparent if one recalls that a standard approach to evaluating divisor sums is approximating the hyperbolic region $\{(x,y) \ : \ x,y \geqslant 0, \ xy \leqslant N\}$ by a union of "small" rectangles. In fact, as we have mentioned, the proof of Theorem 3.2 is based on some ideas of [4].

5.4. *Sato-Tate Conjecture in the "Vertical" Aspect*

We also recall that Theorem 3.2 about the distribution of points on $\mathcal{H}_{a,m}$ on average over a have been used in studying the so-called "vertical" aspect of the Sato-Tate conjecture for Kloosterman sums, see [68]. The relation is provided by the identity $K_m(r,s) = K_m(1,rs)$ which holds if $\gcd(r,m) = 1$. Clearly for the complex conjugated sum we have

$$\overline{K_m(r,s)} = K_m(-r,-s) = K_m(r,s),$$

hence we see that $K_m(r,s)$ is real. Since by the Weil bound, see [39], for any prime p, we have

$$|K_p(r,s)| \leqslant 2\sqrt{p}, \qquad \gcd(r,s,p) = 1,$$

we can now define the angles $\psi_p(r,s)$ by the relations

$$K_p(r,s) = 2\sqrt{p}\cos\psi_p(r,s) \qquad \text{and} \qquad 0 \leqslant \psi_p(r,s) \leqslant \pi.$$

The famous *Sato–Tate* conjecture asserts that, in the "horizontal" aspect, that is, for any fixed non-zero integers r and s, the angles $\psi_p(r,s)$ are distributed according to the *Sato–Tate density*

$$\mu_{ST}(\alpha,\beta) = \frac{2}{\pi}\int_{\alpha}^{\beta} \sin^2\gamma\, d\gamma,$$

see [39, Section 21.2]. More precisely, if $\pi_{r,s}(\alpha,\beta;T)$ denotes the number of primes $p \leqslant T$ with $\alpha \leqslant \psi_p(r,s) \leqslant \beta$, where, as usual $\pi(T)$ denotes the total number of primes $p \leqslant T$, the Sato–Tate conjecture predicts that

$$\pi_{r,s}(\alpha,\beta;T) \sim \mu_{ST}(\alpha,\beta)\pi(T), \qquad T \to \infty, \tag{5.1}$$

for all fixed real $0 \leqslant \alpha < \beta \leqslant \pi$, see [39, Section 21.2]. We remark that for elliptic curves, the Sato-Tate conjecture in its original (and the most difficult) "horizontal" aspect has recently been settled [70], but it still remains open for Kloosterman sums. It is shown in [68] that Theorem 3.2 implies that

$$\frac{1}{4RS}\sum_{0<|r|\leqslant R}\ \sum_{0<|s|\leqslant S} \pi_{r,s}(\alpha,\beta;T) \sim \mu_{ST}(\alpha,\beta)\pi(T)$$

provided $RS \geqslant T^{1+\varepsilon}$ for some fixed $\varepsilon > 0$.

5.5. *Torsion of Elliptic Curves*

A variant of Theorem 3.1 has been obtained in [54] and applied to estimating torsion of elliptic curves. More precisely, let $\mathcal{E}$ be an elliptic curve over an algebraic number field $\mathbb{K}$ of degree d over $\mathbb{Q}$. It is shown in [54] that if $\mathcal{E}$ contains a point of prime order p then

$$p \leqslant d^{3d^2}.$$

5.6. *Approximations by Sums of Two Rationals*

Following [15], we consider the problem of obtaining an upper bound on the approximation of a real α by s rational fractions with denominators at most Q, that is for

$$\delta_{\alpha,s}(Q) = \min_{1\leqslant q_1,\ldots,q_s\leqslant Q}\left|\alpha - \frac{r_1}{q_1} - \ldots - \frac{r_s}{q_s}\right|$$

with positive integers $q_1,\ldots,q_s \leqslant Q$. The question is motivated by the Dirichlet theorem on rational approximations which corresponds to the case $s = 1$.

It is shown in [15] that for $s = 2$ the results on the distribution of points on $\mathcal{H}_{m,a}$ are directly related to this question and imply nontrivial bounds on $\delta_{\alpha,2}(Q)$.

It is natural to expect that there are also links between $\delta_{\alpha,s}(Q)$ and multivariate analogues of the sets $\mathcal{H}_{m,a}$.

5.7. $\mathbf{SL_2(\mathbb{F}_p)}$ *Matrices of Bounded Height*

For a finite field $\mathbb{F}_p$ of p elements and a positive integer $T \leqslant (p-1)/2$, we use $N_p(T)$ to denote the number of matrices

$$\begin{pmatrix} u & v \\ x & y \end{pmatrix} \in \mathrm{SL}_2(\mathbb{F}_p)$$

with $|u|, |v|, |x|, |y| \leqslant T$ (we assume that $\mathbb{F}_p$ is represented by the elements of the set $\{0, \pm 1, \ldots, \pm(p-1)/2\}$).

Clearly

$$N_p(T) = \sum_{a \in \mathbb{F}_p} \#\mathcal{H}_{a+1,p}(\mathcal{T}, \mathcal{T}) \#\mathcal{H}_{a,p}(\mathcal{T}, \mathcal{T})$$

where $\mathcal{T} = \{0, \pm 1, \ldots, \pm T\}$ and also $\#\mathcal{H}_{0,p}(\mathcal{T}, \mathcal{T}) = 4T + 1$.

It has been shown in [1] that using the identity

$$\sum_{a \in \mathbb{F}_p} \#\mathcal{H}_{a,p}(\mathcal{T}, \mathcal{T}) = (2T+1)^2$$

and Theorem 3.2, one can derive that

$$N_p(T) = \frac{(2T+1)^4}{p} + O\left(T^2 p^{o(1)}\right)$$

which is nontrivial if $T \geqslant p^{1/2+\varepsilon}$ for any fixed $\varepsilon > 0$ and sufficiently large p. We remark that this is an $\mathbb{F}_p$ analogue of the results of similar spirit for matrices over $\mathbb{Z}$ and algebraic number fields, see [23,58] and references therein.

5.8. *Computing Discrete Logarithms and Factoring*

It has been discovered in [62] that the distribution of points on hyperbolas $\mathcal{H}_{a,m}$ has direct links with analysis of some discrete logarithm algorithms. Namely, analysis of the algorithm from [62] rests on a weaker version of Theorem 3.2 obtained in [61].

Question 5.1. Study whether Theorem 3.2 can be used to improve some of the results of [62].

A new deterministic integer factorisation algorithm has recently been suggested in [59]. Its analysis depends on a variant of Theorem 3.1.

6. Concluding Remarks

6.1. *Generalisations*

Multidimensional variants of the above problems have also been considered and in principle one can use bounds of multidimensional Kloosterman sums to study the distribution of solutions to the congruence

$$x_1 \dots x_s \equiv a \pmod m \tag{6.1}$$

in the same fashion as in the case of two variables. However, it has turned out that in many cases bounds of multiplicative character sums lead to much stronger results. For examples, in [67], using the bounds of Lemmas 2.2 and 2.3 and the bound (2.4), an improvement is given of some results of [2] on generalised Lehmer problem. Similar ideas have also led in [65] to some other results on distribution of solutions to (6.1). For instance, let $\Delta_{s,a,m}$ denote the discrepancy of the s-dimensional points

$$\left(\frac{x_1}{m}, \dots, \frac{x_s}{m}\right) \in [0,1]^s, \qquad x_1 \dots x_s \equiv a \pmod m, \ 1 \leqslant x_1, \dots, x_s \leqslant m.$$

It is shown in [65] that

$$\Delta_{s,a,m} \leqslant \begin{cases} m^{-1/2+o(1)} & \text{if } s = 3, \\ m^{-1+o(1)} & \text{if } s \geqslant 4. \end{cases}$$

It is worth noticing that one has to be careful with posing multidimensional generalisations, which sometimes lead to rather simple questions for $s \geqslant 3$ (despite that for $s = 2$ they are nontrivial at all). For example, it has been noticed in [65] that often such generalisations do not need any analytic technique used in [78] but follows immediately from a very elementary argument.

Another unexplored line of research is in the direction of function field generalisations, which conceivably should admit results of the same strength or maybe even stronger. For example, a polynomial analogue of the conjecture of [24] could be more accessible.

Question 6.1. Let $\mathbb{F}_q$ be a finite field of q elements. Given an irreducible polynomial $F(X) \in \mathbb{F}_q[X]$ of sufficiently large degree d, show that for any polynomial $A(X) \in \mathbb{F}_q[X]$, relatively prime to $F(X)$, there are two irreducible polynomials $G(X), H(X) \in \mathbb{F}_q[X]$ of degree at most d such that

$$G(X)H(X) \equiv A(X) \pmod{F(X)}.$$

Finally, we believe that it is interesting to explore how much of the theory developed for modular hyperbolas can be extended to modular circles

$$\mathcal{C}_{a,m} = \{(x,y) \ : \ x^2 + y^2 \equiv a \pmod{m}\}.$$

Well known parallels between the properties of integer points on hyperbolas and circles on the Euclidean plane, suggest that many of the results obtained for $\mathcal{H}_{a,m}$ can be extended to $\mathcal{C}_{a,m}$.

6.2. *Further Improvements*

We have seen that bounds of Kloosterman sums play a prominent role in this field. It would be interesting to find some applications of bounds of very short incomplete Kloosterman sums from [11,40,41,46,53,66].

In [76,85] some questions are studied about the distribution of residues modulo m of powers (x^k, y^k) taken over all $(x,y) \in \overline{\mathcal{H}}_{a,m}$. These results can be substantially extended and improved if one uses Lemma 2.1. In particular, Lemma 2.1 immediately implies that the error term in the asymptotic formula of [85] for the quantity $N(k,q)$ can be lowered from $q^{3/4+o(1)}$ to $q^{1/2+o(1)}$. It may also be extended and generalised in various directions. All these results can be obtained by combining Lemma 2.1 with standard arguments like those used in the proof of Theorem 3.1.

References

1. O. Ahmadi and I. E. Shparlinski, Distribution of matrices with restricted entries over finite fields, *Preprint*, 2006.
2. E. Alkan, F. Stan and A. Zaharescu, Lehmer k-tuples, *Proc. Amer. Math. Soc.*, **134** (2006), 2807–2815.
3. A. Ayyad, T. Cochrane and Z. Zheng, The congruence $x_1x_2 \equiv x_3x_4 \pmod{p}$, the equation $x_1x_2 = x_3x_4$ and the mean value of character sums, *J. Number Theory*, **59** (1996), 398–413.
4. W. D. Banks, R. Heath-Brown and I. E. Shparlinski, On the average value of divisor sums in arithmetic progressions, *Intern. Math. Research Notices*, **2005** (2005), 1–25.
5. J. Beck and M. R. Khan, On the uniform distribution of inverses modulo n, *Period. Math. Hung.*, **44** (2002), 147–155.
6. F. P. Boca, On the distribution of angles between geodesic rays associated with hyperbolic lattice points, *Preprint*, 2006.
7. F. P. Boca, C. Cobeli and A. Zaharescu, Distribution of lattice points visible from the origin, *Commun. Math. Phys.*, **213** (2000), 433–470.
8. F. P. Boca and A. Zaharescu, Farey fractions and two-dimensional tori, *Noncommutative Geometry and Number Theory*, Aspects of Mathematics E37, Vieweg Verlag, Wiesbaden, 2006, 57–77.

9. E. Bombieri, On exponential sums in finite fields, *Amer. J. Math.*, **88** (1966), 71–105.
10. E. Bombieri, J. Friedlander and H. Iwaniec, Primes in arithmetic progressions to large moduli, III, *J. Amer. Math. Soc.*, **2** (1989), 215–224.
11. J. Bourgain, More on the sum-product phenomenon in prime fields and its applications, *Intern. J. Number Theory*, **1** (2005), 1–32.
12. J. Bourgain, New encounters in combinatorial number theory: From the Kakeya problem to cryptography, *Perspectives in Analysis*, Mathematical Physics Studies, vol. 27, Springer-Verlag, Berlin, 2005, 17–26.
13. P. Brass, W. Moser and J. Pach, *Research problems in discrete geometry*, Springer, New York, 2005.
14. T. H. Chan, Distribution of difference between inverses of consecutive integers modulo p, *Integers*, **4** (2004), Paper A03, 1–11.
15. T. H. Chan, Approximating reals by sums of two rationals, *Preprint*, 2006.
16. C. Cobeli, S. Gonek and A. Zaharescu, The distribution of patterns of inverses modulo a prime, *J. Number Theory*, **101** (2003), 209–222.
17. C. Cobeli, M. Vâjâitu and A. Zaharescu, Average estimates for the number of tuples of inverses mod$p$ in short intervals, *Bull. Math. Soc. Sci. Math. Roumanie.*, **43** (2000), 155–164.
18. C. Cobeli, M. Vâjâitu and A. Zaharescu, Distribution of gaps between the inverses $\bmod q$, *Proc. Edinb. Math. Soc.*, **46** (2003), 185–203.
19. C. Cobeli and A. Zaharescu, The order of inverses $\bmod q$, *Mathematika*, **47** (2000), 87–108.
20. C. Cobeli and A. Zaharescu, Generalization of a problem of Lehmer, *Manuscr. Math.*, **104** (2001), 301–307.
21. C. Cobeli and A. Zaharescu, On the distribution of the $\mathbb{F}_p$-points on an affine curve in r dimensions, *Acta Arithmetica*, **99** (2001), 321–329.
22. M. Drmota and R. Tichy, *Sequences, discrepancies and applications*, Springer-Verlag, Berlin, 1997.
23. W. Duke, Z. Rudnick and P. Sarnak, Density of integer points on affine homogeneous varieties, *Duke Math. J.*, **71** (1993), 143–179.
24. P. Erdős, A. M. Odlyzko and A. Sárközy, On the residues of products of prime numbers, *Period. Math. Hung.*, **18** (1987), 229–239.
25. K. Ford, The distribution of integers with a divisor in a given interval, *Ann. Math.*, (to appear).
26. K. Ford, M. R. Khan, I. E. Shparlinski and C. L. Yankov, On the maximal difference between an element and its inverse in residue rings, *Proc. Amer. Math. Soc.*, **133** (2005), 3463–3468.
27. J. B. Friedlander and H. Iwaniec, Incomplete Kloosterman sums and a divisor problem, *Ann. of Math.*, **121** (1985), 319–350.
28. J. B. Friedlander and H. Iwaniec, The divisor problem for arithmetic progressions, *Acta Arith.*, **45** (1985), 273–277.
29. A. Fujii and Y. Kitaoka, On plain lattice points whose coordinates are reciprocals modulo a prime, *Nagoya Math. J.*, **147** (1997), 137–146.
30. M. Z. Garaev, Character sums in short intervals and the multiplication table modulo a prime, *Monatsh. Math.*, **148** (2006), 127–138.

31. M. Z. Garaev, On the logarithmic factor in error term estimates in certain additive congruence problems, *Acta Arith.*, **124** (2006), 27–39.
32. M. Z. Garaev and V. Garcia, The equation $x_1x_2 = x_3x_4 + \lambda$ in fields of prime order and applications, *Preprint*, 2007.
33. M. Z. Garaev and A. A. Karatsuba, On character sums and the exceptional set of a congruence problem, *J. Number Theory*, **114** (2005), 182–192.
34. M. Z. Garaev and A. A. Karatsuba, The representation of residue classes by products of small integers, *Proc. Edin. Math. Soc.*, (to appear).
35. M. Z. Garaev and K.-L. Kueh, Distribution of special sequences modulo a large prime, *Int. J. Math. Math. Sci.*, **50** (2003), 3189–3194.
36. A. Granville, I. E. Shparlinski and A. Zaharescu, On the distribution of rational functions along a curve over $\mathbb{F}_p$ and residue races, *J. Number Theory*, **112** (2005), 216–237.
37. R. K. Guy, *Unsolved problems in number theory*, Springer-Verlag, Berlin, 1994.
38. R. Hall and G. Tenenbaum , *Divisors*, Cambridge Tracts in Math. **90**, Cambridge Univ. Press, 1988.
39. H. Iwaniec and E. Kowalski, *Analytic number theory*, Amer. Math. Soc., Providence, RI, 2004.
40. A. A. Karatsuba, Fractional parts of functions of a special form, *Izv. Ross. Akad. Nauk Ser. Mat. (Transl. as Russian Acad. Sci. Izv. Math.)*, (4)**55** (1995), 61–80 (in Russian).
41. A. A. Karatsuba, Analogues of Kloosterman sums, *Izv. Ross. Akad. Nauk Ser. Mat. (Transl. as Russian Acad. Sci. Izv. Math.)*, (5)**55** (1995), 93–102 (in Russian).
42. N. Katz and T. Tao, Recent progress on the Kakeya conjecture, *Proceedings of the 6th International Conference on Harmonic Analysis and Partial Differential Equations* Publ Matem., U. Barcelona, 2002, 161–180.
43. M. R. Khan, Problem 10736: An optimization with a modular constraint, *Amer. Math. Monthly*, **108** (2001), 374–375.
44. M. R. Khan and I. E. Shparlinski, On the maximal difference between an element and its inverse modulo n, *Period. Math. Hung.*, **47** (2003), 111–117.
45. M. R. Khan, I. E. Shparlinski and C. L. Yankov, On the convex closure of the graph of modular inversions, *Preprint*, 2006.
46. M. A. Korolev, Incomplete Kloosterman sums and their applications, *Izv. Ross. Akad. Nauk Ser. Mat. (Transl. as Russian Acad. Sci. Izv. Math.)*, (6)**64** (2000), 41–64.
47. M. Laczkovich, Discrepancy estimates for sets with small boundary, *Studia Sci. Math. Hungar.*, **30** (1995), 105–109.
48. H. N. Liu and W. Zhang, On a problem of D. H. Lehmer, *Acta Math. Sinica*, **22** (2006), 61–68.
49. H. N. Liu and W. Zhang, Hybrid mean value on the difference between a quadratic residue and its inverse modulo p, *Publ. Math. Debrecen*, **69** (2006), 227–243.
50. H. N. Liu and W. Zhang, General Kloosterman sums and the difference between an integer and its inverse modulo q, *Acta Math. Sinica*, **23** (2007),

77–82.
51. H. N. Liu and W. Zhang, Mean value on the difference between a quadratic residue and its inverse modulo p, *Acta Math. Sinica*, **23** (2007), 915–924.
52. S. R. Louboutin, J. Rivat and A. Sárközy, On a problem of D. H. Lehmer, *Proc. Amer. Math. Soc.*, **135** (2007), 969–975.
53. W. Luo, Bounds for incomplete hyper-Kloosterman sums, *J. Number Theory*, **75** (1999), 41–46.
54. L. Merel, Bornes pour la torsion des courbes elliptiques sur les corps de nombres, *Invent. Math.*, **124** (1996), 437–449.
55. N. G. Moshchevitin, On numbers with missing digits: Solvability of the congruences $x_1x_2 \equiv \lambda \pmod p$, *Doklady Akad. Nauk*, **410** (2006), 730–733 (in Russian).
56. H. Niederreiter and J. M. Wills, Diskrepanz und Distanz von Massen bezuglich konvexer und Jordanscher Mengen, *Math. Z.*, **144** (1975), 125–134.
57. A. Rényi and R. Sulanke, Über die Konvexe Hülle von n Zufällig Gewaählten Punkten, *Z. Wahrscheinlichkeitstheorie*, **2** (1963), 75–84.
58. C. Roettger, Counting invertible matrices and uniform distribution, *J. Théorie Nombres Bordeaux*, **17** (2005), 301–322.
59. M. Rubinstein, Hide and seek – A naive factoring algorithm, *Preprint*, 2006.
60. É. Saias, Entiers á Diviseurs Denses 1, *J. Number Theory*, **62** (1997), 163–191.
61. I. A. Semaev, On the number of small solutions of a linear homogeneous congruence, *Mat. Zametki*, **50** (1991), no.4, 102–107, (in Russian).
62. I. A. Semaev, An algorithm for evaluation of discrete logarithms in some nonprime finite fields, *Math. Comp.*, **67** (1998), 1679–1689.
63. I. E. Shparlinski, On exponential sums with sparse polynomials and rational functions, *J. Number Theory*, **60** (1996), 233–244.
64. I. E. Shparlinski, Primitive points on a modular hyperbola, *Bull. Polish Acad. Sci. Math.*, **54** (2006), 193–200.
65. I. E. Shparlinski, On the distribution of points on multidimensional modular hyperbolas, *Proc. Japan Acad. Sci., Ser.A*, **83** (2007), 5–9.
66. I. E. Shparlinski, Bounds of incomplete multiple Kloosterman sums, *J. Number Theory*, (to appear).
67. I. E. Shparlinski, On a generalisation of a Lehmer problem, *Preprint*, 2006.
68. I. E. Shparlinski, Distribution of inverses and multiples of small integers and the Sato–Tate conjecture on average, *Preprint*, 2006.
69. I. E. Shparlinski and A. Winterhof, Distances between the points on modular hyperbolas, *J. Number Theory*, (to appear).
70. R. Taylor, Automorphy for some l-adic lifts of automorphic mod l representations, II, *Preprint*, 2006.
71. G. Tenenbaum, *Introduction to analytic and probabilistic number theory*, Cambridge Univ. Press, 1995.
72. M. Vajaitu and A. Zaharescu, Distribution of values of rational maps on the $\mathbb{F}_p$-points on an affine curve, *Monatsh. Math.*, **136** (2002), 81–86.
73. Y. Weili, On the generalization of the D. H. Lehmer problem and its mean value, *J. Algebra Number Theory and Appl.*, **6** (2006), 479–491.

74. H. Weyl, On the volume of tubes, *Amer. J. Math.*, **61** (1939), 461–472.
75. Z. Xu and W. Zhang, On a problem of D. H. Lehmer over short intervals, *J. Math. Anal. Appl.*, **320** (2006), 756–770.
76. Y. Yi and W. Zhang, On the generalization of a problem of D. H. Lehmer, *Kyushu J. Math.*, **56** (2002), 235–241.
77. A. Zaharescu, The distribution of the values of a rational function modulo a big prime, *J. Theor. Nombres Bordeaux*, bf 15 (2003), 863–872.
78. T. Zhang and W. Zhang, A generalization on the difference between an integer and its inverse modulo q, II, *Proc. Japan Acad. Sci., Ser.A*, **81** (2005), 7–11.
79. W. Zhang, On a problem of D. H. Lehmer and its generalization, *Compos. Math.*, **86** (1993), 307–316.
80. W. Zhang, On a problem of D. H. Lehmer and its generalization, II, *Compos. Math.*, **91** (1994), 47–56.
81. W. Zhang, On the difference between a D. H. Lehmer number and its inverse modulo q, *Acta Arith.*, **68** (1994), 255–263.
82. W. Zhang, On the difference between an integer and its inverse modulo n, *J. Number Theory*, **52** (1995), 1–6.
83. W. Zhang, On the distribution of inverses modulo n, *J. Number Theory*, **61** (1996), 301–310.
84. W. Zhang, On a problem of P. Gallagher, *Acta Math. Hung.*, **78** (1998), 345–357.
85. W. Zhang, On the distribution of inverses modulo p, *Acta Arith.*, **100** (2001), 189–194.
86. W. Zhang, On a problem of D. H. Lehmer and Kloosterman sums, *Monatsh. Math.*, **139** (2003), 247–257.
87. W. Zhang, Z. Xu and Y. Yi, A problem of D. H. Lehmer and its mean square value formula, *J. Number Theory*, **103** (2003), 197–213.
88. Z. Zheng, The distribution of zeros of an irreducible curve over a finite field, *J. Number Theory*, **59** (1996), 106–118.

A SURVEY OF PROBLEMS AND RESULTS ON RESTRICTED SUMSETS

ZHI-WEI SUN

Department of Mathematics, Nanjing University,
Nanjing 210093, China
E-mail: zwsun@nju.edu.cn

Additive number theory is currently an active field related to combinatorics. In this paper we give a survey of problems and results concerning lower bounds for cardinalities of various restricted sumsets with elements in a field or an abelian group.

1. Erdős-Heilbronn conjecture and the polynomial method

Let $A = \{a_1, \ldots, a_k\}$ and $B = \{b_1, \ldots, b_l\}$ be two finite subsets of $\mathbb{Z}$ with $a_1 < \cdots < a_k$ and $b_1 < \cdots < b_l$. Observe that

$$a_1 + b_1 < a_2 + b_1 < \cdots < a_k + b_1 < a_k + b_2 < \cdots < a_k + b_l,$$

whence we see that the sumset

$$A + B = \{a + b\colon a \in A \text{ and } b \in B\}$$

contains at least $k + l - 1$ elements. In particular, $|2A| \geq 2|A| - 1$, where $|A|$ denotes the cardinality of A, and $2A$ stands for $A + A$.

The following fundamental theorem was first proved by A. Cauchy [9] in 1813 and then rediscovered by H. Davenport [11] in 1935.

Cauchy-Davenport Theorem. *Let A and B be non-empty subsets of the field $\mathbb{Z}/p\mathbb{Z}$ where p is a prime. Then*

$$|A + B| \geq \min\{p,\ |A| + |B| - 1\}. \tag{1.1}$$

For lots of important results on sumsets over $\mathbb{Z}$, the reader is referred to the recent book [38] by T. Tao and V. H. Vu. In this paper we mainly focus our attention on restricted sumsets with elements in a field or an abelian group.

In combinatorics, for a finite sequence $\{A_i\}_{i=1}^n$ of sets, a sequence $\{a_i\}_{i=1}^n$ is called a *system of distinct representatives* of $\{A_i\}_{i=1}^n$ if $a_1 \in A_1, \ldots, a_n \in A_n$ and $a_1, \ldots, a_n$ are distinct. A fundamental theorem of P. Hall [17] states that $\{A_i\}_{i=1}^n$ has a system of distinct representatives if and only if $|\bigcup_{i\in I} A_i| \geq |I|$ for all $I \subseteq \{1, \ldots, n\}$. The reader may consult [31] for a simple proof of Hall's theorem.

In 1964 P. Erdős and H. Heilbronn [13] made the following challenging conjecture.

Erdős-Heilbronn Conjecture. *Let p be a prime, and let A be a nonempty subset of the field $\mathbb{Z}/p\mathbb{Z}$. Then $|2^\wedge A| \geq \min\{p, 2|A| - 3\}$, where*

$$2^\wedge A = \{a + b : \ a, b \in A \text{ and } a \neq b\}.$$

This conjecture remained open until it was confirmed by Dias da Silva and Y. Hamidoune [12] thirty years later, with the help of the representation theory of groups.

For a general field F, the additive order of the (multiplicative) identity of F is either infinite or a prime, which we denote by $p(F)$. The *characteristic* of the field F is defined as follows:

$$\mathrm{ch}(F) = \begin{cases} p & \text{if } p(F) \text{ is a prime } p, \\ 0 & \text{if } p(F) = \infty. \end{cases} \tag{1.2}$$

Now we state Dias da Silva and Y. Hamidoune's extension of the Erdős-Heilbronn conjecture.

Dias da Silva–Hamidoune Theorem ([12]). *Let F be a field, and let $n \in \mathbb{Z}^+ = \{1, 2, 3, \ldots\}$. Then, for any finite subset A of F, we have*

$$|n^\wedge A| \geq \min\{p(F),\ n|A| - n^2 + 1\}, \tag{1.3}$$

where $n^\wedge A$ denotes the set of all sums of n distinct elements of A.

If p is a prime, $A \subseteq \mathbb{Z}/p\mathbb{Z}$ and $|A| > \sqrt{4p - 7}$, then by the Dias da Silva–Hamidoune theorem, any element of $\mathbb{Z}/p\mathbb{Z}$ can be written as a sum of $\lfloor |A|/2 \rfloor$ distinct elements of A (see [12]), where $\lfloor \cdot \rfloor$ is the well-known floor function.

In 1995–1996 N. Alon, M. B. Nathanson and I. Z. Ruzsa [4,5] developed a polynomial method rooted in [6] to prove the Erdős-Heilbronn conjecture and some similar results. The method turns out to be very powerful and has many applications in number theory and combinatorics.

Now we introduce the above-mentioned polynomial method. We begin with a lemma.

Lemma 1.1 (Alon, Nathanson and Ruzsa [4,5]). *Let F be a field and let $A_1,\ldots,A_n$ be finite non-empty subsets of F. Let $f(x_1,\ldots,x_n)\in F[x_1,\ldots,x_n]$ have degree smaller than $k_i=|A_i|$ in x_i for each $i=1,\ldots,n$. If $f(a_1,\ldots,a_n)=0$ for all $a_1\in A_1,\ldots,a_n\in A_n$, then $f(x_1,\ldots,x_n)$ is identically zero.*

This lemma can be proved by using induction on n and noting that a non-zero polynomial $P(x)\in F[x]$ of degree smaller than a positive integer k cannot have k distinct zeros in F.

The central part of the polynomial method is the following important principle formulated by Alon in 1999.

Combinatorial Nullstellensatz (Alon [1]). *Let $A_1,\ldots,A_n$ be finite subsets of a field F, and let $f(x_1,\ldots,x_n)\in F[x_1,\ldots,x_n]$.*

(i) *Set $g_i(x)=\prod_{a\in A_i}(x-a)$ for $i=1,\ldots,n$. Then*

$$f(a_1,\ldots,a_n)=0 \quad \text{for all } a_1\in A_1,\ldots,a_n\in A_n \tag{1.4}$$

if and only if there are $h_1(x_1,\ldots,x_n),\ldots,h_n(x_1,\ldots,x_n)\in F[x_1,\ldots,x_n]$ with $\deg h_i\le\deg f-\deg g_i$ for $i=1,\ldots,n$, such that

$$f(x_1,\ldots,x_n)=\sum_{i=1}^n g_i(x_i)h_i(x_1,\ldots,x_n). \tag{1.5}$$

(ii) *Suppose that $\deg f=k_1+\cdots+k_n$ where $0\le k_i<|A_i|$ for $i=1,\ldots,n$. If (1.4) holds then*

$$[x_1^{k_1}\cdots x_n^{k_n}]f(x_1,\ldots,x_n)=0,$$

where $[x_1^{k_1}\cdots x_n^{k_n}]f(x_1,\ldots,x_n)$ denotes the coefficient of $x_1^{k_1}\cdots x_n^{k_n}$ in $f(x_1,\ldots,x_n)$.

Proof. (i) If there are $h_1(x_1,\ldots,x_n),\ldots,h_n(x_1,\ldots,x_n)\in F[x_1,\ldots,x_n]$ such that (1.5) holds, then for any $a_1\in A_1,\ldots,a_n\in A_n$ we have

$$f(a_1,\ldots,a_n)=\sum_{i=1}^n g_i(a_i)h_i(a_1,\ldots,a_n)=0.$$

Now we consider the converse. Write

$$f(x_1,\ldots,x_n)=\sum_{j_1,\ldots,j_n\ge 0} f_{j_1,\ldots,j_n}x_1^{j_1}\ldots x_n^{j_n}$$

and

$$x^j=g_i(x)q_{ij}(x)+r_i^{(j)}(x),$$

where $q_{ij}(x), r_i^{(j)}(x) \in F[x]$ and $\deg r_i^{(j)}(x) < \deg g_i(x) = |A_i|$. Note that both $r_i^{(j)}(x)$ and $g_i(x)q_{ij}(x) = x^j - r_i^{(j)}(x)$ have degree not exceeding j. Clearly

$$\begin{aligned}f(x_1, \ldots, x_n) = & \sum_{\substack{j_1,\ldots,j_n \geqslant 0 \\ j_1+\cdots+j_n \leqslant \deg f}} f_{j_1,\ldots,j_n} \prod_{i=1}^{n} \left(g_i(x_i)q_{ij_i}(x_i) + r_i^{(j_i)}(x_i) \right) \\ = & \bar{f}(x_1, \ldots, x_n) + \sum_{i=1}^{n} g_i(x_i)h_i(x_1, \ldots, x_n),\end{aligned}$$

where

$$\bar{f}(x_1, \ldots, x_n) = \sum_{j_1,\ldots,j_n \geq 0} f_{j_1,\ldots,j_n} \prod_{i=1}^{n} r_i^{(j_i)}(x_i)$$

and each $h_i(x_1, \ldots, x_n)$ is a suitable polynomial over F with $\deg g_i + \deg h_i \leq \deg f$. If $a_1 \in A_1, \ldots, a_n \in A_n$, then

$$\bar{f}(a_1, \ldots, a_n) = \sum_{j_1,\ldots,j_n \geq 0} f_{j_1,\ldots,j_n} \prod_{i=1}^{n} a_i^{j_i} = f(a_1, \ldots, a_n) = 0.$$

Since the degree of $\bar{f}(x_1, \ldots, x_n)$ in x_i is smaller than $|A_i|$, by Lemma 1.1 the polynomial $\bar{f}(x_1, \ldots, x_n)$ is identically zero. Therefore (1.5) holds.

(ii) By part (i) we can write

$$f(x_1, \ldots, x_n) = \sum_{i=1}^{n} g_i(x_i)h_i(x_1, \ldots, x_n)$$

with $h_i(x_1, \ldots, x_n) \in F[x_1, \ldots, x_n]$ and $\deg h_i \leq \deg f - \deg g_i$. Since $k_1 + \cdots + k_n = \deg f$ and $k_i < |A_i|$ for $i = 1, \ldots, n$, we have

$$[x_1^{k_1} \cdots x_n^{k_n}]f(x_1, \ldots, x_n) = \sum_{i=1}^{n} [x_1^{k_1} \cdots x_n^{k_n}] x_i^{|A_i|} h_i(x_1, \ldots, x_n) = 0.$$

This concludes the proof. □

Here is a useful lemma implied by the Combinatorial Nullstellensatz.

ANR Lemma ([5]). *Let $A_1, \ldots, A_n$ be finite subsets of a field F with $k_i = |A_i| > 0$ for $i = 1, \ldots, n$. Let $f(x_1, \ldots, x_n) \in F[x_1, \ldots, x_n] \setminus \{0\}$ and $\deg f \leq \sum_{i=1}^{n}(k_i - 1)$. If*

$$[x_1^{k_1-1} \cdots x_n^{k_n-1}]f(x_1, \ldots, x_n)(x_1 + \cdots + x_n)^{\sum_{i=1}^{n}(k_i-1)-\deg f} \neq 0, \quad (1.6)$$

then

$$|\{a_1+\cdots+a_n\colon a_i\in A_i,\ f(a_1,\ldots,a_n)\neq 0\}|\geq \sum_{i=1}^{n}(k_i-1)-\deg f+1. \quad (1.7)$$

Proof. Assume that $C=\{a_1+\cdots+a_n\colon a_i\in A_i,\ f(a_1,\ldots,a_n)\neq 0\}$ has cardinality not exceeding $K=\sum_{i=1}^n(k_i-1)-\deg f$. Then the polynomial

$$P(x_1,\ldots,x_n):=f(x_1,\ldots,x_n)(x_1+\cdots+x_n)^{K-|C|}\prod_{c\in C}(x_1+\cdots+x_n-c)$$

is of degree $\sum_{i=1}^n(k_i-1)$ with the coefficient of $x_1^{k_1-1}\cdots x_n^{k_n-1}$ non-zero. Applying the second part of the Combinatorial Nullstellensatz, we find that $P(a_1,\ldots,a_n)\neq 0$ for some $a_1\in A_1,\ldots,a_n\in A_n$. This is impossible since $a_1+\cdots+a_n\in C$ if $f(a_1,\ldots,a_n)\neq 0$. □

We remark that a variant of this lemma appeared in Q. H. Hou and Z. W. Sun [18].

Alon-Nathanson-Ruzsa Theorem ([5]). *Let $A_1,\ldots,A_n$ be finite non-empty subsets of a field F with $|A_1|<\cdots<|A_n|$. Then, for the set*

$$A_1\dotplus\cdots\dotplus A_n=\left\{\sum_{i=1}^{n}a_i\colon a_i\in A_i,\ \text{and } a_i\neq a_j \text{ if } i\neq j\right\}, \quad (1.8)$$

we have

$$|A_1\dotplus\cdots\dotplus A_n|\geq \min\left\{p(F),\ \sum_{i=1}^{n}|A_i|-\frac{n(n+1)}{2}+1\right\}. \quad (1.9)$$

This follows from the ANR lemma and the following fact: If $k_1,\ldots,k_n\in\mathbb{Z}^+$, then

$$[x_1^{k_1-1}\cdots x_n^{k_n-1}]\prod_{1\leq i<j\leq n}(x_j-x_i)\times(x_1+\cdots+x_n)^{\sum_{i=1}^n k_i-n(n+1)/2}$$
$$=\frac{(k_1+\cdots+k_n-n(n+1)/2)!}{(k_1-1)!\cdots(k_n-1)!}\prod_{1\leq i<j\leq n}(k_j-k_i). \quad (1.10)$$

The Dias da Silva–Hamidoune theorem can be deduced from the ANR theorem in the following way: Suppose that $|A|=k\geq n$. Let $A_1,\ldots,A_n$ be subsets of A with cardinalities $k-n+1,k-n+2,\ldots,k$ respectively. By the ANR theorem,

$$|A_1\dotplus\cdots\dotplus A_n|\geq\min\left\{p(F),\sum_{i=1}^{n}(|A_i|-i)+1\right\}=\min\{p(F),(k-n)n+1\}.$$

As $n^\wedge A \supseteq A_1 \dotplus \cdots \dotplus A_n$, the desired inequality (1.3) follows.

In addition, the reader may also consult [24], [3] and [38] for the polynomial method, and [34] for its connections with covers of $\mathbb{Z}$ by residue classes and zero-sum problems on abelian p-groups.

2. Various sumsets with polynomial restrictions

By a sophisticated induction argument (cf. [7] and [30]), it can be shown that if $A_1, \ldots, A_n$ are finite subsets of $\mathbb{Z}$ with $|A_1| \le \cdots \le |A_n|$ and $|A_i| \ge i$ for all $i = 1, \ldots, n$, then

$$|A_1 \dotplus \cdots \dotplus A_n| \ge 1 + \sum_{i=1}^{n} \min_{i \le j \le n} (|A_j| - j).$$

Now we state a result on sumsets with linear restrictions over $\mathbb{Z}$.

Theorem 2.1 (Z. W. Sun [30]). *Let $A_1, \ldots, A_n$ be finite subsets of $\mathbb{Z}$, and let V be a finite set of quin-tuples (s, t, μ, ν, w) with $1 \le s, t \le n$, $s \ne t$, $\mu, \nu \in \mathbb{Z} \setminus \{0\}$ and $w \in \mathbb{Z}$. If each $V_i = \{(s, t, \mu, \nu, w) \in V : i \in \{s, t\}\}$ has cardinality smaller than $|A_i|$, then*

$$\begin{aligned}&|\{a_1 + \ldots + a_n : a_i \in A_i, \text{ and } \mu a_i + \nu a_j \ne w \text{ if } (i, j, \mu, \nu, w) \in V\}| \\ &\quad \ge \sum_{i=1}^{n} |A_i| - 2|V| - n + 1 = 1 + \sum_{i=1}^{n} (|A_i| - |V_i| - 1) > 0.\end{aligned} \tag{2.1}$$

Clearly Theorem 2.1 has the following consequence.

Corollary 2.1 (Z. W. Sun [30]). *Let $A_1, \ldots, A_n$ be finite subsets of $\mathbb{Z}$ with $|A_i| \ge 2n - 1$ for all $i = 1, \ldots, n$. Then*

$$\left|\left\{ \sum_{i=1}^{n} a_i : a_i \in A_i,\ a_i \ne \pm a_j \text{ if } i \ne j \right\}\right| \ge \sum_{i=1}^{n} |A_i| - 2n^2 + n + 1. \tag{2.2}$$

All the remaining theorems in this section have been obtained via the polynomial method. Preceding a theorem we usually state a lemma which makes the method applicable.

Lemma 2.1. *Let k, m, n be integers with $m \ge 0$, $n > 1$ and $k > m(n-1)$.*

(i) (Q. H. Hou and Z. W. Sun [18]) *We have*

$$\begin{aligned}&[x_1^{k-1} \cdots x_n^{k-1}] \prod_{1 \leqslant i < j \leqslant n} (x_i - x_j)^{2m} \cdot (x_1 + \cdots + x_n)^{(k-1-m(n-1))n} \\ &= (-1)^{mn(n-1)/2} \frac{((k-1-m(n-1))n)!}{(m!)^n} \prod_{j=1}^{n} \frac{(jm)!}{(k-1-(j-1)m)!}.\end{aligned} \tag{2.3}$$

(ii) (Z. W. Sun and Y. N. Yeh [37]) *If* $m>0$ *then*

$$[x_1^{k-n}\cdots x_n^{k-1}]\prod_{1\le i<j\le n}(x_i-x_j)^{2m-1}\cdot(x_1+\cdots+x_n)^{(k-1-m(n-1))n}$$
$$=(-1)^{(m-1)n(n-1)/2}\frac{((k-1-m(n-1))n)!}{(m!)^n n!}\prod_{j=1}^{n}\frac{(jm)!}{(k-1-(j-1)m)!}. \tag{2.4}$$

Theorem 2.2. *Let* $k,m\in\mathbb{N}=\{0,1,2,\ldots\}$ *and* $n\in\mathbb{Z}^+$. *Let* F *be a field with* $p(F)>\max\{mn,\ (k-1-m(n-1))n\}$, *and let* $A_1,\ldots,A_n$ *be finite subsets of* F *with* $\max_{1\le i\le n}|A_i|=k$. *Set*

$$C=\{a_1+\cdots+a_n\colon a_1\in A_1,\ldots,a_n\in A_n,\ a_i-a_j\notin S_{ij}\ \text{if}\ i<j\},$$

where S_{ij} $(1\le i<j\le n)$ *are subsets of* F.

(i) (Q. H. Hou and Z. W. Sun [18]) *If* $|A_1|=\cdots=|A_n|=k$, *and* $|S_{ij}|\le 2m$ *for all* $1\le i<j\le n$, *then we have* $|C|\ge(k-1-m(n-1))n+1$.

(ii) (Z. W. Sun and Y. N. Yeh [37]) *If* $|A_i|=k-n+i$ *for* $i=1,\ldots,n$, *and* $|S_{ij}|<2m$ *for all* $1\le i<j\le n$, *then* $|C|\ge(k-1-m(n-1))n+1$.

The following conjecture posed by Z. W. Sun in [18] is open even for the rational field.

Conjecture 2.1 (Z. W. Sun, 2002). *Let* $A_1,\ldots,A_n$ *be finite non-empty subsets of a field* F. *For* $1\le i<j\le n$, *let* S_{ij} *and* S_{ji} *be finite subsets of* F *with* $|S_{ij}|\equiv|S_{ji}|\pmod 2$. *Then*

$$|\{a_1+\cdots+a_n\colon a_1\in A_1,\ldots,a_n\in A_n,\ a_i-a_j\notin S_{ij}\ \text{if}\ i\ne j\}|$$
$$\ge\min\Big\{p(F),\ \sum_{i=1}^n|A_i|-\sum_{1\le i<j\le n}(|S_{ij}|+|S_{ji}|)-n+1\Big\}. \tag{2.5}$$

Lemma 2.2 (J. X. Liu and Z. W. Sun [23]). *Let* $k,m,n\in\mathbb{Z}^+$ *with* $k-1\ge m(n-1)$. *Then*

$$[x_1^{k-n}\cdots x_n^{k-1}]\prod_{1\le i<j\le n}(x_j^m-x_i^m)\cdot(x_1+\cdots+x_n)^{(k-1)n-(m+1)\binom n2}$$
$$=(-m)^{\binom n2}\frac{((k-1)n-(m+1)\binom n2)!1!2!\cdots(n-1)!}{(k-1)!(k-1-m)!\cdots(k-1-(n-1)m)!}. \tag{2.6}$$

Theorem 2.3 (J. X. Liu and Z. W. Sun [23]). *Let* $k,m,n\in\mathbb{Z}^+$ *with* $k>m(n-1)$, *and let* $A_1,\ldots,A_n$ *be subsets of a field* F *such that*

$$|A_n|=k\ \text{and}\ |A_{i+1}|-|A_i|\in\{0,1\}\ \text{for}\ i=1,\ldots,n-1.$$

Let $P_1(x), \ldots, P_n(x) \in F[x]$ *be monic and of degree* m. *If* $p(F) > (k-1)n-(m+1)\binom{n}{2}$, *then*

$$|\{a_1 + \cdots + a_n\colon a_i \in A_i, \text{ and } P_i(a_i) \neq P_j(a_j) \text{ if } i \neq j\}| \\ \geq (k-1)n - (m+1)\binom{n}{2} + 1. \tag{2.7}$$

Here we pose the following conjecture.

Conjecture 2.2. *Under the conditions of Theorem 2.3, we have*

$$|\{a_1 + \cdots + a_n\colon a_i \in A_i, \text{ and } P_i(a_i) \neq P_j(a_j) \text{ if } i \neq j\}| \geq p(F)$$

if $p(F) \leq (k-1)n - (m+1)\binom{n}{2}$.

Lemma 2.3 (Z. W. Sun [33]). *Let* R *be a commutative ring with identity. Let* $A = (a_{ij})_{1\leq i,j\leq n}$ *be a matrix over* R, *and let* $\det(A) = |a_{ij}|_{1\leq i,j\leq n}$ *be the determinant of* A. *Let* $k, m_1, \ldots, m_n \in \mathbb{N}$.

(i) *If* $m_1 \leq \cdots \leq m_n \leq k$, *then we have*

$$[x_1^k \cdots x_n^k]|a_{ij}x_j^{m_i}|_{1\leq i,j\leq n}(x_1 + \cdots + x_n)^{kn-\sum_{i=1}^n m_i} \\ = \frac{(kn - \sum_{i=1}^n m_i)!}{\prod_{i=1}^n (k-m_i)!}\det(A). \tag{2.8}$$

(ii) *If* $m_1 < \cdots < m_n \leq k$ *then*

$$[x_1^k \cdots x_n^k]|a_{ij}x_j^{m_i}|_{1\leqslant i,j\leqslant n} \prod_{1\leqslant i<j\leqslant n}(x_j - x_i)\cdot\left(\sum_{s=1}^n x_s\right)^{kn-\binom{n}{2}-\sum_{i=1}^n m_i} \\ = (-1)^{\binom{n}{2}}\frac{(kn-\binom{n}{2}-\sum_{i=1}^n m_i)!}{\prod_{i=1}^n \prod_{\substack{m_i<j\leqslant k \\ j\neq m_{i+1},\ldots,m_n}}(j-m_i)}\operatorname{per}(A), \tag{2.9}$$

where $\operatorname{per}(A)$ *is the permanent* $\|a_{ij}\|_{1\leq i,j\leq n} = \sum_{\sigma\in S_n} a_{1,\sigma(1)}\cdots a_{n,\sigma(n)}$ *and* S_n *is the symmetric group of all the permutations on* $\{1, \ldots, n\}$.

(iii) *Suppose that* $K = kn - \sum_{i=1}^n (l_i + m_i) \geq 0$ *where* $l_1, \ldots, l_n \in \mathbb{N}$. *Then*

$$[x_1^k \cdots x_n^k]|a_{ij}x_j^{l_i}|_{1\leq i,j\leq n}|x_j^{m_i}|_{1\leq i,j\leq n}(x_1 + \cdots + x_n)^K \\ = [x_1^k \cdots x_n^k]|a_{ij}x_j^{m_i}|_{1\leq i,j\leq n}|x_j^{l_i}|_{1\leq i,j\leq n}(x_1 + \cdots + x_n)^K. \tag{2.10}$$

Theorem 2.4 (Z. W. Sun [33]). *Let* $k, m, n \in \mathbb{Z}^+$ *with* k *greater than* $m(n-1)$, *and let* $A_1, \ldots, A_n$ *be subsets of a field* F *with cardinality* k. *Let*

$P_1(x),\ldots,P_n(x)\in F[x]$ have degree m with leading coefficients $b_1,\ldots,b_n$ respectively.

(i) *Suppose that $b_1,\ldots,b_n$ are distinct. If $p(F)>(k-1)n-m\binom{n}{2}$, then*

$$\begin{aligned}\left|\left\{\sum_{i=1}^{n}a_i\colon a_1\in A_1,\ldots,a_n\in A_n,\ \text{and } P_i(a_i)\neq P_j(a_j) \text{ if } i\neq j\right\}\right|&\\ \geq (k-1)n-m\binom{n}{2}+1.&\end{aligned}\tag{2.11}$$

(ii) *Assume that the permanent $\|b_j^{i-1}\|_{1\le i,j\le n}$ does not vanish. If $p(F)$ is greater than $(k-1)n-(m+1)\binom{n}{2}$, then*

$$\begin{aligned}\left|\left\{\sum_{i=1}^{n}a_i\colon a_i\in A_i,\ a_i\neq a_j \text{ and } P_i(a_i)\neq P_j(a_j) \text{ if } i\neq j\right\}\right|&\\ \geq (k-1)n-(m+1)\binom{n}{2}+1.&\end{aligned}\tag{2.12}$$

(iii) *We have $\|b_j^{i-1}\|_{1\le i,j\le n}\neq 0$, if F is the complex field $\mathbb{C}$, $b_1,\ldots,b_n$ are qth roots of unity, and $n!$ does not belong to the set*

$$D(q)=\left\{\sum_{p\mid q}px_p\colon x_p\in\{0,1,2,\ldots\} \text{ for any prime divisor } p \text{ of } q\right\}.$$

Now we raise the following conjecture.

Conjecture 2.3. *When F is a field with $p(F)\le (k-1)n-m\binom{n}{2}$, the right-hand side of the inequality (2.11) in Theorem 2.4(i) should be replaced by $p(F)$. Similarly, when F is a field with $p(F)\le (k-1)n-(m+1)\binom{n}{2}$, the right-hand side of the inequality (2.12) in Theorem 2.4(ii) should be replaced by $p(F)$.*

The following lemma has the same flavor as Lemma 2.3, but it was only recently noted and applied by the author.

Lemma 2.4 (Z. W. Sun [35]). *Let R be a commutative ring with identity. Let $A=(a_{ij})_{1\le i,j\le n}$ be a matrix over R, and let $\operatorname{per}(A)=\|a_{ij}\|_{1\le i,j\le n}$ be the permanent of A. Let $k,m_1,\ldots,m_n\in\mathbb{N}$.*

(i) *If $m_1\le\cdots\le m_n\le k$, then we have*

$$\begin{aligned}&[x_1^k\cdots x_n^k]\|a_{ij}x_j^{m_i}\|_{1\le i,j\le n}(x_1+\cdots+x_n)^{kn-\sum_{i=1}^n m_i}\\ &\quad=\frac{(kn-\sum_{i=1}^n m_i)!}{\prod_{i=1}^n(k-m_i)!}\operatorname{per}(A).\end{aligned}\tag{2.13}$$

(ii) *If $m_1<\cdots<m_n\le k$ then*

$$[x_1^k \cdots x_n^k]\|a_{ij}x_j^{m_i}\|_{1\leqslant i,j\leqslant n} \prod_{1\leqslant i<j\leqslant n}(x_j - x_i) \cdot \left(\sum_{s=1}^n x_s\right)^{kn-\binom{n}{2}-\sum_{i=1}^n m_i}$$
$$= (-1)^{\binom{n}{2}} \frac{(kn - \binom{n}{2} - \sum_{i=1}^n m_i)!}{\prod_{i=1}^n \prod_{\substack{m_i<j\leqslant k \\ j\neq m_{i+1},\ldots,m_n}} (j - m_i)} \det(A). \tag{2.14}$$

(iii) *Suppose that* $K = kn - \sum_{i=1}^n (l_i + m_i) \geq 0$ *where* $l_1, \ldots, l_n \in \mathbb{N}$. *Then*

$$\begin{aligned}&[x_1^k \cdots x_n^k]\|a_{ij}x_j^{l_i}\|_{1\leq i,j\leq n}|x_j^{m_i}|_{1\leq i,j\leq n}(x_1 + \cdots + x_n)^K \\ =&[x_1^k \cdots x_n^k]\|a_{ij}x_j^{m_i}\|_{1\leq i,j\leq n}|x_j^{l_i}|_{1\leq i,j\leq n}(x_1 + \cdots + x_n)^K\end{aligned} \tag{2.15}$$

and also

$$\begin{aligned}&[x_1^k \cdots x_n^k]\|a_{ij}x_j^{l_i}\|_{1\leq i,j\leq n}\|x_j^{m_i}\|_{1\leq i,j\leq n}(x_1 + \cdots + x_n)^K \\ =&[x_1^k \cdots x_n^k]\|a_{ij}x_j^{m_i}\|_{1\leq i,j\leq n}\|x_j^{l_i}\|_{1\leq i,j\leq n}(x_1 + \cdots + x_n)^K.\end{aligned} \tag{2.16}$$

Theorem 2.5 (Z. W. Sun [35]). *Let* $A_1, \ldots, A_n$ *be finite subsets of a field* F *with* $|A_1| = \cdots = |A_n| = k > m(n-1)$ *where* $m \in \mathbb{Z}^+$, *and let* $P_1(x), \ldots, P_n(x) \in F[x]$ *have degree at most* m *with* $[x^m]P_1(x), \ldots, [x^m]P_n(x)$ *distinct. If* $p(F) > (k-1)n - (m+1)\binom{n}{2}$, *then the restricted sumset*

$$\left\{\sum_{i=1}^n a_i : a_i \in A_i,\ a_i \neq a_j \text{ for } i \neq j,\ \text{and } \|P_j(a_j)^{i-1}\|_{1\leq i,j\leq n} \neq 0\right\} \tag{2.17}$$

has cardinality at least $(k-1)n - (m+1)\binom{n}{2} + 1 > (m-1)\binom{n}{2}$.

Conjecture 2.4. *Under the conditions of Theorem 2.5, if* $p(F) \leq (k-1)n - (m+1)\binom{n}{2}$ *then the restricted sumset in (2.17) has cardinality at least* $p(F)$.

Corollary 2.2 (Z. W. Sun [35]). *Let* $A_1, \ldots, A_n$ *and* $B = \{b_1, \ldots, b_n\}$ *be subsets of a field with cardinality* n. *Then there are distinct* $a_1 \in A_1, \ldots, a_n \in A_n$ *such that the permanent* $\|(a_jb_j)^{i-1}\|_{1\leq i,j\leq n}$ *is non-zero.*

Theorem 2.6 (Z. W. Sun [35]). *Let* h, k, l, m, n *be positive integers satisfying*

$$k - 1 \geq m(n-1) \quad \text{and} \quad l - 1 \geq h(n-1).$$

Let F *be a field with* $p(F) > \max\{K, L\}$, *where*

$$K = (k-1)n - (m+1)\binom{n}{2} \text{ and } L = (l-1)n - (h+1)\binom{n}{2}.$$

Assume that $c_1,\ldots,c_n\in F$ are distinct and $A_1,\ldots,A_n,B_1,\ldots,B_n$ are subsets of F with

$$|A_1|=\cdots=|A_n|=k \text{ and } |B_1|=\cdots=|B_n|=l.$$

Let $P_1(x),\ldots,P_n(x),Q_1(x),\ldots,Q_n(x)\in F[x]$ be monic polynomials with $\deg P_i(x)=m$ and $\deg Q_i(x)=h$ for $i=1,\ldots,n$. Then, for any $S,T\subseteq F$ with $|S|\le K$ and $|T|\le L$, there exist $a_1\in A_1,\ldots,a_n\in A_n, b_1\in B_1,\ldots,b_n\in B_n$ such that $a_1+\cdots+a_n\notin S$, $b_1+\cdots+b_n\notin T$, and also

$$a_ib_ic_i\ne a_jb_jc_j,\ P_i(a_i)\ne P_j(a_j),\ Q_i(b_i)\ne Q_j(b_j)\ \text{if}\ 1\le i<j\le n. \tag{2.18}$$

Lemma 2.5 (Z. W. Sun [35]). *Let $k,m,n\in\mathbb{Z}^+$ with $k-1\geqslant m(n-1)$. Then*

$$[x_1^{k-1}\cdots x_n^{k-1}]\prod_{1\leqslant i<j\leqslant n}(x_j-x_i)^{2m-1}(x_jy_j-x_iy_i)\cdot(x_1+\cdots+x_n)^N$$
$$=(-1)^{m\binom{n}{2}}\frac{(mn)!N!}{(m!)^n n!}\prod_{r=0}^{n-1}\frac{(rm)!}{(k-1-rm)!}\cdot\|y_j^{i-1}\|_{1\leqslant i,j\leqslant n}, \tag{2.19}$$

where $N=(k-1-m(n-1))n$.

Theorem 2.7 (Z. W. Sun [35]). *Let k,m,n be positive integers with $k-1\ge m(n-1)$, and let F be a field with $p(F)>\max\{mn,(k-1-m(n-1))n\}$. Assume that $c_1,\ldots,c_n\in F$ are distinct, and $A_1,\ldots,A_n,B_1,\ldots,B_n$ are subsets of F with $|A_1|=\cdots=|A_n|=k$ and $|B_1|=\cdots=|B_n|=n$. Let $S_{ij}\subseteq F$ with $|S_{ij}|<2m$ for all $1\le i<j\le n$. Then there are distinct $b_1\in B_1,\ldots,b_n\in B_n$ such that the restricted sumset*

$$S=\{a_1+\cdots+a_n:\ a_i\in A_i,\ a_i-a_j\notin S_{ij}\ \text{and}\ a_ib_ic_i\ne a_jb_jc_j\ \text{if}\ i<j\} \tag{2.20}$$

has at least $(k-1-m(n-1))n+1$ elements.

3. Snevily's conjecture and additive theorems

Suppose that $\{a_1,\ldots,a_n\}$, $\{b_1,\ldots,b_n\}$ and $\{a_1+b_1,\ldots,a_n+b_n\}$ are complete systems of residues modulo n. Let $\sigma=0+1+\cdots+(n-1)=n(n-1)/2$. Since $\sum_{i=1}^n(a_i+b_i)=\sum_{i=1}^n a_i+\sum_{i=1}^n b_i$, we have $\sigma\equiv\sigma+\sigma \pmod n$ and hence $2\nmid n$.

In 1999 H. S. Snevily [28] made the following interesting conjecture.

Snevily's Conjecture. *Let G be an additive abelian group with $|G|$ odd. Let A and B be subsets of G with cardinality $n>0$. Then there is a numbering $\{a_i\}_{i=1}^n$ of the elements of A and a numbering $\{b_i\}_{i=1}^n$ of the elements of B such that $a_1+b_1,\ldots,a_n+b_n$ are distinct.*

Theorem 3.1. (i) (N. Alon [2]) *Let p be an odd prime and A be a non-empty subset of $\mathbb{Z}/p\mathbb{Z}$ with cardinality $n < p$. For any given $b_1, \ldots, b_n \in \mathbb{Z}/p\mathbb{Z}$, we can find a numbering $\{a_i\}_{i=1}^n$ of the elements of A such that the sums $a_1 + b_1, \cdots, a_n + b_n$ are distinct.*

(ii) (Q. H. Hou and Z. W. Sun [18]) *Let $k \geq n \geq 1$ be integers, and let F be a field with $p(F) > \max\{n, (k-n)n\}$. Let $A_1, \ldots, A_n$ be subsets of F with cardinality k, and let $b_1, \ldots, b_n$ be elements of F. Then the restricted sumset*

$$\{a_1 + \cdots + a_n \colon a_i \in A_i,\ a_i \neq a_j \text{ and } a_i + b_i \neq a_j + b_j \text{ if } i \neq j\}$$

has more than $(k-n)n$ elements.

Note that part (ii) in the case $k = n$ and $A_1 = \cdots = A_n$ yields part (i). In order to get part (i) by the polynomial method, Alon noted that

$$[x_1^{n-1} \cdots x_n^{n-1}] \prod_{1 \leq i < j \leq n} (x_j - x_i)(x_j + b_j - (x_i + b_i)) = (-1)^{n(n-1)/2} n!.$$

Part (ii) is a consequence of Theorem 2.2(i) (due to Hou and Sun) with $m = 1$.

Theorem 3.2 (Dasgupta, Károlyi, Serra and Szegedy [10]). *Snevily's conjecture holds for any cyclic group of odd order.*

Proof. (Dasgupta, Károlyi, Serra and Szegedy). Let $m > 0$ be any odd integer. As $2^{\varphi(m)} \equiv 1 \pmod{m}$ by Euler's theorem, the multiplicative group of the finite field with order $2^{\varphi(m)}$ has a cyclic subgroup of order m. Thus, in view of the Combinatorial Nullstellensatz, Snevily's conjecture for the cyclic group of order m follows from the following statement: *If F is a field of characteristic 2 and $b_1, \ldots, b_n$ are distinct elements of $F^* = F \setminus \{0\}$, then*

$$c := [x_1^{n-1} \cdots x_n^{n-1}] \prod_{1 \leq i < j \leq n} (x_j - x_i)(b_j x_j - b_i x_i) \neq 0.$$

In fact,

$$\begin{aligned} &\prod_{1 \leqslant i < j \leqslant n} (x_j - x_i)(b_j x_j - b_i x_i) \\ &= (-1)^{\binom{n}{2}} |x_j^{n-i}|_{1 \leqslant i,j \leqslant n} |b_j^{i-1} x_j^{i-1}|_{1 \leqslant i,j \leqslant n} \text{ (Vandermonde)} \\ &= (-1)^{\binom{n}{2}} \sum_{\sigma \in S_n} \operatorname{sign}(\sigma) \prod_{i=1}^{n} x_{\sigma(i)}^{n-i} \cdot \sum_{\tau \in S_n} \operatorname{sign}(\tau) \prod_{i=1}^{n} b_{\tau(i)}^{i-1} x_{\tau(i)}^{i-1}, \end{aligned}$$

where sign(σ) (the sign of σ) is 1 or -1 according as $\sigma \in S_n$ is even or odd. Therefore

$$\begin{aligned}(-1)^{\binom{n}{2}}c &= \sum_{\tau\in S_n}\prod_{i=1}^{n} b_{\tau(i)}^{i-1}\\ &= \sum_{\tau\in S_n}\operatorname{sign}(\tau)\prod_{i=1}^{n} b_{\tau(i)}^{i-1} \quad (\text{because } \operatorname{ch}(F)=2)\\ &= |b_j^{i-1}|_{1\leqslant i,j\leqslant n} = \prod_{1\leqslant i<j\leqslant n}(b_j-b_i)\neq 0.\end{aligned}$$

□

It is well known that all finite subgroups of the multiplicative group of a field are cyclic. On the other hand, Z. W. Sun [33] observed that any finitely generated abelian group whose finite subgroups are all cyclic, can be embedded in the unit group of the ring of algebraic integers in a suitable cyclotomic field, which allows us to view it as a subgroup of the multiplicative group $\mathbb{C}^*$ of non-zero complex numbers. So we have the following lemma.

Lemma 3.1. *Let G be a finitely generated abelian group. Then the torsion group*

$$\operatorname{Tor}(G) = \{a \in G:\ a \text{ has a finite order}\} \tag{3.1}$$

is cyclic if and only if there is a field F such that the multiplicative group $F^* = F\setminus\{0\}$ contains a subgroup isomorphic to G.

This lemma, together with Theorem 2.4, enabled the author to establish the following theorem which extends both Theorem 3.1 and Theorem 3.2.

Theorem 3.3 (Z. W. Sun [33]). *Let G be an additive abelian group whose finite subgroups are all cyclic. Let m, n be positive integers and let $b_1, \ldots, b_n$ be elements of G. Assume that $A_1, \ldots, A_n$ are finite subsets of G with cardinality $k > m(n-1)$.*

(i) *If $b_1, \ldots, b_n$ are distinct, then there are at least $(k-1)n - m\binom{n}{2} + 1$ multi-sets $\{a_1, \ldots, a_n\}$ such that $a_i \in A_i$ for $i = 1, \ldots, n$ and all the ma_i+b_i are distinct.*

(ii) *The sets*

$$\{\{a_1, \ldots, a_n\}\colon a_i \in A_i,\ a_i \neq a_j \text{ and } ma_i + b_i \neq ma_j + b_j \text{ if } i \neq j\} \tag{3.2}$$

and

$$\{\{a_1, \ldots, a_n\}\colon a_i \in A_i,\ ma_i \neq ma_j \text{ and } a_i + b_i \neq a_j + b_j \text{ if } i \neq j\} \tag{3.3}$$

have more than $(k-1)n-(m+1)\binom{n}{2} \geq (m-1)\binom{n}{2}$ *elements, provided that* $b_1, \ldots, b_n$ *are distinct and of odd order, or they have finite order and* $n!$ *cannot be written in the form* $\sum_{p \in P} px_p$ *where all the* x_p *are non-negative integers and* P *is the set of primes dividing one of the orders of* $b_1, \ldots, b_n$.

In Snevily's conjecture the abelian group is required to have odd order. For a general abelian group G with cyclic torsion subgroup, what additive properties can we impose on several subsets of G with cardinality n?

Lemma 3.1 and Theorem 2.6 together yield the following result.

Theorem 3.4 (Z. W. Sun [35]). *Let* G *be an additive abelian group with cyclic torsion subgroup. Let* $h, k, l, m, n \in \mathbb{Z}^+$ *with* $k > m(n-1)$ *and* $l > h(n-1)$. *Assume that* $c_1, \ldots, c_n \in G$ *are distinct, and* $A_1, \ldots, A_n, B_1, \ldots, B_n$ *are subsets of* G *with* $|A_1| = \cdots = |A_n| = k$ *and* $|B_1| = \cdots = |B_n| = l$. *Then, for any sets* S *and* T *with* $|S| \leq (k-1)n-(m+1)\binom{n}{2}$ *and* $|T| \leq (l-1)n-(h+1)\binom{n}{2}$, *there are* $a_1 \in A_1, \ldots, a_n \in A_n, b_1 \in B_1, \ldots, b_n \in B_n$ *such that* $\{a_1, \ldots, a_n\} \notin S$, $\{b_1, \ldots, b_n\} \notin T$, *and also*

$$a_i + b_i + c_i \neq a_j + b_j + c_j, \ ma_i \neq ma_j, \ hb_i \neq hb_j \ \text{ if } 1 \leq i < j \leq n. \quad (3.4)$$

Corollary 3.1 (Z. W. Sun [35]). *Let* G *be an additive abelian group with cyclic torsion subgroup, and let* $A_1, \ldots, A_n, B_1, \ldots, B_n$ *and* $C = \{c_1, \ldots, c_n\}$ *be finite subsets of* G *with the same cardinality* $n > 0$. *Then there are distinct* $a_1 \in A_1, \ldots, a_n \in A_n$ *and distinct* $b_1 \in B_1, \ldots, b_n \in B_n$ *such that all the sums* $a_1 + b_1 + c_1, \ldots, a_n + b_n + c_n$ *are distinct.*

Proof. Just apply Theorem 3.4 with $k = l = n$ and $m = h = 1$. □

In contrast with Snevily's conjecture, Corollary 3.1 in the case $A_1 = \cdots = A_n = A$ and $B_1 = \cdots = B_n = B$ is of particular interest. Here we state a general additive theorem.

Theorem 3.5 (Z. W. Sun [35]). *Let* G *be any additive abelian group with cyclic torsion subgroup, and let* $A_1, \ldots, A_m$ *be subsets of* G *with the same cardinality* $n \in \mathbb{Z}^+$. *If* m *is odd or all the elements of* A_m *are of odd order, then the elements of* A_i $(1 \leq i \leq m)$ *can be listed in a suitable order* $a_{i1}, \ldots, a_{in}$, *so that all the sums* $\sum_{i=1}^m a_{ij}$ $(1 \leq j \leq n)$ *are distinct.*

Sun [35] also noted that Theorem 3.5 with m odd cannot be extended to general abelian groups since there are counter-examples for the Klein quaternion group $\mathbb{Z}/2\mathbb{Z} \oplus \mathbb{Z}/2\mathbb{Z}$.

A line of a square, an $n \times n$ matrix, is a row or column of the matrix. We define a line of an $n \times n \times n$ cube in a similar way. A *Latin cube* over a set S of cardinality n is an $n \times n \times n$ cube whose entries come from the set S and no line of which contains a repeated element. A *transversal* of an $n \times n \times n$ cube is a collection of n cells no two of which lie in the same line. A *Latin transversal* is a transversal whose cells contain no repeated element.

Corollary 3.2 (Z. W. Sun [35])). (*Let N be any positive integer. For the $N \times N \times N$ Latin cube over $\mathbb{Z}/N\mathbb{Z}$ formed by the Cayley addition table, each $n \times n \times n$ sub-cube with $n \leq N$ contains a Latin transversal.*

Proof. Just apply Theorem 3.5 with $m = 3$ (or Corollary 3.1) to the cyclic group $\mathbb{Z}/NZ$. □

In contrast, Theorem 3.2 has the following equivalent version observed by Snevily [28]: *Let N be a positive odd integer. For the $N \times N$ Latin square over $\mathbb{Z}/N\mathbb{Z}$ formed by the Cayley addition table, each of its sub-squares contains a Latin transversal.*

Conjecture 3.1 (Z. W. Sun [35]). *Every $n \times n \times n$ Latin cube contains a Latin transversal.*

4. On a conjecture of Lev and related results

Let A and B be finite non-empty subsets of an additive abelian group G. In contrast with the Cauchy-Davenport theorem, J.H.B. Kemperman [21] and P. Scherk [27] proved that

$$|A + B| \geq |A| + |B| - \min_{c \in A+B} \nu_{A,B}(c), \tag{4.1}$$

where

$$\nu_{A,B}(c) = |\{(a, b) \in A \times B: a + b = c\}|; \tag{4.2}$$

in particular, we have $|A + B| \geq |A| + |B| - 1$ if some $c \in A + B$ can be uniquely written as $a + b$ with $a \in A$ and $b \in B$.

Motivated by the Kemperman-Scherk theorem and the Erdős-Heilbronn conjecture, V. F. Lev [22] proposed the following interesting conjecture.

Lev's Conjecture. *Let G be an abelian group, and let A and B be finite non-empty subsets of G. Then we have*

$$|A \dotplus B| \geq |A| + |B| - 2 - \min_{c \in A+B} \nu_{A,B}(c). \tag{4.3}$$

By a sophisticated application of the first part of the Combinatorial Nullstellensatz, H. Pan and Z. W. Sun [26] made the following progress on Lev's conjecture.

Theorem 4.1 (H. Pan and Z. W. Sun [26]). *Let A and B be finite non-empty subsets of a field F. Let $P(x,y) \in F[x,y]$ and*

$$C = \{a+b\colon a \in A,\ b \in B,\ \text{and } P(a,b) \neq 0\}. \tag{4.4}$$

If C is non-empty, then

$$|C| \geq |A| + |B| - \deg P - \min_{c \in C} \nu_{A,B}(c). \tag{4.5}$$

Theorem 4.2 (H. Pan and Z. W. Sun [26]). *Let A and B be finite non-empty subsets of an abelian group G with cyclic torsion subgroup. For $i = 1, \ldots, l$ let m_i and n_i be non-negative integers and let $d_i \in G$. Suppose that*

$$C = \{a+b\colon a \in A,\ b \in B,\ \text{and } m_i a - n_i b \neq d_i \text{ for all } i = 1, \ldots, l\} \tag{4.6}$$

is non-empty. Then

$$|C| \geq |A| + |B| - \sum_{i=1}^{l} (m_i + n_i) - \min_{c \in C} \nu_{A,B}(c). \tag{4.7}$$

The following result on difference-restricted sumsets follows from Theorems 4.1 and 4.2.

Theorem 4.3 (H. Pan and Z. W. Sun [26]). *Let G be an abelian group, and let A, B, S be finite non-empty subsets of G with*

$$C = \{a+b\colon a \in A,\ b \in B,\ \text{and } a - b \notin S\} \neq \emptyset. \tag{4.8}$$

(i) *If G is torsion-free or elementary abelian, then*

$$|C| \geq |A| + |B| - |S| - \min_{c \in C} \nu_{A,B}(c). \tag{4.9}$$

(ii) *If Tor(G) is cyclic, then*

$$|C| \geq |A| + |B| - 2|S| - \min_{c \in C} \nu_{A,B}(c). \tag{4.10}$$

Proof. Without loss of generality we can assume that G is generated by the finite set $A \cup B$.

If $G \cong \mathbb{Z}^n$, then we can simply view G as the ring of algebraic integers in an algebraic number field K with $[K : \mathbb{Q}] = n$. If $G \cong (\mathbb{Z}/p\mathbb{Z})^n$ where

p is a prime, then G is isomorphic to the additive group of the finite field with p^n elements. Thus part (i) follows from Theorem 4.1 with $P(x,y) = \prod_{s\in S}(x-y-s)$.

Let $d_1,\ldots,d_l$ be a list of all the elements of S. Applying Theorem 4.2 with $m_i = n_i = 1$ for all $i = 1,\ldots,l$, we immediately get part (ii), completing the proof. □

Given two finite subsets A and B of a field F and a general $P(x,y) \in F[x,y]$, what can we say about the cardinality of the restricted sumset $\{a+b\colon a\in A,\ b\in B, \text{ and } P(a,b)\neq 0\}$? In 2002 H. Pan and Z. W. Sun [25] made progress in this direction by relaxing (to some extent) the limitations of the polynomial method, their approach allows one to draw conclusions even if no coefficients in question are explicitly known.

Lemma 4.1 (H. Pan and Z. W. Sun [25]). *Let $P(x)$ be a polynomial over a field F. Let $\bar{F}$ be the algebraic closure of the field F and $m_P(\alpha)$ be the multiplicity of $\alpha \in \bar{F}$ as a root of $P(x) = 0$ over $\bar{F}$. Suppose that there exist non-negative integers $k < l$ such that $[x^i]P(x) = 0$ for all i with $k < i < l$. Then either $x^l \mid P(x)$, or $\deg P(x) \le k$, or $N_q(P) \ge l-k$ for some $q \in \mathcal{P}(p) = \{1,p,p^2,\ldots\}$, where $p = \mathrm{ch}(F)$,*

$$N_q(P) = q|\{\alpha\in\bar{F}\setminus\{0\}\colon m_P(\alpha)\ge q\}| - \sum_{\alpha\in\bar{F}\setminus\{0\}}\{m_P(\alpha)\}_q \tag{4.11}$$

and $\{m\}_q$ denotes the least non-negative residue of $m\in\mathbb{Z}$ modulo q.

We remark that $N_1(P)$ is the number of distinct roots in $\bar{F}\setminus\{0\}$ of the equation $P(x)=0$ over $\bar{F}$.

Theorem 4.4 (H. Pan and Z. W. Sun [25]). *Let A and B be two finite non-empty subsets of a field F. Furthermore, let $P(x,y)$ be a polynomial over F of degree $d = \deg P(x,y)$ such that for some $i < |A|$ and $j < |B|$ we have $[x^iy^{d-i}]P(x,y)\neq 0$ and $[x^{d-j}y^j]P(x,y)\neq 0$. Define $P_0(x,y)$ to be the homogeneous polynomial of degree d such that $P(x,y) = P_0(x,y) + R(x,y)$ for some $R(x,y)\in F[x,y]$ with $\deg R(x,y) < d$, and put $P^*(x) = P_0(x,1)$. For any α in the algebraic closure $\bar{F}$ of F, let $m_{P^*}(\alpha)$ denote the multiplicity of α as a zero of $P^*(x)\in\bar{F}[x]$. Then*

$$\begin{aligned}&|\{a+b\colon a\in A,\ b\in B,\ \text{and } P(a,b)\neq 0\}|\\ &\qquad\ge \min\{p(F)-m_{P^*}(-1),\ |A|+|B|-1-d-N(P^*)\},\end{aligned} \tag{4.12}$$

where

$$N(P^*) = \max_{q\in\mathcal{P}(\mathrm{ch}(F))} q|\{\alpha\in\bar{F}\setminus\{0,-1\} : m_{P^*}(\alpha)\ge q\}|. \tag{4.13}$$

For the sake of clarity, here we state a consequence of Theorem 4.4.

Corollary 4.1 (H. Pan and Z. W. Sun [25]). *Let F be a field with $p = \mathrm{ch}(F) \neq 2$, and let A, B and S be finite non-empty subsets of F. Then*

$$|\{a+b\colon a \in A,\ b \in B,\ \text{and}\ a-b \notin S\}| \geq \min\{p(F),\ |A|+|B|-|S|-q-1\}, \tag{4.14}$$

where q is the largest element of $\mathcal{P}(p)$ not exceeding $|S|$.

5. Working with general abelian groups

Theorem 5.1 (Kneser's Theorem). *Let G be an additive abelian group. Let A and B be finite non-empty subsets of G, and let $H = H(A+B)$ be the stabilizer $\{g \in G : g+A+B = A+B\}$. If $|A+B| \leq |A|+|B|-1$, then*

$$|A+B| = |A+H| + |B+H| - |H|. \tag{5.1}$$

The following consequence is an extension of the Cauchy-Davenport theorem.

Corollary 5.1. *Let G be an additive abelian group. Let $p(G) = +\infty$ if G is torsion-free, otherwise we let $p(G)$ be the least order of a non-zero element of G. Then, for any finite non-empty subsets A and B of G, we have*

$$|A+B| \geq \min\{p(G),\ |A|+|B|-1\}. \tag{5.2}$$

Proof. Suppose that $|A+B| < |A|+|B|-1$. Then $H = H(A+B) \neq \{0\}$ by Kneser's theorem. Therefore $|H| \geq p(G)$ and hence

$$|A+B| = |A+H| + |B+H| - |H| \geq |A+H| \geq |H| \geq p(G).$$

We are done. □

G. Károlyi [19,20] extended the Erdős-Heilbronn conjecture to general abelian groups.

Theorem 5.2. *Let G be an additive abelian group and let A be a finite non-empty subset of G.*

(i) (G. Károlyi [19]) *We have*

$$|2^\wedge A| \geq \min\{p(G),\ 2|A|-3\}. \tag{5.3}$$

(ii) (G. Károlyi [20]) *When $|A| \geq 5$ and $p(G) > 2|A|-3$, the equality $|2^\wedge A| = 2|A|-3$ holds if and only if A is an arithmetic progression.*

Using the fact that any finitely generated abelian group can be written as the direct sum of some cyclic groups of infinite or prime power order, Károlyi proved Theorem 5.2 in two steps. First, he showed that Theorem 5.2 is true for any cyclic group G of infinite or prime power order; then, he proved that those abelian groups possessing the required property are closed under direct sum. In the first step for Theorem 5.2(i), he actually obtained the following more general result.

Theorem 5.3 (G. Károlyi [19]). *Let $\emptyset \neq A, B \subseteq \mathbb{Z}/p^{\alpha}\mathbb{Z}$, where p is a prime and $\alpha \in \mathbb{Z}^+$. Then*

$$|A \dotplus B| \geq \min\{p,\ |A| + |B| - 3\}. \tag{5.4}$$

For non-empty subsets A and B of $\mathbb{Z}/p\mathbb{Z}$ with p a prime, if $|A| \neq |B|$ then we have

$$|A \dotplus B| \geq \min\left\{p,\ |A| + |B| - \frac{2(2+1)}{2} + 1\right\} = \min\{p,\ |A| + |B| - 2\}$$

by the ANR theorem; if $|A| = |B|$ then

$$|A \dotplus B| \geq |(A \setminus \{a_0\}) \dotplus B)| \geq \min\{p,\ (|A| - 1) + |B| - 2\},$$

where a_0 is any fixed element of A.

When $q = p^{\alpha}$ is not a prime, $\mathbb{Z}/q\mathbb{Z}$ is not a subgroup of the additive group of a field but Károlyi considered it as the group of qth roots of unity (up to isomorphism) which can be viewed as a subgroup of the multiplicative group $\mathbb{C}^*$ of non-zero complex numbers.

Lemma 5.1 (Z. W. Sun [29,32]). *Let $\lambda_1, \ldots, \lambda_k$ be qth roots of unity, and let $c_1, \ldots, c_k$ be non-negative integers with $c_1\lambda_1 + \cdots + c_k\lambda_k = 0$. Then $c_1 + \cdots + c_k \in D(q)$, where $D(q)$ is as in Theorem 2.4(iii).*

Proof of Theorem 5.3. Since $\mathbb{Z}/q\mathbb{Z}$ is isomorphic to the multiplicative group C_q of qth roots of unity, we may view A and B as subsets of C_q. If $|A| + |B| - 3 > p$, then we can choose $\emptyset \neq A' \subseteq A$ and $\emptyset \neq B' \subseteq B$ so that $|A'| + |B'| - 3 = p$. Thus, without loss of generality, we may assume that $k + l - 3 \leq p$ where $k = |A|$ and $l = |B|$.

Suppose that $|C| \ngeq \min\{p,\ k + l - 3\} = k + l - 3$, where

$$C = \{ab : a \in A,\ b \in B \text{ and } a \neq b\}.$$

If

$$c_0 := [x^{k-1}y^{l-1}](xy - 1)\prod_{c \in C}(x - cy) \times (x - y)^{k+l-4-|C|} \neq 0,$$

then by the polynomial method, there exist $a \in A$ and $b^{-1} \in B^{-1}$ such that $ab^{-1} \neq 1$ and $a \neq cb^{-1}$ for all $c \in C$, which leads to a contradiction since $a \neq b$ and $ab \in C$. Thus, it suffices to show $c_0 \neq 0$.

Observe that

$$c_0 = [x^{k-2}y^{l-2}] \prod_{s=1}^{k+l-4} (x - \rho_s y) = (-1)^{l-2} \sum_{1 \le i_1 < \cdots < i_{l-2} \le k+l-4} \rho_{i_1} \cdots \rho_{i_{l-2}},$$

where $\rho_1, \ldots, \rho_{k+l-4}$ are suitable qth roots of unity. Since $\binom{k+l-4}{l-2} \notin D(q) = \{pn : n \in \mathbb{N}\}$, we have $c_0 \neq 0$ by Lemma 5.1.

Now we mention a celebrated theorem of M. Hall which was conjectured by G. Cramer for cyclic groups.

Theorem 5.4 (M. Hall [16]). *Let $G = \{b_1, \ldots, b_n\}$ be an additive abelian group of order n, and let $a_1, \ldots, a_n$ be (not necessarily distinct) elements of G. Then $a_1 + \cdots + a_n = 0$ if and only if $\{a_i + b_{\sigma(i)} : i = 1, \ldots, n\} = G$ for some $\sigma \in S_n$.*

Z. W. Sun and Y. N. Yeh [37] observed that Hall's theorem implies the following conjecture of Parker (cf. [14]): For integers $a_1, \ldots, a_n$ with $a_1 + \cdots + a_n \equiv 0 \pmod{n} + 1$, there are $\sigma, \tau \in S_n$ such that $a_i \equiv \sigma(i) + \tau(i) \pmod{n} + 1$ for all $i = 1, \ldots, n$.

In contrast with Snevily's conjecture, we have the following consequence of Theorem 5.4.

Corollary 5.2. *Let G be a finite abelian group, and let $a_1, \ldots, a_n \in G$ with $n < |G|$. Then there are distinct $b_1, \ldots, b_n \in G$ such that the sums $a_1 + b_1, \ldots, a_n + b_n$ are distinct.*

Proof. Write $G = \{c_1, \ldots, c_m\}$ with $m = |G|$. Set $a_{n+1} = -(a_1 + \cdots + a_n)$ and $a_k = 0$ for $n + 1 < k \le m$. As $a_1 + \cdots + a_m = 0$, by Theorem 5.4, for some $\sigma \in S_m$ we have $\{a_i + c_{\sigma(i)} : i = 1, \ldots, m\} = G$. Let $b_i = c_{\sigma(i)}$ for $i = 1, \ldots, n$. Then $b_1, \ldots, b_n$ are distinct, and so are the sums $a_1+b_1, \ldots, a_n+b_n$. We are done. □

Let us conclude this paper with a new open problem.

Problem 5.1. *Let G be a finite abelian group, and let n be a positive integer smaller than $|G|$. Determine the smallest positive integer $m \le |G|$ such that whenever $a_1, \ldots, a_n \in G$ are distinct and $B \subseteq G$ with $|B| \ge m$ there are distinct $b_1, \ldots, b_n \in B$ such that all the sums $a_1 + b_1, \ldots, a_n + b_n$ are distinct.*

6. On value sets of polynomials

Let p be a prime, and let $c_1, \dots, c_n$ be non-zero elements of the field $F_p = \mathbb{Z}/p\mathbb{Z}$. In 1959 Chowla, Mann and Straus (cf. Theorem 2.8 of [24]) used Vosper's theorem (cf. pp. 52-57 of [24]) to deduce that if $p > 3$, $1 < k < (p-1)/2$ and $k \mid p-1$, then

$$|\{c_1x_1^k + \cdots + c_nx_n^k : x_1, \dots, x_n \in F_p\}| \geq \min\left\{p,\ (2n-1)\frac{p-1}{k} + 1\right\}.$$

In 1956 Carlitz [8] proved that whenever $n \geq k \geq 1$, $k \mid p-1$ and $g(x_1, \dots, x_n) \in F_p[x_1, \dots, x_n]$ with $\deg g < k$, the equation

$$c_1x_1^k + \cdots + c_nx_n^k + g(x_1, \dots, x_n) = 0$$

has a solution with $x_1, \dots, x_n \in F_p$. In 2006 Felszeghy [15] extended this result by showing that

$$\{c_1x_1^k + \cdots + c_nx_n^k + g(x_1, \dots, x_n) :\ x_1, \dots, x_n \in F_p\} = F_p$$

if $k \in \{1, \dots, p-1\}$ and $n \geq (p-1)/\lfloor (p-1)/k \rfloor$.

With the help of the Combinatorial Nullstellensatz, recently Z. W. Sun established the following result on value sets of polynomials.

Theorem 6.1 (Z. W. Sun [36]). *Let F be a field, and let*

$$f(x_1, \dots, x_n) = c_1x_1^k + \cdots + c_nx_n^k + g(x_1, \dots, x_n) \in F(x_1, \dots, x_n), \quad (6.1)$$

where

$$c_1, \dots, c_n \in F^* = F \setminus \{0\} \ \text{ and } \ \deg g < k \in \mathbb{Z}^+. \quad (6.2)$$

Then, for any finite non-empty subsets $A_1, \dots, A_n$ of F, we have

$$|\{f(a_1, \dots, a_n) : a_1 \in A_1, \dots, a_n \in A_n\}| \geq \min\left\{p(F),\ \sum_{i=1}^{n} \left\lfloor \frac{|A_i| - 1}{k} \right\rfloor + 1\right\}. \quad (6.3)$$

Note that Theorem 6.1 in the case $f(x_1, \dots, x_n) = x_1 + \cdots + x_n$ yields the Cauchy-Davenport theorem. Felszeghy's result is also a special case of Theorem 6.1.

Here is another result of [36] obtained by the Combinatorial Nullstellensatz.

Theorem 6.2 (Z. W. Sun [36]). *Let $f(x_1, \dots, x_n)$ be a polynomial over a field F given by (6.1) and (6.2) with $n \leq k = \deg f$. And let $A_1, \dots, A_n$*

be finite subsets of F *with* $|A_i| \geq i$ *for* $i = 1, \ldots, n$. *Then, for the restricted value set*

$$V = \{f(a_1, \ldots, a_n) : \ a_1 \in A_1, \ldots, a_n \in A_n, \ \text{and } a_i \neq a_j \text{ if } i \neq j\}, \quad (6.4)$$

we have

$$|V| \geq \min\left\{p(F), \sum_{i=1}^{n} \left\lfloor \frac{|A_i| - i}{k} \right\rfloor + 1\right\}. \quad (6.5)$$

Theorem 6.2 has the following consequence for the case $n \geq k$.

Corollary 6.1 (Z. W. Sun [36]). *Let* A *be a finite subset of a field* F, *and let* $f(x_1, \ldots, x_n)$ *be a polynomial given by (6.1) and (6.2). If* $n \geq k$, *then we have*

$$\begin{aligned} &|\{f(a_1, \ldots, a_n) : a_1, \ldots, a_n \in A, \text{ and } a_i \neq a_j \text{ if } i \neq j\}| \\ &\qquad \geq \min\{p(F), \ |A| - n + 1\}. \end{aligned} \quad (6.6)$$

Let us conclude this paper with a conjecture raised in [36].

Conjecture 6.1 (Z. W. Sun [36]). *Let* $f(x_1, \ldots, x_n)$ *be a polynomial over a field* F *given by (6.1) and (6.2), and let* A *be any finite subset of* F. *Provided* $n > k$, *we have*

$$\begin{aligned} &|\{f(a_1, \ldots, a_n) : a_1, \ldots, a_n \in A, \ \text{and } a_i \neq a_j \text{ if } i \neq j\}| \\ &\quad \geq \min\left\{p(F) - \delta, \ \frac{n(|A| - n)}{k} - k\left\{\frac{n}{k}\right\}\left\{\frac{|A| - n}{k}\right\} + 1\right\}, \end{aligned} \quad (6.7)$$

where $\{\alpha\}$ *denotes the fractional part* $\alpha - \lfloor \alpha \rfloor$ *of a real number* α, *and*

$$\delta = \begin{cases} 1 & \text{if } n = 2 \text{ and } c_1 = -c_2, \\ 0 & \text{otherwise.} \end{cases} \quad (6.8)$$

By Corollary 3 of [25], this conjecture holds when $n = 2$. Note also that the Dias da Silva–Hamidoune theorem is a special case of Conjecture 6.1 with $k = 1$.

References

1. N. Alon, Combinatorial Nullstellensatz, *Combin. Prob. Comput.*, **8** (1999), 7–29.
2. N. Alon, Additive Latin transversals, *Israel J. Math.*, **117** (2000), 125–130.
3. N. Alon, Discrete mathematics: methods and challenges, *Proceedings of the International Congress of Mathematicians*, Vol. I, (Beijing, 2002), Higher Ed. Press, Beijing, 2002, 119–135.

4. N. Alon, M. B. Nathanson and I. Z. Ruzsa, Adding distinct congruence classes modulo a prime, *Amer. Math. Monthly*, **102** (1995), 250–255.
5. N. Alon, M. B. Nathanson and I. Z. Ruzsa, The polynomial method and restricted sums of congruence classes, *J. Number Theory*, **56** (1996), 404–417.
6. N. Alon and M. Tarsi, A nowhere-zero point in linear mappings, *Combinatorica*, **9** (1989), 393–395.
7. H. Q. Cao and Z. W. Sun, On sums of distinct representatives, *Acta Arith.*, **87** (1998), 159–169.
8. L. Carlitz, Solvablity of certain equations in a finite field, *Quart. J. Math.*, **7** (1956), 3–4.
9. A. Cauchy, Recherches sur les nombres, *Jour. Ecole Polytechn.*, **9** (1813), 99–116.
10. S. Dasgupta, G. Károlyi, O. Serra and B. Szegedy, Transversals of additive Latin squares, *Israel J. Math.*, **126** (2001), 17–28.
11. H. Davenport, On the addition of residue classes, *J. London Math. Soc.*, **10** (1935), 30–32.
12. J. A. Dias da Silva and Y. O. Hamidoune, Cyclic spaces for Grassmann derivatives and additive theory, *Bull. London Math. Soc.*, **26** (1994), 140–146.
13. P. Erdős and H. Heilbronn, On the addition of residue classes modulo p, *Acta Arith.*, **9** (1964), 149–159.
14. R. K. Guy, Parker's permutation problem involves the Catalan numbers, *Amer. Math. Monthly*, **100** (1993), 287–289.
15. B. Felzeghy, On the solvability of some special equations over finite fields, *Publ. Math. Debrecen*, **68** (2006), 15–23.
16. M. Hall, A combinatorial problem on abelian groups, *Proc. Amer. Math. Soc.*, **3** (1952), 584–587.
17. P. Hall, On representatives of subsets, *J. London Math. Soc.*, **10** (1935), 26–30.
18. Q. H. Hou and Z. W. Sun, Restricted sums in a field, *Acta Arith.*, **102** (2002), 239–249.
19. G. Károlyi, The Erdős-Heilbronn problem in abelian groups, *Israel J. Math.*, **139** (2004), 349–359.
20. G. Károlyi, An inverse theorem for the restricted set addition in abelian groups, *J. Algebra*, **290** (2005), 557–593.
21. J. H. B. Kemperman, On small sumsets in an abelian group, *Acta Math.*, **103** (1960), 63–88.
22. V. F. Lev, Restricted set addition in Abelian groups: results and conjectures, *J. Théor. Nombres Bordeaux*, **17** (2005), 181–193.
23. J. X. Liu and Z. W. Sun, Sums of subsets with polynomial restrictions, *J. Number Theory*, **97** (2002), 301–304.
24. M. B. Nathanson, *Additive Number Theory: Inverse Problems and the Geometry of Sumsets* (Graduated texts in mathematics; 165), Springer, New York, 1996.
25. H. Pan and Z. W. Sun, A lower bound for $|\{a+b\colon a \in A,\ b \in B,\ P(a,b) \neq 0\}|$,

J. Combin. Theory Ser. A, **100** (2002), 387–393.

26. H. Pan and Z. W. Sun, Restricted sumsets and a conjecture of Lev, *Israel J. Math.*, **154** (2006), 21–28.
27. P. Scherk, Distinct elements in a set of sums, *Amer. Math. Monthly*, **62** (1955), 46–47.
28. H. S. Snevily, The Cayley addition table of $\mathbb{Z}_n$, *Amer. Math. Monthly*, **106** (1999), 584–585.
29. Z. W. Sun, Covering the integers by arithmetic sequences II, *Trans. Amer. Math. Soc.*, **348** (1996), 4279–4320.
30. Z. W. Sun, Restricted sums of subsets of $\mathbb{Z}$, *Acta Arith.*, **99** (2001), 41–60.
31. Z. W. Sun, Hall's theorem revisited, *Proc. Amer. Math. Soc.*, **129** (2001), 3129–3131.
32. Z. W. Sun, On the function $w(x) = |\{1 \leq s \leq k : x \equiv a_s \pmod{n_s}\}|$, *Combinatorica*, **23** (2003), 681–691.
33. Z. W. Sun, On Snevily's conjecture and restricted sumsets, *J. Combin. Theory* Ser. A, **103** (2003), 291–304.
34. Z. W. Sun, Unification of zero-sum problems, subset sums and covers of $\mathbb{Z}$, *Electron. Res. Announc. Amer. Math. Soc.*, **9** (2003), 51–60.
35. Z. W. Sun, *An additive theorem and restricted sumsets*, preprint, 2006. On-line version: `http://arxiv.org/abs/math.CO/0610981`.
36. Z. W. Sun, *On value sets of polynomials over a field*, preprint, 2007. On-line version: `http://arxiv.org/abs/math.NT/0703180`.
37. Z. W. Sun and Y. N. Yeh, On various restricted sumsets, *J. Number Theory*, **114** (2005), 209–220.
38. T. Tao and V. H. Vu, Additive Combinatorics, Cambridge Univ. Press, Cambridge, 2006.

A GENERAL MODULAR RELATION IN ANALYTIC NUMBER THEORY

HARUO TSUKADA

Department of Information and Computer Sciences,
School of Humanity-Oriented Science and Engineering,
University of Kinki, Iizuka, Fukuoka 820-8555, Japan
E-mail: tsukada@fuk.kindai.ac.jp

In this paper, we consider Dirichlet series $\sum\limits_{k=1}^{\infty} \frac{\alpha_k}{\lambda_k^s}$ and associated series of type

$$\sum_{k=1}^{\infty} \alpha_k H_{p+P,q+Q}^{m+M,n+N}\left(z\lambda_k \,\middle|\, \begin{matrix} \{(a_j - A_j s, A_j)\}_{j=1}^{n},\ \{(c_j, C_j)\}_{j=1}^{N}, \\ \{(b_j - B_j s, B_j)\}_{j=1}^{m},\ \{(d_j, D_j)\}_{j=1}^{M}, \end{matrix} \right.$$
$$\left. \begin{matrix} \{(a_j - A_j s, A_j)\}_{j=n+1}^{p},\ \{(c_j, C_j)\}_{j=N+1}^{P} \\ \{(b_j - B_j s, B_j)\}_{j=m+1}^{q},\ \{(d_j, D_j)\}_{j=M+1}^{Q} \end{matrix} \right),$$

where H denotes the Fox H-function. Assuming a functional equation with gamma factors between the Dirichlet series, we prove a modular relation of a general form for the corresponding H-function series. This formula enables us to treat wide varieties of similar formulas in analytic number theory in a systematic way. For example, if the Dirichlet series satisfy a functional equation of Hecke type, the modular relation includes Bochner's formula, the Riesz sum, a K-Bessel expansion, the incomplete gamma expansion, etc. as special cases, and we present examples associated to the Riemann zeta-function. This paper is an expanded version of the preprint [28].

1. Preliminaries

In this section, we fix some notation related to Fox H-functions. Let $\mathbb{C}$ be the set of complex numbers, and let $\mathbb{R}^+$ be the set of positive real numbers. For any set A, the set of finite sequences of elements of A is denoted by $A^{(\infty)}$. We set $\Omega = \left((\mathbb{C} \times \mathbb{R}^+)^{(\infty)}\right)^4$ and for an element

$$\Delta = \begin{pmatrix} \{(a_j, A_j)\}_{j=1}^{n}; \{(a_j, A_j)\}_{j=n+1}^{p} \\ \{(b_j, B_j)\}_{j=1}^{m}; \{(b_j, B_j)\}_{j=m+1}^{q} \end{pmatrix} \tag{1.1}$$

of Ω, we define

$$\Delta^+ = \begin{pmatrix} \{(a_j, A_j)\}_{j=1}^n ; \{(a_j, A_j)\}_{j=n+1}^p \\ - \qquad ; \{(b_j, B_j)\}_{j=m+1}^q \end{pmatrix} \in \Omega \tag{1.2}$$

and

$$\Delta^- = \begin{pmatrix} - \qquad ; \{(a_j, A_j)\}_{j=n+1}^p \\ \{(b_j, B_j)\}_{j=1}^m ; \{(b_j, B_j)\}_{j=m+1}^q \end{pmatrix} \in \Omega, \tag{1.3}$$

where "–" means the empty sequence. We define the Gamma factor associated to Δ by

$$\Delta(s) = \frac{\prod_{j=1}^{m} \Gamma(b_j + B_j s) \prod_{j=1}^{n} \Gamma(1 - a_j - A_j s)}{\prod_{j=n+1}^{p} \Gamma(a_j + A_j s) \prod_{j=m+1}^{q} \Gamma(1 - b_j - B_j s)}. \tag{1.4}$$

Finally, we define the Fox H-function [12,20,24,25] associated to Δ by

$$\begin{aligned} H(z \mid \Delta) &= H_{p,q}^{m,n}\left(z \,\middle|\, \begin{matrix} \{(a_j, A_j)\}_{j=1}^n, \{(a_j, A_j)\}_{j=n+1}^p \\ \{(b_j, B_j)\}_{j=1}^m, \{(b_j, B_j)\}_{j=m+1}^q \end{matrix} \right) \\ &= \frac{1}{2\pi i} \int_L \Delta(s)\, z^{-s}\, \mathrm{d}s, \end{aligned} \tag{1.5}$$

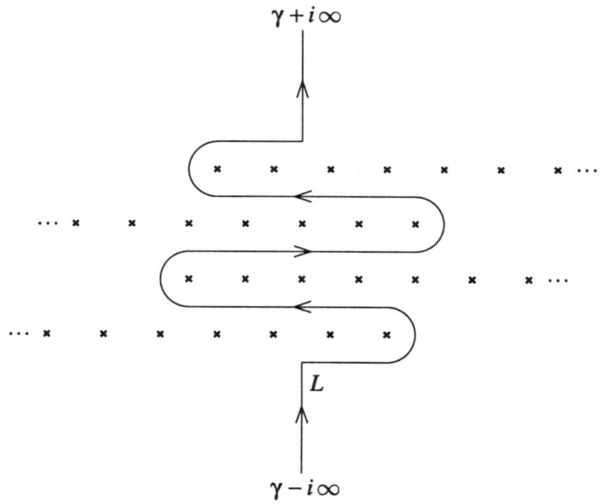

Fig. 1. the path L

provided that the integral converges absolutely, where the path $L : \gamma - i\infty \to \gamma + i\infty$ (γ : real) is taken so that the poles of $\Delta^+(s)$ lie to the right of L, and those of $\Delta^-(s)$ lie to the left of L (cf. Fig. 1).

We note that the Meijer G-functions [21,22] are special cases of H-functions.

$$\begin{aligned} &G^{m,n}_{p,q}\left(z \,\middle|\, \begin{matrix} \{a_j\}_{j=1}^n, \{a_j\}_{j=n+1}^p \\ \{b_j\}_{j=1}^m, \{b_j\}_{j=m+1}^q \end{matrix}\right) \\ &= H^{m,n}_{p,q}\left(z \,\middle|\, \begin{matrix} \{(a_j,1)\}_{j=1}^n, \{(a_j,1)\}_{j=n+1}^p \\ \{(b_j,1)\}_{j=1}^m, \{(b_j,1)\}_{j=m+1}^q \end{matrix}\right) \end{aligned} \tag{1.6}$$

For $c \in \mathbb{C}$, we define

$$\Delta + c = \begin{pmatrix} \{(a_j + A_j c, A_j)\}_{j=1}^n ; \{(a_j + A_j c, A_j)\}_{j=n+1}^p \\ \{(b_j + B_j c, B_j)\}_{j=1}^m ; \{(b_j + B_j c, B_j)\}_{j=m+1}^q \end{pmatrix} \in \Omega, \tag{1.7}$$

and for $C \in \mathbb{R}^+$, we define

$$C \cdot \Delta = \begin{pmatrix} \{(a_j, CA_j)\}_{j=1}^n ; \{(a_j, CA_j)\}_{j=n+1}^p \\ \{(b_j, CB_j)\}_{j=1}^m ; \{(b_j, CB_j)\}_{j=m+1}^q \end{pmatrix} \in \Omega. \tag{1.8}$$

We also set

$$\Delta^* = \begin{pmatrix} \{(1-b_j, B_j)\}_{j=1}^m ; \{(1-b_j, B_j)\}_{j=m+1}^q \\ \{(1-a_j, A_j)\}_{j=1}^n ; \{(1-a_j, A_j)\}_{j=n+1}^p \end{pmatrix} \in \Omega, \tag{1.9}$$

and for another element $\Delta' = \begin{pmatrix} \{(a'_j, A'_j)\}_{j=1}^{n'} ; \{(a'_j, A'_j)\}_{j=n'+1}^{p'} \\ \{(b'_j, B'_j)\}_{j=1}^{m'} ; \{(b'_j, B'_j)\}_{j=m'+1}^{q'} \end{pmatrix}$ of Ω, we define

$$\begin{aligned} &\Delta \oplus \Delta' \\ &= \begin{pmatrix} \{(a_j, A_j)\}_{j=1}^n, \{(a'_j, A'_j)\}_{j=1}^{n'} ; \{(a_j, A_j)\}_{j=n+1}^p, \{(a'_j, A'_j)\}_{j=n'+1}^{p'} \\ \{(b_j, B_j)\}_{j=1}^m, \{(b'_j, B'_j)\}_{j=1}^{m'} ; \{(b_j, B_j)\}_{j=m+1}^q, \{(b'_j, B'_j)\}_{j=m'+1}^{q'} \end{pmatrix} \in \Omega. \end{aligned} \tag{1.10}$$

Then we have the following relations between the Gamma factors:

$$\begin{aligned} (\Delta + c)(s) &= \Delta(s + c) \quad (c \in \mathbb{C}), \\ (C \cdot \Delta)(s) &= \Delta(C\,s) \quad (C \in \mathbb{R}^+), \\ \Delta^*(s) &= \Delta(-s), \\ (\Delta \oplus \Delta')(s) &= \Delta(s)\,\Delta'(s). \end{aligned}$$

Furthermore, we have the following relations between the H-functions:

$$H(z\,|\,\Delta+c) = z^c\,H(z\,|\,\Delta) \quad (c\in\mathbb{C}), \tag{1.11}$$

$$H\left(z\,\middle|\,\frac{1}{C}\cdot\Delta\right) = C\,H(z^C|\,\Delta) \quad (C\in\mathbb{R}^+), \tag{1.12}$$

$$H(z\,|\,\Delta^*) = H\left(\frac{1}{z}\,\middle|\,\Delta\right), \tag{1.13}$$

$$\int_0^\infty H(t\,z\,|\,\Delta)\,H\left(\frac{1}{t\,z'}\,\middle|\,\Delta'\right)\frac{\mathrm{d}t}{t} = H\left(\frac{z}{z'}\,\middle|\,\Delta\oplus\Delta'\right). \tag{1.14}$$

2. Assumptions

In this section, we state the assumptions in our theorem. Let H and I be two natural numbers. Let $\{\lambda_k^{(h)}\}_{k=1}^\infty$ $(1 \leqq h \leqq H)$, $\{\mu_k^{(i)}\}_{k=1}^\infty$ $(1 \leqq i \leqq I)$ be increasing sequences of positive real numbers tending to ∞, and let $\{\alpha_k^{(h)}\}_{k=1}^\infty$ $(1 \leqq h \leqq H)$, $\{\beta_k^{(i)}\}_{k=1}^\infty$ $(1 \leqq i \leqq I)$ be complex sequences. We assume that the Dirichlet series

$$\varphi_h(s) = \sum_{k=1}^\infty \frac{\alpha_k^{(h)}}{\lambda_k^{(h)s}} \quad (1 \leqq h \leqq H) \tag{2.1}$$

and

$$\psi_i(s) = \sum_{k=1}^\infty \frac{\beta_k^{(i)}}{\mu_k^{(i)s}} \quad (1 \leqq i \leqq I) \tag{2.2}$$

have finite abscissas of absolute convergence $\sigma_{\varphi_h}(1 \leqq h \leqq H)$ and $\sigma_{\psi_i}(1 \leqq i \leqq I)$, respectively. We also assume the existence of a meromorphic function χ satisfying the functional equation

$$\chi(s) = \begin{cases} \displaystyle\sum_{h=1}^H \Delta_1^{(h)}(s)\,\varphi_h(s) & \operatorname{Re}(s) > \displaystyle\max_{1\leqq h\leqq H}(\sigma_{\varphi_h}), \\ \displaystyle\sum_{i=1}^I \Delta_2^{(i)*}(r-s)\,\psi_i(r-s) & \operatorname{Re}(s) < \displaystyle\min_{1\leqq i\leqq I}(r-\sigma_{\psi_i}), \end{cases} \tag{2.3}$$

where $\Delta_1^{(h)}(1 \leqq h \leqq H)$, $\Delta_2^{(i)}(1 \leqq i \leqq I)$ are elements of Ω, and r is a real number. We further assume that among the poles of $\chi(s)$, only finitely many distinct s_k $(1 \leqq k \leqq L)$ are neither a pole of $\Delta_1^{(h)+}(s)$ nor a pole of $\left(\Delta_2^{(i)} - r\right)^-(s)$. We denote the set of such poles by $S = \{s_k \mid 1 \leqq k \leqq L\}$.

Now, we introduce a gamma factor associated to $\Delta \in \Omega$ and suppose that for any pair of real numbers $u_1, u_2\,(u_1 < u_2)$, we have a convergence

$\lim_{|v|\to\infty} (\Delta - s)(u+iv)\,\chi(u+iv) = 0$ uniformly in $u_1 \leqq u \leqq u_2$. Let

$$L_1 = L_1(s) : \gamma_1 - i\infty \to \gamma_1 + i\infty,$$
$$L_2 = L_2(s) : \gamma_2 - i\infty \to \gamma_2 + i\infty,$$

(γ_1, γ_2 : real, $\gamma_2 < \gamma_1$) be two non-intersecting paths in the w-plane depending on s. We assume that there exists a real number Y such that L_1 and L_2 coincide with the two lines $\mathrm{Re}(w) = \gamma_1$ and $\mathrm{Re}(w) = \gamma_2$, respectively, for $|\mathrm{Im}(w)| > Y$, and that all the poles of $\big((\Delta - s) \oplus \Delta_1^{(h)}\big)^+(w)$ lie to the right of L_1, and those of $\big((\Delta - s) \oplus \Delta_1^{(h)}\big)^-(w)$ and S lie to the left of L_1, and all the poles of $\big((\Delta - s) \oplus \big(\Delta_2^{(i)} - r\big)\big)^+(w)$ and S lie to the right of L_2, and those of $\big((\Delta - s) \oplus \big(\Delta_2^{(i)} - r\big)\big)^-(w)$ lie to the left of L_2 (cf. Fig. 2).

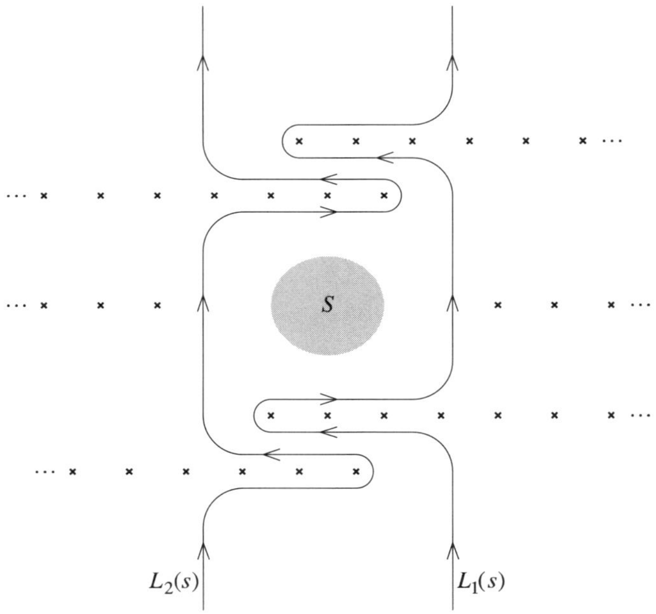

Fig. 2. the paths L_1 and L_2

Under these assumptions, we define a function $X(z, s, \Delta)$ by

$$X(z, s, \Delta) = \frac{1}{2\pi i} \int_{L_1} \Delta(w - s)\,\chi(w)\,z^{-w}\,\mathrm{d}w. \tag{2.4}$$

Then, by the Cauchy residue theorem, we have

$$\begin{aligned} X(z,s,\Delta) &= \frac{1}{2\pi i}\int_{L_2} \Delta(w-s)\,\chi(w)\,z^{-w}\,\mathrm{d}w \\ &\quad + \sum_{k=1}^{L} \operatorname{Res}\big(\Delta(w-s)\,\chi(w)\,z^{-w}, w=s_k\big)\,. \end{aligned} \tag{2.5}$$

3. Theorem

In this section we state and prove our theorem.

Theorem 3.1. *For the H-function series*

$$\Phi_h(z,s,\Delta) = \sum_{k=1}^{\infty} \alpha_k^{(h)} H\Big(z\lambda_k^{(h)} \;\Big|(\Delta - s)\oplus \Delta_1^{(h)}\Big) \tag{3.1}$$

and

$$\Psi_i(z,s,\Delta) = \sum_{k=1}^{\infty} \beta_k^{(i)} H\Big(z\mu_k^{(i)} \;\Big|(\Delta - s)\oplus \Delta_2^{(i)*}\Big), \tag{3.2}$$

the following modular relation holds:

$$X(z,s,\Delta) = \begin{cases} \displaystyle\sum_{h=1}^{H} \Phi_h(z,s,\Delta) & \text{if } L_1 \text{ can be taken to the right of } \displaystyle\max_{1\leqq h\leqq H}(\sigma_{\varphi_h}), \\ \displaystyle z^{-r}\sum_{i=1}^{I}\Psi_i\Big(\frac{1}{z}, r-s, \Delta^*\Big) + \sum_{k=1}^{L}\operatorname{Res}\big(\Delta(w-s)\,\chi(w)\,z^{-w}, w=s_k\big) & \\ \qquad\qquad \text{if } L_2 \text{ can be taken to the left of } \displaystyle\min_{1\leqq i\leqq I}(r-\sigma_{\psi_i}), & \end{cases} \tag{3.3}$$

for Δ and z such that the H-functions on the right-hand side converge absolutely.

Proof. If L_1 can be taken to the right of $\max_{1\leqq h\leqq H}(\sigma_{\varphi_h})$ (cf. Fig. 3), by using (2.4), (2.3) and (2.1) successively, we have

$$\begin{aligned} X(z,s,\Delta) &= \frac{1}{2\pi i}\int_{L_1} \Delta(w-s)\,\chi(w)\,z^{-w}\,\mathrm{d}w \\ &= \frac{1}{2\pi i}\int_{L_1} \Delta(w-s)\,\sum_{h=1}^{H}\Delta_1^{(h)}(w)\,\varphi_h(w)\,z^{-w}\,\mathrm{d}w \end{aligned}$$

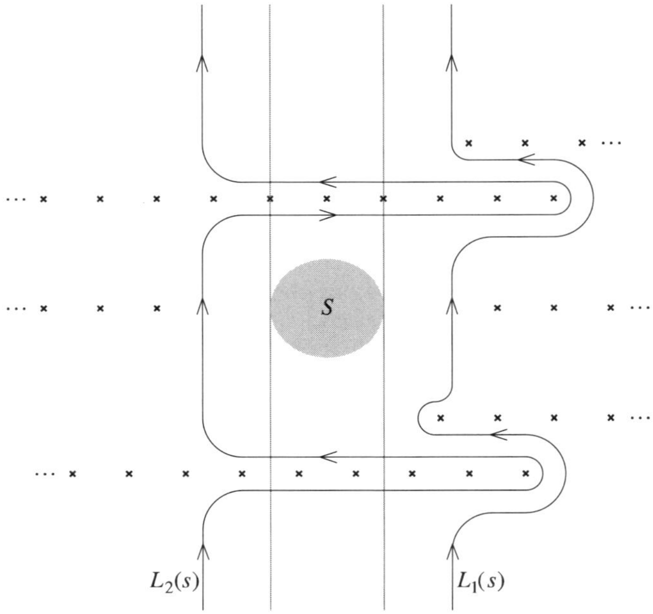

Fig. 3. the paths L_1 and L_2

$$= \sum_{h=1}^{H} \frac{1}{2\pi i} \int_{L_1} \Delta(w-s)\, \Delta_1^{(h)}(w) \sum_{k=1}^{\infty} \frac{\alpha_k^{(h)}}{\left(z\lambda_k^{(h)}\right)^w} \, dw$$

$$= \sum_{h=1}^{H} \sum_{k=1}^{\infty} \alpha_k^{(h)} \frac{1}{2\pi i} \int_{L_1} (\Delta - s)(w)\, \Delta_1^{(h)}(w) \left(z\lambda_k^{(h)}\right)^{-w} dw,$$

where the exchange of the integration and the summation is permitted by absolute convergence. Now the resulting integral is an H-function, and by (3.1), we have

$$\begin{aligned} X(z,s,\Delta) &= \sum_{h=1}^{H} \sum_{k=1}^{\infty} \alpha_k^{(h)} H\left(z\lambda_k^{(h)} \,\middle|\, (\Delta - s) \oplus \Delta_1^{(h)} \right) \\ &= \sum_{h=1}^{H} \Phi_h(z,s,\Delta)\,. \end{aligned}$$

On the other hand, if L_2 can be taken to the left of $\min_{1\leqq i\leqq I}(r-\sigma_{\psi_i})$ (cf. Fig. 4), by using (2.5), (2.3) and (2.2) successively, we have

$$\begin{aligned}
X(z,s,\Delta) &- \sum_{k=1}^{L} \operatorname{Res}\bigl(\Delta(w-s)\,\chi(w)\,z^{-w}, w=s_k\bigr) \\
&= \frac{1}{2\pi i}\int_{L_2} \Delta(w-s)\,\chi(w)\,z^{-w}\,\mathrm{d}w \\
&= \frac{1}{2\pi i}\int_{L_2} \Delta(w-s)\sum_{i=1}^{I}\Delta_2^{(i)*}(r-w)\,\psi_i(r-w)\,z^{-w}\,\mathrm{d}w \\
&= \sum_{i=1}^{I}\frac{1}{2\pi i}\int_{L_2}\Delta(w-s)\,\Delta_2^{(i)}(w-r)\sum_{k=1}^{\infty}\frac{\beta_k^{(i)}}{\left(\frac{\mu_k^{(i)}}{z}\right)^{-w}}\,\mathrm{d}w\left(\frac{1}{\mu_k^{(i)}}\right)^{r} \\
&= \sum_{i=1}^{I}\sum_{k=1}^{\infty}\beta_k^{(i)}\frac{1}{2\pi i}\int_{L_2}(\Delta-s)(w)\bigl(\Delta_2^{(i)}-r\bigr)(w)\left(\frac{z}{\mu_k^{(i)}}\right)^{-w}\mathrm{d}w\left(\frac{1}{\mu_k^{(i)}}\right)^{r},
\end{aligned}$$

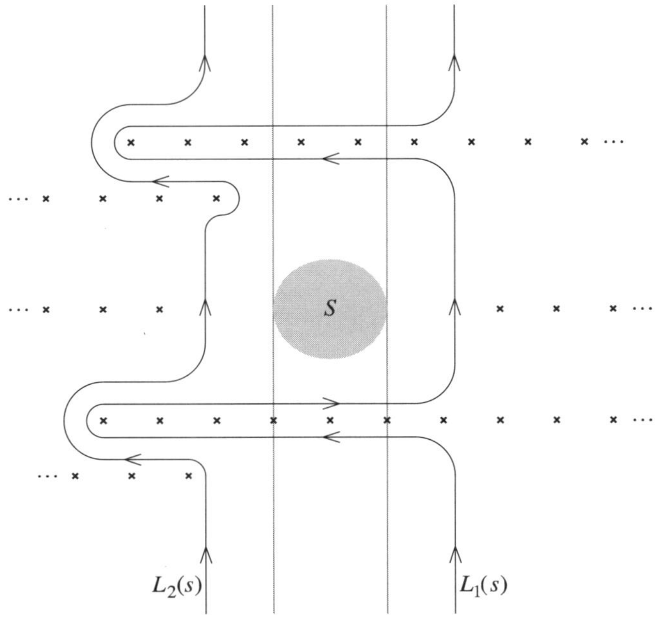

Fig. 4. the paths L_1 and L_2

where we changed the order of the integration and the summation. Now the resulting integral is an H-function which can be transformed as follows by the properties (1.11) and (1.13):

$$\begin{aligned}
&\frac{1}{2\pi i}\int_{L_2}(\Delta - s)(w)(\Delta_2^{(i)} - r)(w)\left(\frac{z}{\mu_k^{(i)}}\right)^{-w} dw \left(\frac{1}{\mu_k^{(i)}}\right)^{r} \\
&= z^{-r}\, H\left(\frac{z}{\mu_k^{(i)}}\,\middle|\,(\Delta - s)\oplus(\Delta_2^{(i)} - r)\right)\left(\frac{z}{\mu_k^{(i)}}\right)^{r} \\
&= z^{-r}\, H\left(\frac{z}{\mu_k^{(i)}}\,\middle|\,(\Delta + (r - s))\oplus \Delta_2^{(i)}\right) \\
&= z^{-r}\, H\left(\frac{\mu_k^{(i)}}{z}\,\middle|\,(\Delta^* - (r - s))\oplus \Delta_2^{(i)*}\right).
\end{aligned}$$

Therefore, we have, by (3.2),

$$\begin{aligned}
X(z, s, \Delta) &- \sum_{k=1}^{L} \operatorname{Res}\big(\Delta(w - s)\,\chi(w)\,z^{-w}, w = s_k\big) \\
&= z^{-r}\sum_{i=1}^{I}\sum_{k=1}^{\infty}\beta_k^{(i)} H\left(\frac{\mu_k^{(i)}}{z}\,\middle|\,(\Delta^* - (r - s))\oplus \Delta_2^{(i)*}\right) \\
&= z^{-r}\sum_{i=1}^{I}\Psi_i\left(\frac{1}{z}, r - s, \Delta^*\right).
\end{aligned}$$

This completes the proof of (3.3). □

Remark 3.1. Let $p : \Omega \to \Xi = \left(\mathbb{R}^{+(\infty)}\right)^4$ be a map defined by

$$\begin{pmatrix}\{(a_j, A_j)\}_{j=1}^{n};\{(a_j, A_j)\}_{j=n+1}^{p} \\ \{(b_j, B_j)\}_{j=1}^{m};\{(b_j, B_j)\}_{j=m+1}^{q}\end{pmatrix} \mapsto \begin{pmatrix}\{A_j\}_{j=1}^{n};\{A_j\}_{j=n+1}^{p} \\ \{B_j\}_{j=1}^{m};\{B_j\}_{j=m+1}^{q}\end{pmatrix},$$

then for each element $\Lambda \in \Xi$, the inverse image $p^{-1}(\Lambda)$ is a finite-dimensional complex vector space, and we can regard the space $\Omega = \coprod_{\Lambda\in\Xi} p^{-1}(\Lambda)$ as a disjoint union. In this sense, we may regard $\Delta \in \Omega$ as a "complex variable", and then the use of variable s in Theorem 3.1 is useful but unnecessary, since we have $\Phi_h(z, s, \Delta) = \Phi_h(z, 0, \Delta - s)$ and $\Psi_i(z, s, \Delta) = \Psi_i(z, 0, \Delta - s)$.

4. A General Modular Relation associated to the Riemann Zeta-function

Even in the simplest case of Hecke type

$$\chi(s) = \begin{cases} \Gamma(s)\,\varphi(s) & \mathrm{Re}(s) > \sigma_\varphi, \\ \Gamma(r-s)\,\psi(r-s) & \mathrm{Re}(s) < r - \sigma_\psi, \end{cases} \tag{4.1}$$

(i. e. $H = I = 1$, $\Delta_1 = \Delta_2^* = \left(\begin{smallmatrix} - & ; & - \\ (0,1) & ; & - \end{smallmatrix}\right)$), the above modular relation (3.3) includes Bochner's formula [6], the Riesz sum [8], a K-Bessel expansion [5], the incomplete Gamma expansion [9,11,17,27], and another Bochner's formula [7], to name a few. Examples of non-Hecke type were studied in [1–4], and also in [14,15] using our formalism. Numerous concrete examples will appear in the forthcoming book [16].

In this section, we consider the general modular relation associated to the Riemann zeta-function defined by the Dirichlet series

$$\zeta(s) = \sum_{k=1}^{\infty} \frac{1}{k^s}, \quad \mathrm{Re}(s) > 1. \tag{4.2}$$

If we set

$$Z(s) = \frac{\zeta(2s)}{\pi^s} = \sum_{k=1}^{\infty} \frac{1}{\left(\pi k^2\right)^s}, \tag{4.3}$$

then it satisfies the functional equation of Hecke type:

$$\Gamma(s)\,Z(s) = \Gamma\left(\tfrac{1}{2} - s\right) Z\left(\tfrac{1}{2} - s\right), \tag{4.4}$$

after analytic continuation.

By Theorem 3.1, we have the following modular relation:

$$\begin{aligned} &\sum_{k=1}^{\infty} H^{m+1,n}_{p,q+1}\left(z\pi k^2 \,\middle|\, \begin{matrix} \{(1-a_j-A_js, A_j)\}_{j=1}^{n}, \{(a_j-A_js, A_j)\}_{j=n+1}^{p} \\ (0,1), \{(b_j-B_js, B_j)\}_{j=1}^{m}, \{(1-b_j-B_js, B_j)\}_{j=m+1}^{q} \end{matrix} \right) \\ &\qquad - \mathrm{Res}\left(\Gamma(w)Z(w)\Delta(w-s)\,z^{-w}, w=0\right) \\ &= z^{-\frac{1}{2}} \sum_{k=1}^{\infty} H^{n+1,m}_{q,p+1}\left(\frac{\pi k^2}{z} \,\middle|\, \begin{matrix} \{(1-b_j-B_j(\tfrac{1}{2}-s), B_j)\}_{j=1}^{m}, \{(b_j-B_j(\tfrac{1}{2}-s), B_j)\}_{j=m+1}^{q} \\ (0,1), \{(a_j-A_j(\tfrac{1}{2}-s), A_j)\}_{j=1}^{n}, \{(1-a_j-A_j(\tfrac{1}{2}-s), A_j)\}_{j=n+1}^{p} \end{matrix} \right) \\ &\qquad + \mathrm{Res}\left(\Gamma(w)Z(w)\Delta(w-s)\,z^{-w}, w=\tfrac{1}{2}\right), \end{aligned} \tag{4.5}$$

which can be rewritten as

$$\begin{aligned}&\frac{1}{\pi^s}\sum_{k=1}^{\infty}\frac{1}{k^{2s}}\,H^{m+1,n}_{p,q+1}\left(z\pi k^2\,\middle|\,\begin{matrix}\{(1-a_j,A_j)\}_{j=1}^{n},\ \{(a_j,A_j)\}_{j=n+1}^{p}\\(s,1),\{(b_j,B_j)\}_{j=1}^{m},\{(1-b_j,B_j)\}_{j=m+1}^{q}\end{matrix}\right)\\&\qquad+\operatorname{Res}\Bigl(\Gamma(w)Z(w)\Delta\bigl(-w-s+\tfrac{1}{2}\bigr)\,z^{s-\frac{1}{2}+w},w=\tfrac{1}{2}\Bigr)\\&=\frac{1}{\pi^{\frac{1}{2}-s}}\sum_{k=1}^{\infty}\frac{1}{k^{1-2s}}\\&\quad\times H^{n+1,m}_{q,p+1}\left(\frac{\pi k^2}{z}\,\middle|\,\begin{matrix}\{(1-b_j,B_j)\}_{j=1}^{m},\ \{(b_j,B_j)\}_{j=m+1}^{q}\\(\frac{1}{2}-s,1),\{(a_j,A_j)\}_{j=1}^{n},\{(1-a_j,A_j)\}_{j=n+1}^{p}\end{matrix}\right)\\&\qquad+\operatorname{Res}\bigl(\Gamma(w)Z(w)\Delta(w-s)\,z^{s-w},w=\tfrac{1}{2}\bigr),\end{aligned}\tag{4.6}$$

and in this form, the modular relation is given by the following simultaneous exchange of the parameters:

$$\boxed{\begin{array}{c}s\leftrightarrow\frac{1}{2}-s\\[4pt] z\leftrightarrow\frac{1}{z}\\[4pt] \bigl\{(a_j,A_j)\bigr\}_{j=1}^{n}\leftrightarrow\bigl\{(b_j,B_j)\bigr\}_{j=1}^{m}\\[4pt] \bigl\{(a_j,A_j)\bigr\}_{j=n+1}^{p}\leftrightarrow\bigl\{(b_j,B_j)\bigr\}_{j=m+1}^{q}\\[4pt] \bigl(\Delta\leftrightarrow\Delta^*\bigr)\end{array}}\tag{4.7}$$

Example 4.1. (Bochner's formula [6]). As the simplest example of (4.5), we have

$$\begin{aligned}&\sum_{k=1}^{\infty}H^{1,0}_{0,1}\left(z\pi k^2\,\middle|\,\begin{matrix}-\\(0,1)\end{matrix}\right)+\frac{1}{2}\\&\qquad=z^{-\frac{1}{2}}\sum_{k=1}^{\infty}H^{1,0}_{0,1}\left(\frac{\pi k^2}{z}\,\middle|\,\begin{matrix}-\\(0,1)\end{matrix}\right)+\frac{1}{2}\,z^{-\frac{1}{2}}.\end{aligned}$$

Since

$$H^{1,0}_{0,1}\left(z\,\middle|\,\begin{matrix}-\\(0,1)\end{matrix}\right)=e^{-z},\tag{4.8}$$

we get Bochner's formula:

$$\sum_{k=1}^{\infty}e^{-z\pi k^2}+\frac{1}{2}=z^{-\frac{1}{2}}\left(\sum_{k=1}^{\infty}e^{-\frac{\pi k^2}{z}}+\frac{1}{2}\right),$$

which can be rewritten as the modular relation of the theta function:

$$\theta(z)=\frac{1}{\sqrt{z}}\,\theta\Big(\frac{1}{z}\Big),\quad \theta(z)=\sum_{k\in\mathbf{Z}}e^{-z\pi k^2}, \tag{4.9}$$

where $\mathrm{Re}(z)>0$.

Example 4.2. (Riesz sum [8]). As an example of (4.6), we have

$$\begin{aligned}&\frac{1}{\pi^s}\sum_{k=1}^{\infty}\frac{1}{k^{2s}}\,H_{1,1}^{1,0}\Big(z\pi k^2\,\Big|\begin{matrix}(a,1)\\(s,1)\end{matrix}\Big)+\mathrm{Res}\Big(\frac{\Gamma(w)Z(w)}{\Gamma(a-w-s+\frac{1}{2})}\,z^{s-\frac{1}{2}+w},w=\tfrac{1}{2}\Big)\\&=\frac{1}{\pi^{\frac{1}{2}-s}}\sum_{k=1}^{\infty}\frac{1}{k^{1-2s}}\,H_{0,2}^{1,0}\Big(\frac{\pi k^2}{z}\,\Big|\begin{matrix}-\\(\frac{1}{2}-s,1),(1-a,1)\end{matrix}\Big)\\&\qquad\qquad+\mathrm{Res}\Big(\frac{\Gamma(w)Z(w)}{\Gamma(a+w-s)}\,z^{s-w},w=\tfrac{1}{2}\Big).\end{aligned}$$

Setting $a=1$, and using

$$H_{1,1}^{1,0}\Big(z\,\Big|\begin{matrix}(1,1)\\(s,1)\end{matrix}\Big)=\begin{cases}\dfrac{1}{\Gamma(1-s)}\Big(\dfrac{z}{1-z}\Big)^s & |z|<1,\\ 0 & |z|>1,\end{cases} \tag{4.10}$$

and

$$H_{0,2}^{1,0}\Big(z\,\Big|\begin{matrix}-\\(s,1),(0,1)\end{matrix}\Big)=z^{\frac{s}{2}}J_s\big(2\sqrt{z}\big), \tag{4.11}$$

where J_s denotes the Bessel function of the first kind, we get the Riesz sum formula:

$$\begin{aligned}&\frac{1}{\Gamma(1-s)}\sum_{\substack{-w<k<w\\k\in\mathbf{Z}}}\frac{1}{(w^2-k^2)^s}\\&\quad=2\,\pi^s w^{\frac{1}{2}-s}\sum_{k=1}^{\infty}\frac{1}{k^{\frac{1}{2}-s}}\,J_{\frac{1}{2}-s}(2w\pi k)+\frac{1}{\Gamma(\frac{3}{2}-s)}\,\pi^{\frac{1}{2}}w^{1-2s},\end{aligned} \tag{4.12}$$

where $w=\frac{1}{\sqrt{z\pi}}>0$, $w\notin\mathbf{Z}$, $s<1$.

Example 4.3. (K-Bessel expansion [5,30]). As the second example of (4.6), we have

$$\begin{aligned}\frac{1}{\pi^s}\sum_{k=1}^{\infty}\frac{1}{k^{2s}}\,&H_{0,2}^{2,0}\left(z\pi k^2\,\middle|\,\begin{matrix}-\\(s,1),(a,1)\end{matrix}\right)\\&+\operatorname{Res}\left(\Gamma(w)Z(w)\Gamma\left(a-w-s+\tfrac{1}{2}\right)z^{s-\frac{1}{2}+w},w=\tfrac{1}{2}\right)\\=\frac{1}{\pi^{\frac{1}{2}-s}}\sum_{k=1}^{\infty}\frac{1}{k^{1-2s}}\,&H_{1,1}^{1,1}\left(\frac{\pi k^2}{z}\,\middle|\,\begin{matrix}(1-a,1)\\(\frac{1}{2}-s,1)\end{matrix}\right)\\&+\operatorname{Res}\left(\Gamma(w)Z(w)\Gamma(a+w-s)\,z^{s-w},w=\tfrac{1}{2}\right).\end{aligned}$$

Setting $a=0$, and using

$$H_{0,2}^{2,0}\left(z\,\middle|\,\begin{matrix}-\\(s,1),(0,1)\end{matrix}\right)=2\,z^{\frac{s}{2}}K_s(2\sqrt{z}) \tag{4.13}$$

where K_s denotes the modified Bessel function of the second kind, and

$$H_{1,1}^{1,1}\left(z\,\middle|\,\begin{matrix}(1,1)\\(s,1)\end{matrix}\right)=\Gamma(s)\left(\frac{z}{z+1}\right)^{s}, \tag{4.14}$$

we get a K-Bessel expansion:

$$\begin{aligned}\frac{2\,z^{\frac{s}{2}}}{\pi^{\frac{s}{2}}}\sum_{k=1}^{\infty}\frac{1}{k^s}\,K_s(2\sqrt{z\pi}\,k)+\frac{1}{2}\Gamma(-s)\,z^s\\=\frac{1}{2}\Gamma\left(\tfrac{1}{2}-s\right)\sum_{k\in\mathbf{Z}}\frac{1}{(\pi k^2+z)^{\frac{1}{2}-s}}.\end{aligned} \tag{4.15}$$

We note that Formula (4.15) leads to the Chowla-Selberg formula [10,26].

In the special case of $s=-\frac{1}{2}$, we have

$$H_{0,2}^{2,0}\left(z\,\middle|\,\begin{matrix}-\\(-\frac{1}{2},1),(0,1)\end{matrix}\right)=\frac{\sqrt{\pi}}{\sqrt{z}}\,e^{-2\sqrt{z}}, \tag{4.16}$$

and therefore

$$\frac{\sqrt{\pi}}{\sqrt{z}}\sum_{k\in\mathbf{Z}}e^{-2\sqrt{z\pi}\,|k|}=\sum_{k\in\mathbf{Z}}\frac{1}{\pi k^2+z},$$

which is the partial fraction expansion of the hyperbolic cotangent:

$$\frac{\coth(\pi w)}{w}=\frac{1}{\pi}\sum_{k\in\mathbf{Z}}\frac{1}{k^2+w^2}, \tag{4.17}$$

where $w=\frac{\sqrt{z}}{\sqrt{\pi}}$.

Example 4.4. (Incomplete Gamma expansion [9,11,27], Bochner's formula [7]). As the third example of (4.6), we have

$$\frac{1}{\pi^s}\sum_{k=1}^{\infty}\frac{1}{k^{2s}}H^{2,0}_{1,2}\left(z\pi k^2\,\middle|\,\begin{matrix}(a,A)\\(s,1),(b,B)\end{matrix}\right)$$
$$+\operatorname{Res}\left(\Gamma(w)Z(w)\frac{\Gamma(b-Bw-Bs+\frac{B}{2})}{\Gamma(a-Aw-As+\frac{A}{2})}z^{s-\frac12+w}, w=\tfrac12\right)$$
$$=\frac{1}{\pi^{\frac12-s}}\sum_{k=1}^{\infty}\frac{1}{k^{1-2s}}H^{1,1}_{1,2}\left(\frac{\pi k^2}{z}\,\middle|\,\begin{matrix}(1-b,B)\\(\frac12-s,1),(1-a,A)\end{matrix}\right)$$
$$+\operatorname{Res}\left(\Gamma(w)Z(w)\frac{\Gamma(b+Bw-Bs)}{\Gamma(a+Aw-As)}z^{s-w}, w=\tfrac12\right).$$

Setting $a=1$, $A=1$, $b=0$, $B=1$, and using

$$H^{2,0}_{1,2}\left(z\,\middle|\,\begin{matrix}(1,1)\\(s,1),(0,1)\end{matrix}\right)=\Gamma(s,z) \tag{4.18}$$

and

$$H^{1,1}_{1,2}\left(z\,\middle|\,\begin{matrix}(1,1)\\(s,1),(0,1)\end{matrix}\right)=\Gamma(s)-\Gamma(s,z), \tag{4.19}$$

we get an incomplete Gamma expansion:

$$\frac{1}{\pi^s}\sum_{k=1}^{\infty}\frac{1}{k^{2s}}\Gamma(s,z\pi k^2)-\frac12\,\frac1s\,z^s$$
$$=\frac{1}{\pi^{\frac12-s}}\sum_{k=1}^{\infty}\frac{1}{k^{1-2s}}\left(\Gamma(\tfrac12-s)-\Gamma\left(\tfrac12-s,\frac{\pi k^2}{z}\right)\right)-\frac12\,\frac{1}{s-\frac12}\,z^{s-\frac12}. \tag{4.20}$$

We note that Formula (4.20) is often referred to as the Ewald expansion [11,27] in the disciplines.

On the other hand, setting $a=s$, $A=1$, $b=0$, and using

$$H^{2,0}_{1,2}\left(z\,\middle|\,\begin{matrix}(s,1)\\(s,1),(0,B)\end{matrix}\right)=H^{1,0}_{0,1}\left(z\,\middle|\,\begin{matrix}-\\(0,B)\end{matrix}\right)=\frac1B e^{-\sqrt[B]{z}} \tag{4.21}$$

and

$$H^{1,1}_{1,2}\left(z\,\middle|\,\begin{matrix}(1,B)\\(s,1),(s+\frac12,1)\end{matrix}\right)$$
$$=z^s\,{}_1\Psi_1\left(\begin{matrix}(Bs,B)\\(\frac12,1)\end{matrix};-z\right)=z^s\sum_{k=0}^{\infty}\frac{\Gamma(Bs+Bk)}{\Gamma(\frac12+k)}\frac{(-z)^k}{k!}, \tag{4.22}$$

where ${}_1\Psi_1$ denotes the Ψ-function, we get another formula of Bochner:

$$\frac{1}{B}\,\frac{1}{\pi^s}\sum_{k=1}^{\infty}\frac{1}{k^{2s}}\,e^{-\sqrt[B]{z\pi k^2}} = \frac{1}{2}\,z^{s-\frac{1}{2}}\sum_{k\in\mathbf{Z}}{}_1\Psi_1\left(\begin{matrix}\left(\frac{B}{2}-Bs,B\right)\\ \left(\frac{1}{2},1\right)\end{matrix};-\frac{\pi k^2}{z}\right). \tag{4.23}$$

In particular, for $B=1$, we have

$$H^{1,1}_{1,2}\left(z\,\middle|\,\begin{matrix}(1,1)\\ (s,1),\left(s+\frac{1}{2},1\right)\end{matrix}\right) = \frac{\Gamma(s)}{\sqrt{\pi}}\,z^s\,{}_1F_1\left(\begin{matrix}s\\ \frac{1}{2}\end{matrix};-z\right), \tag{4.24}$$

where ${}_1F_1$ denotes the hypergeometric function, and

$$\frac{1}{\pi^s}\sum_{k=1}^{\infty}\frac{1}{k^{2s}}\,e^{-z\pi k^2} = \frac{1}{2}\,\frac{\Gamma\left(\frac{1}{2}-s\right)}{\sqrt{\pi}}\,z^{s-\frac{1}{2}}\sum_{k\in\mathbf{Z}}{}_1F_1\left(\begin{matrix}\frac{1}{2}-s\\ \frac{1}{2}\end{matrix};-\frac{\pi k^2}{z}\right). \tag{4.25}$$

Example 4.5. (Hurwitz's formula [19,29]). As the last example of (4.6), we have

$$\begin{aligned}
&\frac{1}{\pi^s}\sum_{k=1}^{\infty}\frac{1}{k^{2s}}\,H^{2,2}_{2,2}\left(z\pi k^2\,\middle|\,\begin{matrix}(1-a,1),(1-b,1)\\ (s,1),(c,1)\end{matrix}\right)\\
&\quad+\operatorname{Res}\Big(\Gamma(w)Z(w)\Gamma\left(a+w+s-\tfrac{1}{2}\right)\Gamma\left(b+w+s-\tfrac{1}{2}\right)\\
&\qquad\qquad\qquad\qquad\Gamma\left(c-w-s+\tfrac{1}{2}\right)z^{s-\frac{1}{2}+w},\,w=\tfrac{1}{2}\Big)\\
&=\frac{1}{\pi^{\frac{1}{2}-s}}\sum_{k=1}^{\infty}\frac{1}{k^{1-2s}}\,H^{3,1}_{1,3}\left(\frac{\pi k^2}{z}\,\middle|\,\begin{matrix}(1-c,1)\\ \left(\frac{1}{2}-s,1\right),(a,1),(b,1)\end{matrix}\right)\\
&\quad+\operatorname{Res}\left(\Gamma(w)Z(w)\Gamma(a-w+s)\Gamma(b-w+s)\Gamma(c+w-s)\,z^{s-w},\,w=\tfrac{1}{2}\right).
\end{aligned}$$

Setting $a=\frac{1}{2}$, $b=0$, $c=s+\frac{1}{2}$, and using

$$H^{2,2}_{2,2}\left(z\,\middle|\,\begin{matrix}\left(\frac{1}{2},1\right),(1,1)\\ (s,1),\left(s+\frac{1}{2},1\right)\end{matrix}\right) = 2^{1-2s}\pi\,\Gamma(2s)\,\frac{z^s}{\left(1+\sqrt{z}\right)^{2s}} \tag{4.26}$$

and

$$\begin{aligned}
&H^{3,1}_{1,3}\left(z\,\middle|\,\begin{matrix}(s,1)\\ (s,1),\left(\frac{1}{2},1\right),(0,1)\end{matrix}\right)\\
&=\sqrt{z}\,\Gamma(1-2s)\Big(e^{-s\pi i+2\sqrt{z}\,i}\,\Gamma\left(2s,2\sqrt{z}\,i\right)+e^{s\pi i-2\sqrt{z}\,i}\,\Gamma\left(2s,-2\sqrt{z}\,i\right)\Big),
\end{aligned} \tag{4.27}$$

we get a formula (cf. Ueno and Nishizawa [29])

$$\zeta(2s,w) = \frac{1}{w^{2s-1}} \sum_{k=1}^{\infty} \left(\frac{e^{-2\pi k w i}}{\left(2\pi k w\, e^{-\frac{\pi i}{2}}\right)^{1-2s}} \Gamma(1-2s, -2\pi k w\, i) + \frac{e^{2\pi k w i}}{\left(2\pi k w\, e^{\frac{\pi i}{2}}\right)^{1-2s}} \Gamma(1-2s, 2\pi k w\, i) \right) + \frac{1}{2\, w^{2s}} + \frac{1}{w^{2s-1}} \frac{1}{2s-1}, \tag{4.28}$$

where $0 < w = \frac{1}{\sqrt{\pi z}} < 1$, and

$$\zeta(s,w) = \sum_{k=0}^{\infty} \frac{1}{(k+w)^s}, \quad \mathrm{Re}(s) > 1, \tag{4.29}$$

is the Hurwitz zeta-function. We note that Formula (4.28) leads to Hurwitz's formula:

$$\zeta(2s,w) = \frac{\Gamma(1-2s)}{(2\pi)^{1-2s}} \left(e^{\frac{1-2s}{2}\pi i}\, l_{1-2s}(-w) + e^{-\frac{1-2s}{2}\pi i}\, l_{1-2s}(w) \right), \tag{4.30}$$

where

$$l_s(w) = \sum_{k=1}^{\infty} \frac{e^{2\pi i k w}}{k^s}, \quad \mathrm{Re}(s) > 1, \tag{4.31}$$

is the Lerch zeta-function or the polylogarithm function.

5. A General Modular Relation associated to a product of the Riemann Zeta-function

In this section, we consider a product of the Riemann zeta-function

$$\begin{aligned} Z\left(s+\tfrac{\nu}{2}\right) Z\left(s-\tfrac{\nu}{2}\right) &= \frac{\zeta(2s+\nu)\,\zeta(2s-\nu)}{\pi^{2s}} \\ &= \sum_{l=1}^{\infty} \sum_{k=1}^{\infty} \frac{l^{\nu}\, k^{-\nu}}{\left(\pi^2\, l^2\, k^2\right)^s} = \sum_{k=1}^{\infty} \frac{\sigma_{2\nu}(k)}{k^{\nu} \left(\pi^2\, k^2\right)^s}, \end{aligned} \tag{5.1}$$

where $\sigma_t(k) = \sum_{j|k} j^t$.

By the functional equation (4.4), we have

$$\begin{aligned} &\Gamma\left(s+\tfrac{\nu}{2}\right) \Gamma\left(s-\tfrac{\nu}{2}\right) Z\left(s+\tfrac{\nu}{2}\right) Z\left(s-\tfrac{\nu}{2}\right) \\ &= \Gamma\left(\tfrac{1}{2}-s+\tfrac{\nu}{2}\right) \Gamma\left(\tfrac{1}{2}-s-\tfrac{\nu}{2}\right) Z\left(\tfrac{1}{2}-s+\tfrac{\nu}{2}\right) Z\left(\tfrac{1}{2}-s-\tfrac{\nu}{2}\right). \end{aligned} \tag{5.2}$$

Therefore, by Theorem 3.1, we have the following modular relation:

$$\sum_{k=1}^{\infty}\frac{\sigma_{2\nu}(k)}{k^{\nu}}\,H_{p,q+2}^{m+2,n}\left(z\pi^2k^2\,\middle|\,\begin{matrix}\{(1-a_j-A_js,A_j)\}_{j=1}^{n},\\ (\frac{\nu}{2},1),(-\frac{\nu}{2},1),\{(b_j-B_js,B_j)\}_{j=1}^{m},\end{matrix}\begin{matrix}\{(a_j-A_js,A_j)\}_{j=n+1}^{p}\\ \{(1-b_j-B_js,B_j)\}_{j=m+1}^{q}\end{matrix}\right)$$
$$=z^{-\frac12}\sum_{k=1}^{\infty}\frac{\sigma_{2\nu}(k)}{k^{\nu}}\,H_{q,p+2}^{n+2,m}\left(\frac{\pi^2k^2}{z}\,\middle|\,\begin{matrix}\{(1-b_j-B_j(\frac12-s),B_j)\}_{j=1}^{m},\\ (\frac{\nu}{2},1),(-\frac{\nu}{2},1),\{(a_j-A_j(\frac12-s),A_j)\}_{j=1}^{n},\end{matrix}\begin{matrix}\{(b_j-B_j(\frac12-s),B_j)\}_{j=m+1}^{q}\\ \{(1-a_j-A_j(\frac12-s),A_j)\}_{j=n+1}^{p}\end{matrix}\right)$$
$$+\sum_{v\in S_\nu}\mathrm{Res}\Big(\Gamma(w+\tfrac{\nu}{2})\,\Gamma(w-\tfrac{\nu}{2})\,Z(w+\tfrac{\nu}{2})\,Z(w-\tfrac{\nu}{2})\,\Delta(w-s)\,z^{-w},w=v\Big),\tag{5.3}$$

which can be rewritten as

$$\frac{1}{\pi^{2s}}\sum_{k=1}^{\infty}\frac{\sigma_{2\nu}(k)}{k^{2s+\nu}}\,H_{p,q+2}^{m+2,n}\left(z\pi^2k^2\,\middle|\,\begin{matrix}\{(1-a_j,A_j)\}_{j=1}^{n},\\ (s+\frac{\nu}{2},1),(s-\frac{\nu}{2},1),\{(b_j,B_j)\}_{j=1}^{m},\end{matrix}\begin{matrix}\{(a_j,A_j)\}_{j=n+1}^{p}\\ \{(1-b_j,B_j)\}_{j=m+1}^{q}\end{matrix}\right)$$
$$=\frac{1}{\pi^{1-2s}}\sum_{k=1}^{\infty}\frac{\sigma_{2\nu}(k)}{k^{1-2s+\nu}}$$
$$\times H_{q,p+2}^{n+2,m}\left(\frac{\pi^2k^2}{z}\,\middle|\,\begin{matrix}\{(1-b_j,B_j)\}_{j=1}^{m},\\ (\frac12-s+\frac{\nu}{2},1),(\frac12-s-\frac{\nu}{2},1),\{(a_j,A_j)\}_{j=1}^{n},\end{matrix}\begin{matrix}\{(b_j,B_j)\}_{j=m+1}^{q}\\ \{(1-a_j,A_j)\}_{j=n+1}^{p}\end{matrix}\right)$$
$$+\sum_{v\in S_\nu}\mathrm{Res}\Big(\Gamma(w+\tfrac{\nu}{2})\,\Gamma(w-\tfrac{\nu}{2})\,Z(w+\tfrac{\nu}{2})\,Z(w-\tfrac{\nu}{2})\,\Delta(w-s)\,z^{s-w},w=v\Big),\tag{5.4}$$

where

$$S_\nu=\begin{cases}\{0.\frac12\} & \nu=0,\\ \{-\frac14,\frac14,\frac34\} & \nu=\pm\frac12,\\ \{\pm\frac{\nu}{2},\pm\frac{\nu}{2}+\frac12\} & \nu\neq0,\pm\frac12,\end{cases}\tag{5.5}$$

is the set of poles.

Example 5.1. As the simplest example of (5.3), we have

$$\sum_{k=1}^{\infty} \frac{\sigma_{2\nu}(k)}{k^{\nu}} H_{0,2}^{2,0}\left(z\pi^2 k^2 \,\middle|\, \begin{matrix} - \\ (\frac{\nu}{2},1),(-\frac{\nu}{2},1) \end{matrix} \right)$$
$$= z^{-\frac{1}{2}} \sum_{k=1}^{\infty} \frac{\sigma_{2\nu}(k)}{k^{\nu}} H_{0,2}^{2,0}\left(\frac{\pi^2 k^2}{z} \,\middle|\, \begin{matrix} - \\ (\frac{\nu}{2},1),(-\frac{\nu}{2},1) \end{matrix} \right)$$
$$+ \sum_{v\in S_\nu} \operatorname{Res}\left(\Gamma(w+\tfrac{\nu}{2})\,\Gamma(w-\tfrac{\nu}{2})\,Z(w+\tfrac{\nu}{2})\,Z(w-\tfrac{\nu}{2})\,z^{-w}, w=v\right).$$

Since

$$H_{0,2}^{2,0}\left(z \,\middle|\, \begin{matrix} - \\ (\frac{\nu}{2},1),(-\frac{\nu}{2},1) \end{matrix} \right) = 2\,K_\nu(2\sqrt{z}\,), \tag{5.6}$$

we get

$$2\sum_{k=1}^{\infty} \frac{\sigma_{2\nu}(k)}{k^{\nu}} K_\nu(2\sqrt{z}\,\pi k)$$
$$= 2\,z^{-\frac{1}{2}} \sum_{k=1}^{\infty} \frac{\sigma_{2\nu}(k)}{k^{\nu}} K_\nu\left(\frac{2\,\pi k}{\sqrt{z}}\right)$$
$$+ \sum_{v\in S_\nu} \operatorname{Res}\left(\Gamma(w+\tfrac{\nu}{2})\,\Gamma(w-\tfrac{\nu}{2})\,Z(w+\tfrac{\nu}{2})\,Z(w-\tfrac{\nu}{2})\,z^{-w}, w=v\right). \tag{5.7}$$

In particular, for $\nu = \frac{1}{2}$, we have

$$H_{0,2}^{2,0}\left(z \,\middle|\, \begin{matrix} - \\ (\frac{1}{4},1),(-\frac{1}{4},1) \end{matrix} \right) = \frac{\sqrt{\pi}}{\sqrt[4]{z}}\, e^{-2\sqrt{z}}, \tag{5.8}$$

and by using the relation

$$\begin{aligned} \sum_{k=1}^{\infty} \frac{\sigma_1(k)}{k} e^{-kz} &= \sum_{n=1}^{\infty}\sum_{m=1}^{\infty} \frac{1}{m} e^{-nmz} \\ &= -\sum_{n=1}^{\infty} \log\left(1-e^{-nz}\right) = -\log\left(\eta\left(\frac{iz}{2\pi}\right)\right) - \frac{z}{24}, \end{aligned} \tag{5.9}$$

where

$$\eta(\tau) = e^{\frac{\pi i}{12}\tau} \prod_{n=1}^{\infty}\left(1-e^{2\pi i n\tau}\right) \tag{5.10}$$

is the eta function, we get its transformation formula

$$\log(\eta(\tau)) = \log\left(\eta\left(-\frac{1}{\tau}\right)\right) - \frac{1}{2}\log(-\,i\tau)$$

or

$$\eta\Big(-\frac{1}{\tau}\Big) = \sqrt{-i\tau}\,\eta(\tau). \tag{5.11}$$

Example 5.2. (Oppenheim's formula [23]). As the second example of (5.3), we have

$$\begin{aligned}
&\sum_{k=1}^{\infty} \frac{\sigma_{2\nu}(k)}{k^{\nu}}\, H_{2,2}^{2,0}\Big(z\pi^2k^2 \,\Big|\, \begin{matrix}(a-s,1),(b-s,1)\\ (\frac{\nu}{2},1),(-\frac{\nu}{2},1)\end{matrix}\Big)\\
&= z^{-\frac12} \sum_{k=1}^{\infty} \frac{\sigma_{2\nu}(k)}{k^{\nu}}\\
&\qquad \times H_{0,4}^{2,0}\Big(\frac{\pi^2k^2}{z} \,\Big|\, \begin{matrix}-\\ (\frac{\nu}{2},1),(-\frac{\nu}{2},1),(\frac12-a+s,1),(\frac12-b+s,1)\end{matrix}\Big)\\
&\qquad + \sum_{v\in S_\nu} \mathrm{Res}\Big(\frac{\Gamma(w+\frac{\nu}{2})\,\Gamma(w-\frac{\nu}{2})\,Z(w+\frac{\nu}{2})\,Z(w-\frac{\nu}{2})}{\Gamma(a+w-s)\,\Gamma(b+w-s)}\, z^{-w}, w=v\Big).
\end{aligned}$$

Setting $a = s+\frac{\nu}{2}+1$, $b = s-\frac{\nu}{2}$, and using

$$\begin{aligned}
&H_{2,2}^{2,0}\Big(z \,\Big|\, \begin{matrix}(\frac{\nu}{2}+1,1),(-\frac{\nu}{2},1)\\ (\frac{\nu}{2},1),(-\frac{\nu}{2},1)\end{matrix}\Big)\\
&= H_{1,1}^{1,0}\Big(z \,\Big|\, \begin{matrix}(\frac{\nu}{2}+1,1)\\ (\frac{\nu}{2},1)\end{matrix}\Big) = \begin{cases} z^{\frac{\nu}{2}} & |z|<1,\\ 0 & |z|>1,\end{cases}
\end{aligned} \tag{5.12}$$

and

$$\begin{aligned}
&H_{0,4}^{2,0}\Big(z \,\Big|\, \begin{matrix}-\\ (\frac{\nu}{2},1),(-\frac{\nu}{2},1),(-\frac12-\frac{\nu}{2},1),(\frac12+\frac{\nu}{2},1)\end{matrix}\Big)\\
&= z^{-\frac14}\, F_{2\nu+1}\big(4\sqrt[4]{z}\,\big),
\end{aligned} \tag{5.13}$$

where

$$F_\nu(z) = -\frac{2}{\pi}\sin\Big(\frac{\nu}{2}\,\pi\Big)\,K_\nu(z) - \sin\Big(\frac{\nu}{2}\,\pi\Big)\,Y_\nu(z) + \cos\Big(\frac{\nu}{2}\,\pi\Big)\,J_\nu(z), \tag{5.14}$$

and Y_ν denotes the Bessel function of the second kind, we get

$$\begin{aligned}
z^{-\nu}\sum_{k<z}\sigma_{2\nu}(k) &= z^{\frac12}\sum_{k=1}^{\infty}\frac{\sigma_{2\nu}(k)}{k^{\nu+\frac12}}\,F_{2\nu+1}\Big(4\pi\sqrt{zk}\Big)\\
&\qquad + \sum_{v\in S_\nu}\mathrm{Res}\Big(\frac{\zeta(2w+\nu)\,\zeta(2w-\nu)}{w+\frac{\nu}{2}}\,z^{2w}, w=v\Big),
\end{aligned} \tag{5.15}$$

where we have replaced z by $\frac{1}{\pi^2z^2}$.

Example 5.3. (Wilton's formula [31]). As another example of (5.3), we have

$$\sum_{k=1}^{\infty} \frac{\sigma_{2\nu}(k)}{k^{\nu}} H_{4,4}^{2,2}\left(z\pi^2 k^2 \,\middle|\, \begin{matrix} (1-a-s,1),(1-b-s,1),(c-s,1),(d-s,1) \\ (\frac{\nu}{2},1),(-\frac{\nu}{2},1),(1-e-s,1),(1-f-s,1) \end{matrix} \right)$$

$$= z^{-\frac{1}{2}} \sum_{k=1}^{\infty} \frac{\sigma_{2\nu}(k)}{k^{\nu}} H_{2,6}^{4,0}\left(\frac{\pi^2 k^2}{z} \,\middle|\, \begin{matrix} - \\ (\frac{\nu}{2},1),(-\frac{\nu}{2},1),(a-\frac{1}{2}+s,1),(b-\frac{1}{2}+s,1), \\ (e-\frac{1}{2}+s,1),(f-\frac{1}{2}+s,1) \\ (\frac{1}{2}-c+s,1),(\frac{1}{2}-d+s,1) \end{matrix} \right)$$

$$+ \sum_{v\in S_\nu} \operatorname{Res}\left(\frac{\Gamma(w+\frac{\nu}{2})\,\Gamma(w-\frac{\nu}{2})\,Z(w+\frac{\nu}{2})\,Z(w-\frac{\nu}{2})\,\Gamma(a+s-w)\Gamma(b+s-w)}{\Gamma(c-s+w)\,\Gamma(d-s+w)\,\Gamma(e+s-w)\,\Gamma(f+s-w)} z^{-w}, w=v \right).$$

Let $\lambda = 0,1,2,\ldots$. Setting $a=-s-\frac{\nu}{2}-\frac{\lambda}{2}+\frac{1}{2}$, $b=-s-\frac{\nu}{2}-\frac{\lambda}{2}$, $c=s-\frac{\nu}{2}$, $d=s+\frac{\nu}{2}+\lambda+1$, $e=-s-\frac{\nu}{2}+\frac{1}{2}$, $f=-s-\frac{\nu}{2}-\lambda$, and using

$$H_{4,4}^{2,2}\left(z \,\middle|\, \begin{matrix} (\frac{\nu}{2}+\frac{\lambda}{2}+\frac{1}{2},1),(\frac{\nu}{2}+\frac{\lambda}{2}+1,1),(-\frac{\nu}{2},1),(\frac{\nu}{2}+\lambda+1,1) \\ (\frac{\nu}{2},1),(-\frac{\nu}{2},1),(\frac{\nu}{2}+\frac{1}{2},1),(\frac{\nu}{2}+\lambda+1,1) \end{matrix} \right)$$

$$= H_{3,3}^{1,2}\left(z \,\middle|\, \begin{matrix} (\frac{\nu}{2}+\frac{\lambda}{2}+\frac{1}{2},1),(\frac{\nu}{2}+\frac{\lambda}{2}+1,1),(\frac{\nu}{2}+\lambda+1,1) \\ (\frac{\nu}{2},1),(\frac{\nu}{2}+\frac{1}{2},1),(\frac{\nu}{2}+\lambda+1,1) \end{matrix} \right)$$

$$= H_{4,4}^{2,2}\left(z \,\middle|\, \begin{matrix} (\frac{\nu}{2}+\frac{\lambda}{2}+\frac{1}{2},1),(\frac{\nu}{2}+\frac{\lambda}{2}+1,1),(\frac{\nu}{2}+\frac{1}{2},1),(\frac{\nu}{2}+\lambda+1,1) \\ (\frac{\nu}{2},1),(\frac{\nu}{2}+\frac{1}{2},1),(\frac{\nu}{2}+\frac{1}{2},1),(\frac{\nu}{2}+\lambda+1,1) \end{matrix} \right)$$

$$= 2^{\lambda} H_{1,1}^{1,0}\left(\sqrt{z} \,\middle|\, \begin{matrix} (\nu+\lambda+1,1) \\ (\nu,1) \end{matrix} \right) = \begin{cases} \dfrac{2^{\lambda}}{\Gamma(\lambda+1)} z^{\frac{\nu}{2}} (1-\sqrt{z})^{\lambda} & |z|<1, \\ 0 & |z|>1, \end{cases} \tag{5.16}$$

and

$$H_{2,6}^{4,0}\left(z \,\middle|\, \begin{matrix} - \\ (\frac{\nu}{2},1),(-\frac{\nu}{2},1),(-\frac{\nu}{2}-\frac{\lambda}{2},1),(-\frac{\nu}{2}-\frac{\lambda}{2}-\frac{1}{2},1), \\ (-\frac{\nu}{2},1),(-\frac{\nu}{2}-\lambda-\frac{1}{2},1) \\ (\frac{\nu}{2}+\frac{1}{2},1),(-\frac{\nu}{2}-\lambda-\frac{1}{2},1) \end{matrix} \right)$$

$$= H_{1,5}^{3,0}\left(z \,\middle|\, \begin{matrix} - \\ (\frac{\nu}{2},1),(-\frac{\nu}{2}-\frac{\lambda}{2},1),(-\frac{\nu}{2}-\frac{\lambda}{2}-\frac{1}{2},1), \\ (-\frac{\nu}{2}-\lambda-\frac{1}{2},1) \\ (\frac{\nu}{2}+\frac{1}{2},1),(-\frac{\nu}{2}-\lambda-\frac{1}{2},1) \end{matrix} \right)$$

$$= H^{4,0}_{2,6}\left(z \,\middle|\, \begin{matrix} - \\ (\frac{\nu}{2},1),(\frac{\nu}{2}+\frac{1}{2},1),(-\frac{\nu}{2}-\frac{\lambda}{2},1),(-\frac{\nu}{2}-\frac{\lambda}{2}-\frac{1}{2},1), \\ (\frac{\nu}{2}+\frac{1}{2},1),(-\frac{\nu}{2}-\lambda-\frac{1}{2},1) \\ (\frac{\nu}{2}+\frac{1}{2},1),(-\frac{\nu}{2}-\lambda-\frac{1}{2},1) \end{matrix} \right)$$

$$= z^{-\frac{\lambda}{4}-\frac{1}{4}}\, G^{\lambda}_{2\nu+\lambda+1}\left(4\sqrt[4]{z}\right), \tag{5.17}$$

where

$$\begin{aligned} G^{\lambda}_{\nu}(z) = &-(-1)^{\lambda}\,\frac{2}{\pi}\sin\left(\frac{\nu-\lambda}{2}\pi\right)K_{\nu}(z) \\ &-\sin\left(\frac{\nu-\lambda}{2}\pi\right)Y_{\nu}(z)+\cos\left(\frac{\nu-\lambda}{2}\pi\right)J_{\nu}(z), \end{aligned} \tag{5.18}$$

we get

$$\begin{aligned} &\frac{2^{\lambda}}{\Gamma(\lambda+1)}\sum_{k<z}\sigma_{2\nu}(k)\left(z-k\right)^{\lambda} \\ &= \pi^{-\lambda}\, z^{\nu+\frac{\lambda}{2}+\frac{1}{2}}\sum_{k=1}^{\infty}\frac{\sigma_{2\nu}(k)}{k^{\nu+\frac{\lambda}{2}+\frac{1}{2}}}\, G^{\lambda}_{2\nu+\lambda+1}\left(4\pi\sqrt{kz}\right) \\ &+\sum_{v\in S_{\nu}}\operatorname{Res}\left(\frac{\Gamma\left(w+\frac{\nu}{2}\right)\zeta(2w+\nu)\zeta(2w-\nu)\,\Gamma\left(-\frac{\nu}{2}-\frac{\lambda}{2}+\frac{1}{2}-w\right)\Gamma\left(-\frac{\nu}{2}-\frac{\lambda}{2}-w\right)}{\Gamma\left(w+\frac{\nu}{2}+\lambda+1\right)\Gamma\left(-\frac{\nu}{2}+\frac{1}{2}-w\right)\Gamma\left(-\frac{\nu}{2}-\lambda-w\right)} \right. \\ &\qquad\qquad\left. z^{\nu+\lambda+2w}, w=v\right), \end{aligned} \tag{5.19}$$

where we have replaced z by $\frac{1}{\pi^2 z^2}$. This is the Wilton's formula.

Acknowledgements

The author would like to thank Professor S. Kanemitsu for introducing him to the world of analytic number theory and special functions [13,18]. This work was not possible without his continuous encouragement and ongoing friendship as well as his extensive and meticulous knowledge on the subjects. The author would also like to thank the organizers of the conference, especially Professor J. Liu for unforgettable excellent experience he had in the beautiful city of Weihai.

References

1. B. C. Berndt, Identities involving the coefficients of a class of Dirichlet series I, *Trans. Amer. Math. Soc.*, **137** (1969), 345–359.

2. B. C. Berndt, Identities involving the coefficients of a class of Dirichlet series II, *Trans. Amer. Math. Soc.*, **137** (1969), 361–374.
3. B. C. Berndt, Identities involving the coefficients of a class of Dirichlet series III, *Trans. Amer. Math. Soc.*, **146** (1969), 323–348.
4. B. C. Berndt, Identities involving the coefficients of a class of Dirichlet series IV, *Trans. Amer. Math. Soc.*, **149** (1970), 179–185.
5. B. C. Berndt, Identities involving the coefficients of a class of Dirichlet series VI, *Trans. Amer. Math. Soc.*, **160** (1971), 157–167.
6. S. Bochner, Some properties of modular relations, *Ann. Math.*, **53** (1951), 332–363.
7. S. Bochner, On Riemann's functional equation with multiple gamma factors, *Ann. Math.*, **67** (1958), 29–41.
8. K. Chandrasekharan and R. Narasimhan, Hecke's functional equation and arithmetical identities, *Ann. Math.*, **74** (1961), 1–23.
9. M. A. Chaudhry and S. M. Zubair, *On a Class of Incomplete Gamma Functions with Applications*, Chapman & Hall/CRC, 2001.
10. S. Chowla and A. Selberg, On Epstein's zeta-function (I), *Proc. Nat. Acad. Sci. U.S.A.*, **35** (1949), 371–374.
11. P. P. Ewald, Die Berechnung optischer und elektrostatischer Gitterpotentiale, *Ann. d. Phys.*, **64** (1921), 253–287.
12. C. Fox, The G and H functions as symmetrical Fourier Kernels, *Trans. Amer. Math. Soc.*, **98** (1961), 395–429.
13. S. Kanemitsu, *Hypergeometric functions as the special functions of the 21st century*, private communication, 2003.
14. S. Kanemitsu, Y. Tanigawa and H. Tsukada, A generalization of Bochner's formula, *Hardy-Ramanujan J.*, **27** (2004), 28–46.
15. S. Kanemitsu, Y. Tanigawa and H. Tsukada, Some number theoretic applications of a general modular relation, *International J. of Number Theory*, (4)**2** (2006), 599–615.
16. S. Kanemitsu, Y. Tanigawa and H. Tsukada, *Contributions to the theory of zeta-functions: the modular relation supremacy*, in preparation.
17. S. Kanemitsu, Y. Tanigawa, H. Tsukada and M. Yoshimoto, Crystal symmetry viewed as zeta symmetry, *Proceedings of Kinki University Symposium "Zeta Functions, Topology and Quantum Physics 2003"* (Developments in Mathematics, Vol.14) 91–129, Springer, 2005.
18. S. Kanemitsu, Y. Tanigawa and M. Yoshimoto, On rapidly convergent series for the Riemann-zeta values via the modular relation, *Abh. Math. Sem. Univ. Hamburg*, **72** (2002), 187–206.
19. A. F. Lavrik, An approximate functional equation for the Dirichlet L-function, *Trady Moskov. Math. Obv sv c.*, **18** (1968), 91–104 = *Trans. Moskow Math. Soc.* **18** (1968), 101–115.
20. A. M. Mathai and R. K. Saxena, *The H-function with applications in statistics and other disciplines*, Wiley, New Delhi, 1978.
21. C. S. Meijer, Über Whittakersche bezw. Besselsche Funktionen und deren Produkte, *Nieuw Archief voor. Wiskunde*, (2)**18** (4tes Heft) (1936), 10–39.
22. C. S. Meijer, On the G-function I-VIII, *Proc. Nederl. Akad. Wetensch.*, **49**

(1946), 227–237, 344–356, 457–469, 632–641, 765–772, 936-943, 1063–1072, 1165–1175.

23. A. Oppenheim, Some identities in the theory of numbers, *Proc. London Math. Soc.*, (2)**26** (1927), 295–350.
24. R. B. Paris and D. Kaminski, *Asymptotics and Mellin-Barnes Integrals*, Cambridge University Press, 2001.
25. A. P. Prudnikov, Yu. A. Bychkov and O. I. Marichev, *Integrals and series, supplementary chapters*, Izd. Nauka, Moscow, 1986.
26. A. Selberg and S. Chowla, On Epstein's zeta-function, *J. Reine Angew. Math.*, **227** (1967), 86–110.
27. A. Terras, *Harmonic analysis on symmetric spaces and applications I, II*, Springer, New York, 1985.
28. H. Tsukada, *A general functional equation in analytic number theory*, privately circulated preprint, 2003.
29. K. Ueno and M. Nishizawa, Quantum groups and zeta-functions, in Quantum Groups: Formalism and Applications, *Proc. of the Thirtieth Karpacz Winter School (Karpacz, 1994)* (J. Lukierski et al., Editors), 115–126, Polish Sci. Publ. PWN, Warsaw, 1995.
30. G. N. Watson, Some self-reciprocal functions, *Quart. J. Math. Oxford Ser.*, **2** (1931), 298–309.
31. J. R. Wilton, An extended form of Dirichlet's divisor problem, *Proc. London Math. Soc.*, (2)**36** (1933), 391–426.

L-FUNCTIONS OF FUNCTION FIELDS

DAQING WAN

Department of Mathematics, University of California,
Irvine, CA 92697-3875, U.S.A.
E-mail: dwan@math.uci.edu

This is a short expository paper on L-functions of function fields, based on the author's lecture given at the Weihai China-Japan Seminar on number theory.

1. Introduction

Let $\mathbb{F}_q$ be a finite field of q elements with characteristic p. Let C be a smooth projective geometrically connected curve defined over $\mathbb{F}_q$ with function field K. Let $j : U \hookrightarrow C$ be a Zariski open dense subset of C. Let $G_K = \mathrm{Gal}(K^{\mathrm{sep}}/K)$ denote the absolute Galois group of K. For example, we can take $C = \mathbb{P}^1$, $U = \mathbb{P}^1 - \{0, \infty\} = \mathbb{G}_m$ and $K = \mathbb{F}_q(t)$.

Let $\pi_1^{\mathrm{arith}}(U)$ denote the arithmetic fundamental group of U. That is,

$$\pi_1^{\mathrm{arith}}(U) = G_K/\langle I_x \rangle_{x \in |U|},$$

where the denominator denotes the closed normal subgroup generated by the inertial subgroups I_x as x runs over the closed points $|U|$ of U. Let D_x denote the decomposition group of G_K at x. One has the following exact sequence

$$1 \to I_x \to D_x \to \mathrm{Gal}(\bar{k}_x/k_x) \to 1,$$

where k_x denotes the residue field of K at x. The Galois group $\mathrm{Gal}(\bar{k}_x/k_x)$ is topologically generated by the geometric Frobenius element Frob_x:

$$\mathrm{Frob}_x^{-1} : \alpha \to \alpha^{\# k_x}.$$

Let P_x denote the p-Sylow subgroup of I_x. Then we have the following exact sequence

$$1 \to P_x \to I_x \to I_x^{\mathrm{tame}} = \prod_{\ell \neq p} \mathbb{Z}_\ell(1) \to 1.$$

Let F_ℓ be a finite extension of $\mathbb{Q}_\ell$, where ℓ is a prime number. Let R_ℓ be the ring of integers in F_ℓ. Let V be a finite dimensional vector space over F_ℓ. Let

$$\rho : G_K \longrightarrow GL(V)$$

be a continuous representation of G_K unramified on U. Equivalently,

$$\rho : \pi_1^{\mathrm{arith}}(U) \longrightarrow GL(V)$$

is a continuous representation of $\pi_1^{\mathrm{arith}}(U)$. The L-function of ρ on U is defined by

$$L(U, \rho) = \prod_{x \in |U|} \frac{1}{\det(I - \rho(\mathrm{Frob}_x)T^{\deg(x)}|V)} \in 1 + TR_\ell[[T]].$$

Similarly, the complete L-function of ρ on C is defined by

$$L(C, \rho) = \prod_{x \in |C|} \frac{1}{\det(I - \rho(\mathrm{Frob}_x)T^{\deg(x)}|V^{I_x})} \in 1 + TR_\ell[[T]],$$

where V^{I_x} is the subspace of V which I_x acts trivially. These two L-functions just differ by a finite number of Euler factors. We are interested in analytic properties of these L-functions, especially for those representations which come from geometry, that is, arising as the relative ètale cohomology with compact support of a family of varieties $f : X \to U$.

2. ℓ-adic cae: $\ell \neq p$

Since $\ell \neq p$, the restriction of the ℓ-adic representation ρ to P_x is of finite order and thus the representation ρ is almost tame. In fact, by class field theory, ρ itself has finite order up to a twist if ρ has rank one. Thus, there are not too many such ℓ-adic representations. The L-function $L(U, \rho)$ is always a rational function. This follows from Grothendieck's trace formula [7]:

Theorem 2.1. *Let $\mathcal{F}_\rho$ denote the lisse ℓ-adic sheaf on U associated with ρ. Then, there are finite dimensional vector spaces $H_c^i(U \otimes \bar{\mathbb{F}}_q, \mathcal{F}_\rho)$ $(i = 0, 1, 2)$ over F_ℓ such that*

$$L(U, \rho) = \prod_{i=0}^{2} \det(I - \mathrm{Frob}_q T | H_c^i(U \otimes \bar{\mathbb{F}}_q, \mathcal{F}_\rho))^{(-1)^{i-1}} \in F_\ell(T).$$

If U is affine, then $H_c^0 = 0$. If ρ does not contain geometrically trivial component, then $H_c^2 = 0$. Thus, in most cases, it is H_c^1 that is most interesting.

Corollary 2.1. *Let U be affine. Assume that ρ does not contain geometrically trivial component. Then, the L-function*

$$L(U,\rho) = \det(I - \mathrm{Frob}_q T | H_c^1(U \otimes \bar{\mathbb{F}}_q, \mathcal{F}_\rho))$$

is a polynomial.

This is the ℓ-adic function field analogue of Artin's entireness conjecture.

Fix an embedding $\iota : \bar{\mathbb{Q}}_\ell \hookrightarrow \mathbb{C}$. A representation ρ is called ι-pure of weight $w \in \mathbb{R}$ if each eigenvalue of Frob_x acting on V has absolute value $q^{\deg(x)w/2}$ for all $x \in |U|$. A representation ρ is called ι-mixed of weights at most w if each irreducible component of ρ is ι-pure of weights at most w. If ρ is ι-pure of weight w for every embedding ι, then ρ is called pure of weight w. Similarly, if ρ is ι-mixed of weights at most w for every ι, then ρ is called mixed of weights at most w. The fundamental theorem of Deligne [2] on the Weil conjectures implies

Theorem 2.2. *If ρ is geometric, then ρ is mixed with integral weights. Furthermore, if ρ is mixed of weights at most w, then $H_c^i(U \otimes \bar{\mathbb{F}}_q, \mathcal{F}_\rho)$ is mixed of weights at most $w + i$.*

The ℓ-adic function field Langlands as established by Lafforgue [9] implies

Theorem 2.3. *If ρ is irreducible, then ρ is geometric up to a twist and hence pure of some weight.*

Thus, in the ℓ-adic case with $\ell \neq p$, essentially all ℓ-adic representations are geometric from L-function point of view.

3. p-adic case

In the case $\ell = p$, the restriction of the p-adic representation ρ to P_x can be infinite and thus ρ can be very wildly ramified. The L-function $L(U,\rho)$ is naturally more complicated and cannot be rational in general. One can ask for its p-adic meromorphic continuation. The function $L(U,\rho)$ is trivially p-adic analytic in the open unit disc $|T|_p < 1$ as the coefficients are in the ring R_p. It was shown in [11] that $L(U,\rho)$ is not p-adic meromorphic in general, disproving a conjecture of Katz [8]. However, one can show that $L(U,\rho)$ is p-adic meromorphic on the closed unit disc $|T|_p \leq 1$. Its zeros and poles on the closed unit disc are explained by p-adic ètale cohomology of ρ. This was proved by Emerton-Kisin [4], confirming a conjecture of Katz [8]. That is,

Theorem 3.1. *For any p-adic representation ρ of $\pi_1^{\mathrm{arith}}(U)$, the quotient*

$$\frac{L(U,\rho)}{\prod_{i=0}^{2}\det(I-\mathrm{Frob}_q T|H_c^i(U\otimes\bar{\mathbb{F}}_q,\mathcal{F}_\rho))^{(-1)^{i-1}}}$$

has no zeros and poles on the closed unit disc $|T|_p \leq 1$.

In the case that ρ has rank one, this was first proved by Crew [1].

For geometric p-adic representation, the following meromorphic continuation was conjectured by Dwork [3] and proved by Wan [12,13].

Theorem 3.2. *If the p-adic representation ρ is geometric, then the L-function $L(U,\rho)$ is p-adic meromorphic everywhere.*

Unlike the ℓ-adic case, most p-adic representations are not geometric. It seems very difficult to classify geometric p-adic representations, even in the rank one case. This can be viewed as the p-adic function field Langlands program, which is still wide open even for rank one case.

We remark that the Artin entireness conjecture fails for L-function $L(U,\rho)$ of geometric p-adic representation, even for non-trivial rank one ρ. From the proof in [12], one can understand why this is the case. The geometric p-adic representation ρ is an effective object in the category of geometric p-adic representations, but it is really not effective in the larger category of L-functions of infinite rank nuclear overconvergent σ-modules, where the L-function is known to be p-adic meromorphic. It is an virtual object!

For example, if the p-adic representation ρ is of rank one and ρ is the slope zero piece of a rank n geometric ℓ-adic representation ψ_ℓ of $\pi_1^{\mathrm{arith}}(U)$, then

$$L(U,\rho)=\prod_{j=0}^{n}\lim_{k\to\infty}L(U,\mathrm{Sym}^{1+(p-1)p^k-i}\psi_\ell\otimes\wedge^i\psi_\ell)^{(-1)^{i-1}i}.$$

Each of the above limit exists as a formal p-adic power series and they can be viewed as the L-function of some infinite rank ℓ-adic representation with p-adic values. These L-functions of infinite symmetric powers are p-adic entire (Artin conjecture holds), not just p-adic meromorphic. For two important examples with much greater details, see the recent papers by Fu-Wan [5,6] for the n-dimensional Kloosterman case, and by Roja-Leon and Wan [10] for the n-dimensional Calabi-Yau hypersurface case.

References

1. R. Crew, L-functions of p-adic characters and geometric Iwasawa theory, *Invent. Math.*, **88** (1987), 395–403.
2. P. Deligne, La Conjecture de Weil II, *Inst. Hautes Études Sci. Publ. Math.*, **52** (1980) 137–252.
3. B. Dwork, Normalized period matrices II, *Ann. Math.*, **98** (1973), 1–57.
4. M. Emerton and M. Kisin, Unit L-functions and a conjecture of Katz, *Ann. Math.*, (2)**153** (2001), 329–354.
5. L. Fu and D. Wan, L-functions for symmetric products of Kloosterman sums, *J. Reine Angew. Math.*, **589** (2005), 79–103.
6. L. Fu and D. Wan, *Trivial factors for L-functions of symmetric products of Kloosterman sheaves*, Compositio Math., to appear.
7. A. Grothendieck, *Formule de Lefschetz et rationalité des fonctions L* Séminaire Bourbaki, exposé 279(1964/65).
8. N. Katz, Travaux de Dwork, *Séminaire Bourbaki*, exposé 409(1971/72), Lecture Notes in Math., **317** (1973), 167–200.
9. L. Lafforgue, Chtoucas de Drinfeld et Correspondance de Langlands, *Invent. Math.*, (1)**147** (2002), 1–241.
10. A. Rojas-Leon and D. Wan, *Moment zeta functions for toric Calabi-Yau hypersurfaces over finite fields*, arXiv:math.AG/0702679.
11. D. Wan, Meromorphic continuation of L-functions of p-adic representations, *Ann. Math.*, **143** (1996), 469–498.
12. D. Wan, Rank one case of Dwork's conjecture, *J. Amer. Math. Soc.*, **13** (2000), 853–908.
13. D. Wan, Higher rank case of Dwork's conjecture, *J. Amer. Math. Soc.*, **13** (2000), 807–852.

INDEX